AF361891

Equine Respiratory Medicine and Cardiology

Equine Respiratory Medicine and Cardiology

Editors

Francesco Ferrucci
Chiara Maria Lo Feudo
Luca Stucchi

Basel • Beijing • Wuhan • Barcelona • Belgrade • Novi Sad • Cluj • Manchester

Editors

Francesco Ferrucci
Department of Veterinary
Medicine and Animal
Sciences
University of Milan
Lodi, Italy

Chiara Maria Lo Feudo
Department of Veterinary
Medicine and Animal
Sciences
University of Milan
Lodi, Italy

Luca Stucchi
Department of Veterinary
Medicine
University of Sassari
Sassari, Italy

Editorial Office
MDPI
St. Alban-Anlage 66
4052 Basel, Switzerland

This is a reprint of articles from the Special Issue published online in the open access journal *Animals* (ISSN 2076-2615) (available at: https://www.mdpi.com/journal/animals/special_issues/equine_respiratory_medicine_cardiology).

For citation purposes, cite each article independently as indicated on the article page online and as indicated below:

Lastname, A.A.; Lastname, B.B. Article Title. *Journal Name* **Year**, *Volume Number*, Page Range.

ISBN 978-3-0365-9794-2 (Hbk)
ISBN 978-3-0365-9795-9 (PDF)
doi.org/10.3390/books978-3-0365-9795-9

Contents

Luca Stucchi, Chiara Maria Lo Feudo and Francesco Ferrucci
Equine Respiratory Medicine and Cardiology
Reprinted from: *Animals* **2023**, *13*, 2952, doi:10.3390/ani13182952 . . . . **1**

Luca Stucchi, Chiara Maria Lo Feudo, Giovanni Stancari, Bianca Conturba and Francesco Ferrucci
Effect of the Administration of a Nutraceutical Supplement in Racehorses with Lower Airway Inflammation
Reprinted from: *Animals* **2022**, *12*, 2479, doi:10.3390/ani12182479 . . . . **5**

Mohamed Marzok, Adel I. Almubarak, Zakriya Al Mohamad, Mohamed Salem, Alshimaa M. Farag, Hussam M. Ibrahim, et al.
Reference Values and Repeatability of Pulsed Wave Doppler Echocardiography Parameters in Normal Donkeys
Reprinted from: *Animals* **2022**, *12*, 2296, doi:10.3390/ani12172296 . . . . **15**

John Klier, Sebastian Fuchs, Gerhard Winter and Heidrun Gehlen
Inhalative Nanoparticulate CpG Immunotherapy in Severe Equine Asthma: An Innovative Therapeutic Concept and Potential Animal Model for Human Asthma Treatment
Reprinted from: *Animals* **2022**, *12*, 2087, doi:10.3390/ani12162087 . . . . **29**

Chiara Maria Lo Feudo, Giovanni Stancari, Federica Collavo, Luca Stucchi, Bianca Conturba, Enrica Zucca and Francesco Ferrucci
Upper and Lower Airways Evaluation and Its Relationship with Dynamic Upper Airway Obstruction in Racehorses
Reprinted from: *Animals* **2022**, *12*, 1563, doi:10.3390/ani12121563 . . . . **53**

Ilaria Basano, Alessandra Romolo, Giulia Iamone, Giulia Memoli, Barbara Riccio, Jean-Pierre Lavoie, et al.
Giant Multinucleated Cells Are Associated with Mastocytic Inflammatory Signature Equine Asthma
Reprinted from: *Animals* **2022**, *12*, 1070, doi:10.3390/ani12091070 . . . . **69**

Joana Simões, Mariana Batista and Paula Tilley
The Immune Mechanisms of Severe Equine Asthma—Current Understanding and What Is Missing
Reprinted from: *Animals* **2022**, *12*, 744, doi:10.3390/ani12060744 . . . . **83**

Sophie Mainguy-Seers, Mouhamadou Diaw and Jean-Pierre Lavoie
Lung Function Variation during the Estrus Cycle of Mares Affected by Severe Asthma
Reprinted from: *Animals* **2022**, *12*, 494, doi:10.3390/ani12040494 . . . . **109**

Chiara Maria Lo Feudo, Luca Stucchi, Giovanni Stancari, Elena Alberti, Bianca Conturba, Enrica Zucca and Francesco Ferrucci
Associations between Exercise-Induced Pulmonary Hemorrhage (EIPH) and Fitness Parameters Measured by Incremental Treadmill Test in Standardbred Racehorses
Reprinted from: *Animals* **2022**, *12*, 449, doi:10.3390/ani12040449 . . . . **121**

Natalia Kozłowska, Małgorzata Wierzbicka, Tomasz Jasiński and Małgorzata Domino
Advances in the Diagnosis of Equine Respiratory Diseases: A Review of Novel Imaging and Functional Techniques
Reprinted from: *Animals* **2022**, *12*, 381, doi:10.3390/ani12030381 . . . . **133**

Luca Stucchi, Francesco Ferrucci, Michela Bullone, Raffaele L. Dellacà and Jean Pierre Lavoie
Within-Breath Oscillatory Mechanics in Horses Affected by Severe Equine Asthma in Exacerbation and in Remission of the Disease
Reprinted from: *Animals* **2022**, *12*, 4, doi:10.3390/ani12010004 . . . . **153**

Editorial

Equine Respiratory Medicine and Cardiology

Luca Stucchi [1], Chiara Maria Lo Feudo [2,*] and Francesco Ferrucci [2]

[1] Department of Veterinary Medicine, University of Sassari, Via Vienna 2, 07100 Sassari, Italy; lstucchi@uniss.it
[2] Department of Veterinary Medicine and Animal Sciences, University of Milan, Via Università 6, 26900 Lodi, Italy; francesco.ferrucci@unimi.it
* Correspondence: chiara.lofeudo@unimi.it

Citation: Stucchi, L.; Lo Feudo, C.M.; Ferrucci, F. Equine Respiratory Medicine and Cardiology. *Animals* **2023**, *13*, 2952. https://doi.org/10.3390/ani13182952

Received: 16 August 2023
Accepted: 7 September 2023
Published: 18 September 2023

In the last few years, the attention regarding the health of the lungs and heart of equine patients has been continuously growing. Indeed, if we search for studies categorized in the PubMed database on the topic "equine respiratory medicine", we can find that 980 articles have been published in the last decade [1]. Interestingly, although the first description of heaves as an "asthma-like" condition dates back to the 1960s [2], the recent revision of the nomenclature for chronic inflammatory diseases affecting the equine lower airways [3] brought about a rapid expansion of the knowledge about these conditions. Similarly, the progressive increase in the life expectancy of horses led to a rise in the diagnosis of cardiac problems. Consequently, the rapid advance of research in the cardiology field led to the availability of new technologic tools, such as three-dimensional electro-anatomical mapping of the equine heart [4]. These tools allow innovative diagnostic and therapeutic approaches that, until a few years ago, were considered impossible. Finally, the growing awareness and sensitivity of owners and trainers about the equine welfare played a fundamental role in this positive and rapid evolution.

The content of this Special Issue aims to reflect the current direction of research, dealing with a wider understanding of the pathophysiological mechanisms underlying respiratory and cardiac diseases of horses, as well as the development of novel diagnostic and therapeutic approaches that are essential to improve their management.

For instance, the review by Kozlowska et al. [5] highlights the relevance of new imaging techniques, such as computer tomography and magnetic resonance, in the diagnosis of upper respiratory tract diseases. Moreover, the authors explore the most recent advances in lung function testing techniques, describing the usefulness of electrical impedance tomography and the impulse oscillometry system (IOS). On this last topic, the original article by Stucchi et al. [6] reports the data of IOS measurements in severely asthmatic horses in different stages of the disease, describing for the first time the phenomenon of the expiratory airflow limitation during the exacerbation phase. Mainguy-Seers et al. [7] performed IOS measurements on asthmatic mares in their study, identifying a positive effect of the luteal phase of the estrus cycle on lung function parameters and hypothesizing a role of sex hormones in equine asthma pathophysiology.

Another key theme in the current literature is represented by the complex immunological pathways involved in equine asthma (EA). The review by Simões et al. [8] widely explores the different phenotypes and endotypes of EA, and the current knowledge concerning the cytokines, the inflammatory biomarkers, and the microbiome involved in the development of the disease. The review by Klier et al. [9] proposes a novel immunomodulatory treatment for EA, based on the inhalation of nanoparticulate cytosine–phosphate–guanine oligodeoxynucleotides, which may also find application in asthmatic human patients. Finally, the paper by Basano et al. [10] reports for the first time a positive association between the presence of giant multinucleated cells and mast cells in the broncho-alveolar lavage fluid of EA-affected horses.

This Special Issue also investigates the different respiratory diseases commonly observed in racehorses, such as dynamic upper airway obstructions (DUAOs), exercise-induced pulmonary hemorrhage (EIPH), and mild–moderate equine asthma (MEA). In the first of their two retrospective studies, Lo Feudo et al. [11] identified no association between the presence of DUAOs and lower airway inflammation in Thoroughbred and Standardbred horses, suggesting that disorders of the upper and lower airways follow independent pathological pathways. Moreover, while they did not detect a role of the size of the epiglottis in the development of DUAOs, they reported an association between its flaccid appearance observed during resting endoscopy and the development of dorsal displacement of the soft palate during exercise. In the second article by Lo Feudo et al. [12], the effects of EIPH on fitness parameters, measured through a standardized treadmill test, of Standardbred trotters were investigated. The study concluded that EIPH does not affect the athletic capacity of racehorses, and its role in decreased performance quality may follow a different pathway. Furthermore, the prospective case–control study by Stucchi et al. [13] evaluates the efficacy of the administration of a nutraceutical supplement, composed of different herbal extracts with antioxidant properties, in Thoroughbred racehorses affected by MEA. The results show a reduction in tracheal mucus accumulation and clinical score in the treated patients.

Lastly, the focus of the paper of Marzok et al. [14] is a species that has gained more and more popularity as a companion animal in recent years: the donkey. The authors studied the reference values and repeatability of pulsed-wave Doppler echocardiographic variables, improving the limited availability of data reported in the literature for these animals.

In conclusion, we believe that the papers published in this Special Issue could enrich the current knowledge of equine cardio-respiratory medicine with innovative, interesting, and useful data, and we hope that readers will be positively impressed.

Acknowledgments: The authors would like to thank all the colleagues who contributed with their work to the realization of this Special Issue.

Conflicts of Interest: The authors declare no conflict of interest for this Editorial paper.

References

1. PubMed. Available online: https://pubmed.ncbi.nlm.nih.gov/ (accessed on 4 August 2023).
2. Lowell, F.C. Observations On Heaves. An Asthma-Like Syndrome In The Horse. *J. Allergy* **1964**, *35*, 322–330. [CrossRef] [PubMed]
3. Pirie, R.S.; Couëtil, L.L.; Robinson, N.E.; Lavoie, J.P. Equine asthma: An appropriate, translational and comprehendible terminology? *Equine Vet. J.* **2016**, *48*, 403–405. [CrossRef] [PubMed]
4. Buschmann, E.; Van Steenkiste, G.; Boussy, T.; Vernemmen, I.; Schauvliege, S.; Decloedt, A.; van Loon, G. Three-dimensional electro-anatomical mapping and radiofrequency ablation as a novel treatment for atrioventricular accessory pathway in a horse: A case report. *J. Vet. Intern. Med.* **2023**, *37*, 728–734. [CrossRef] [PubMed]
5. Kozłowska, N.; Wierzbicka, M.; Jasiński, T.; Domino, M. Advances in the Diagnosis of Equine Respiratory Diseases: A Review of Novel Imaging and Functional Techniques. *Animals* **2022**, *12*, 381. [CrossRef] [PubMed]
6. Stucchi, L.; Ferrucci, F.; Bullone, M.; Dellacà, R.L.; Lavoie, J.P. Within-Breath Oscillatory Mechanics in Horses Affected by Severe Equine Asthma in Exacerbation and in Remission of the Disease. *Animals* **2022**, *12*, 4. [CrossRef] [PubMed]
7. Mainguy-Seers, S.; Diaw, M.; Lavoie, J.-P. Lung Function Variation during the Estrus Cycle of Mares Affected by Severe Asthma. *Animals* **2022**, *12*, 494. [CrossRef] [PubMed]
8. Simões, J.; Batista, M.; Tilley, P. The Immune Mechanisms of Severe Equine Asthma—Current Understanding and What Is Missing. *Animals* **2022**, *12*, 744. [CrossRef] [PubMed]
9. Klier, J.; Fuchs, S.; Winter, G.; Gehlen, H. Inhalative Nanoparticulate CpG Immunotherapy in Severe Equine Asthma: An Innovative Therapeutic Concept and Potential Animal Model for Human Asthma Treatment. *Animals* **2022**, *12*, 2087. [CrossRef]
10. Basano, I.; Romolo, A.; Iamone, G.; Memoli, G.; Riccio, B.; Lavoie, J.-P.; Miniscalco, B.; Bullone, M. Giant Multinucleated Cells Are Associated with Mastocytic Inflammatory Signature Equine Asthma. *Animals* **2022**, *12*, 1070. [CrossRef] [PubMed]
11. Lo Feudo, C.M.; Stancari, G.; Collavo, F.; Stucchi, L.; Conturba, B.; Zucca, E.; Ferrucci, F. Upper and Lower Airways Evaluation and Its Relationship with Dynamic Upper Airway Obstruction in Racehorses. *Animals* **2022**, *12*, 1563. [CrossRef] [PubMed]
12. Lo Feudo, C.M.; Stucchi, L.; Stancari, G.; Alberti, E.; Conturba, B.; Zucca, E.; Ferrucci, F. Associations between Exercise-Induced Pulmonary Hemorrhage (EIPH) and Fitness Parameters Measured by Incremental Treadmill Test in Standardbred Racehorses. *Animals* **2022**, *12*, 449. [CrossRef] [PubMed]

13. Stucchi, L.; Lo Feudo, C.M.; Stancari, G.; Conturba, B.; Ferrucci, F. Effect of the Administration of a Nutraceutical Supplement in Racehorses with Lower Airway Inflammation. *Animals* **2022**, *12*, 2479. [CrossRef] [PubMed]
14. Marzok, M.; Almubarak, A.I.; Al Mohamad, Z.; Salem, M.; Farag, A.M.; Ibrahim, H.M.; El-Ashker, M.R.; El-khodery, S. Reference Values and Repeatability of Pulsed Wave Doppler Echocardiography Parameters in Normal Donkeys. *Animals* **2022**, *12*, 2296. [CrossRef] [PubMed]

Article

Effect of the Administration of a Nutraceutical Supplement in Racehorses with Lower Airway Inflammation

Luca Stucchi [1,*], Chiara Maria Lo Feudo [2], Giovanni Stancari [1], Bianca Conturba [1] and Francesco Ferrucci [2,*]

[1] Veterinary Teaching Hospital, Department of Veterinary Medicine and Animal Sciences, Università degli Studi di Milano, 26900 Lodi, Italy
[2] Equine Sports Medicine Laboratory "Franco Tradati", Department of Veterinary Medicine and Animal Sciences, Università degli Studi di Milano, 26900 Lodi, Italy
* Correspondence: luca.stucchi@unimi.it (L.S.); francesco.ferrucci@unimi.it (F.F.)

Simple Summary: The study of non-pharmacological products that could have a beneficial action on the respiratory system of the horse may have a great impact on the welfare and performance of equine athletes. As a matter of fact, young racehorses can be affected by a mild–moderate form of equine asthma (MEA), that can reach a prevalence of 80%, and pharmacological treatments during racing days are almost always forbidden. This study aimed to evaluate the efficacy of the administration of a feed supplement, composed of several nutraceutical herbs, in the treatment of lower airway inflammation. This product was administered to seven racehorses with MEA, while five horses were used as control. After 21 days, horses treated with the supplement showed a lower degree of clinical signs and lower mucus accumulation in the trachea, compared to controls. For this reason, the supplement was shown to be effective in controlling lower airway inflammation, and its use as an adjunctive treatment for MEA may deserve to be evaluated.

Abstract: Mild–moderate equine asthma (MEA) is a chronic inflammatory disorder of the lower airways of the horse, characterized by tracheal mucus accumulation, cough and poor performance. The therapeutic approach is based on pharmacological treatment and environmental management. Moreover, the efficacy of the administration of antioxidant molecules has been reported. The aim of the present study was to evaluate the effect of the administration of a commercial nutraceutical supplement, composed of several herbal extracts, on lower airway inflammation in racehorses. Twelve Thoroughbreds affected by MEA were selected. All horses underwent a clinical examination with assignment of a clinical score, airway endoscopy and cytological examination of bronchoalveolar lavage fluid. In seven horses, the supplement was administered for 21 days in association with environmental changes, while in five horses only environmental changes were performed. All procedures were repeated at the end of the study. Data concerning the clinical score, the endoscopic scores and the cytology at the beginning and at the end of the study were statistically compared. Data showed a significant reduction ($p < 0.0156$) of the clinical score and a significant reduction ($p < 0.0156$) of the tracheal mucus score. The results showed the beneficial effect of the supplement on mild–moderate lower airway inflammation, probably due to its antioxidant activity.

Keywords: equine asthma; racehorse; bronchoalveolar lavage; tracheal mucus; cough

Citation: Stucchi, L.; Lo Feudo, C.M.; Stancari, G.; Conturba, B.; Ferrucci, F. Effect of the Administration of a Nutraceutical Supplement in Racehorses with Lower Airway Inflammation. *Animals* **2022**, *12*, 2479. https://doi.org/10.3390/ani12182479

Academic Editor: Francisco Javier Mendoza

Received: 13 August 2022
Accepted: 16 September 2022
Published: 19 September 2022

Publisher's Note: MDPI stays neutral with regard to jurisdictional claims in published maps and institutional affiliations.

1. Introduction

Young racehorses are often affected by an inflammatory disorder of the lower airways that is characterized by the presence of cough, mucus accumulation in the trachea and reduced athletic performance [1–3]. The diagnosis, that could be only presumptive on the basis of clinical symptoms and airway endoscopy, relies on cytological examination of the bronchoalveolar lavage fluid (BALf), where a pathological accumulation of neutrophils and/or eosinophils and/or mast cells can be detected [2,4]. This condition, that may affect up to 80% of Thoroughbred racehorses in training, has been recently classified as a mild or moderate form of equine asthma (MEA), due to its similarities with human asthma [5–8]. The treatment of this disease is based, similarly to human asthma, on the administration of bronchodilators and corticosteroids, both systemic and by inhalation, in association with changes in the environmental management, aimed to reduce the respirable dust [9–11].

Nevertheless, the presence of pharmacological molecules in the horse organism is forbidden during the racing day for most of the European Racing Association [12]. Therefore, the research on non-pharmacological molecules that may have beneficial effects on the respiratory system of the horse is of primary interest.

Recently, a complementary feed supplement has been introduced to the market, as a nutraceutical support for the respiratory system of the horse. It contains maltodextrin, calcium carbonate, *Arthrospira platensis* (12%) and fermented pineapple (5%), plus several herbal extracts and antioxidant substances. This supplement aims to reduce airway inflammation, primarily modulating the oxidative stress. In fact, it has been reported that oxidative status plays a major role in sustaining equine asthma [13–15]; in particular, neutrophils are responsible for the production of reactive oxidant species (ROS) and their accumulation in the lower airway can result in a high level of oxidative stress, leading to an increase in the inflammation [16,17]. Moreover, it has been demonstrated that the administration of antioxidant substances such as Omega-3 fatty acids, in association with environmental changes, was effective in reducing airway inflammation in horses affected by equine asthma [18].

The hypothesis for this work was that the nutraceutical substances contained in the supplement may have an effective anti-inflammatory action on the horses' respiratory system. Thus, the aim of the study was to evaluate the efficacy of the administration of this supplement in the control of clinical symptoms, improvement of endoscopic findings and reduction of inflammatory cells in the lower airways of Thoroughbred racehorses in training affected by MEA.

2. Materials and Methods

2.1. Sample Selection

To perform the study, 12 client-owned Thoroughbred racehorses in training, referred to the Equine Unit of the Veterinary Teaching Hospital (University of Milan, Italy), were selected. Horses from three different training centers located in the North of Italy were enrolled; the study was performed at the stables where the horses usually lived, from May to September 2021. All horses were stabled in boxes bedded with straw, fed hay and concentrate on the basis of their nutritional requirements, and trained once daily. Horses with a history of decreased performance and symptoms consistent with MEA were included. At the beginning of the study (T0), all horses underwent a clinical examination with the attribution of an adapted clinical score: 1 point for the presence of cough, 1 point for the presence of nasal discharge, from 0 to 2 points for respiratory rate increase and from 0 to 3 points for the severity of the sounds at lung auscultation [19]. Then, all horses underwent a diagnostic protocol that consisted of upper and lower airway endoscopy, BALf collection and cytological examination of the BALf.

All procedures performed on the horses were approved by the University of Milan Animal Welfare Organization (Protocol Number OPBA_39_2021) and included informed owner consent.

2.2. Airway Endoscopy and Attribution of Endoscopic Scores

Horses were contained with a twitch and sedated with detomidine hydrochloride (0.01 mg/kg IV; Domosedan; Vetoquinol, Italy). A flexible videoendoscope (EC-530WL-P, Fujifilm, Tokyo, Japan) was passed through the left nostril, and the upper and lower tracts of the respiratory system were visualized and recorded. Then, the recordings were blind reviewed by an expert veterinarian (L.S.) who assigned a score from 0 to 4 for pharyngeal lymphoid hyperplasia (PLH) [20], from 0 to 5 for tracheal mucus accumulation [21] and from 0 to 4 for the edema of the tracheal bifurcation [22].

2.3. BALf Collection and Cytological Examination

During the airway endoscopy, a BALf sample was obtained. To perform the BALf, 60 mL of a 0.5% lidocaine hydrochloride solution was sprayed at the level of the tracheal bifurcation in order to inhibit the coughing reflex; then, the endoscope was passed into the bronchial tree until it was wedged firmly within a segmental bronchus. Here, a 300 mL pre-warmed sterile saline 0.9% was instilled, and the fluid was immediately aspirated. The BALf sample was stored in sterile ethylenediaminetetraacetic acid (EDTA) tubes and processed within 90 min. To perform the cytological examination, a few drops of pooled BALf were cytocentrifugated (Rotofix 32, Hettich Cyto System, Tuttlingen, Germany) at 26 *g* for 5 min. The slides were air dried, stained with May-Grünwald Giemsa and Perl's Prussian blue, and observed under a light microscope at 400× and 1000× for 400-cell leukocyte differential counting [23]. In order to be included in the study, the cytological examination of BALf had to be consistent with MEA, presenting a percentage of neutrophils > 5%, and/or mast cells > 2% and/or eosinophils > 1% [24].

2.4. Treatment

Once the diagnosis of MEA was made, horses were divided casually by coin-flip into two groups:

- The Treatment group, composed of 7 horses. Horses in this group underwent changes in the management aimed to reduce respirable dust, i.e., bedding with wood shaving and feeding wet hay [25]. Moreover, they were administered with the supplement (BURAN Candioli, Acel Pharma s.r.l., Beinasco, Italy) at the dosage recommended by the company (35 gr/daily, orally) for 21 days. The composition of the supplement is displayed in Table 1.
- The Control group, composed of 5 horses, that underwent the same environmental changes of the Treatment group, but did not receive the nutritional supplement.

Table 1. Herbal extracts and vitamins contained in the supplement.

Additives per kg	
Allium sativum	37.500 mg
Glycyrrhiza glabra	27.500 mg
Thymus vulgaris	18.000 mg
Hedera helix	7.500 mg
Vitamin C	50.000 mg
Vitamin E	30.000 mg

No other changes of stabling, feeding or training were performed during the study. After 21 days (T1), all horses underwent the same clinical protocol performed at T0, including clinical examination and attribution of the clinical score, airway endoscopy with attribution of endoscopic scores, and BALf cytology. The operator assigning scores (L.S.) remained blinded to the group distribution.

2.5. Statistical Analysis

Data were reported in an electronic sheet and analyzed using a commercially available statistical software package (Prism Graphpad 9.1.0 for MacOs; San Diego, CA, USA). Data distribution was evaluated for normality using a Shapiro–Wilk test. The values of age, the clinical score, the endoscopic scores (PLH, tracheal mucus and edema of the tracheal bifurcation), and the results of the differential cell count of the BALf at T0 were compared between groups by means of unpaired t-test or Mann–Whitney test, on the basis of data distribution. The clinical score, the endoscopic scores, and the cytological results of the BALf obtained at T0 and T1 were compared within both groups by means of the paired t-test, if normally distributed, or Wilcoxon test, if not normally distributed. Statistical significance was set at $p < 0.05$.

3. Results

Data are reported as mean ± standard deviation if normally distributed, or as median and interquartile range (I.Q.R.) if not normally distributed.

3.1. Horses

The study population consisted of nine colts, two fillies and one gelding. The median age of the Treatment group was 3, I.Q.R. 2–3, while the age of the Control group was 3, I.Q.R. 2–4. The groups were age-matched as no statistical difference was found between the ages of the groups.

3.2. Clinical Score

The results of clinical score in the two groups at T0 and T1 are reported in Table 2. A significant difference was detected ($p = 0.0189$) for the clinical score between the two groups at T0, with higher values in Treatment group. At T1, a significant difference ($p = 0.0156$) (Figure 1) was found in the Treatment group, while no significant difference was observed in the Control group.

Table 2. Results of the clinical score at T0 and T1 in the Treatment and Control groups.

		T0				T1		
TREATMENT	**Cough**	**Nasal Discharge**	**Respiratory Rate**	**Lung Sound**	**Cough**	**Nasal Discharge**	**Respiratory Rate**	**Lung Sound**
Horse 1	0	0	1	2	0	0	1	0
Horse 2	0	0	0	2	0	0	0	0
Horse 3	1	1	1	0	1	0	1	0
Horse 4	0	1	1	2	0	0	1	0
Horse 5	1	0	1	2	0	0	1	2
Horse 6	1	0	1	2	0	0	1	2
Horse 7	0	0	1	2	0	0	1	0
CONTROL								
Horse 1	0	0	0	0	0	0	0	0
Horse 2	0	0	0	1	0	0	0	0
Horse 3	0	0	0	2	0	0	0	0
Horse 4	0	0	1	0	1	0	0	2
Horse 5	0	0	1	2	1	0	1	2

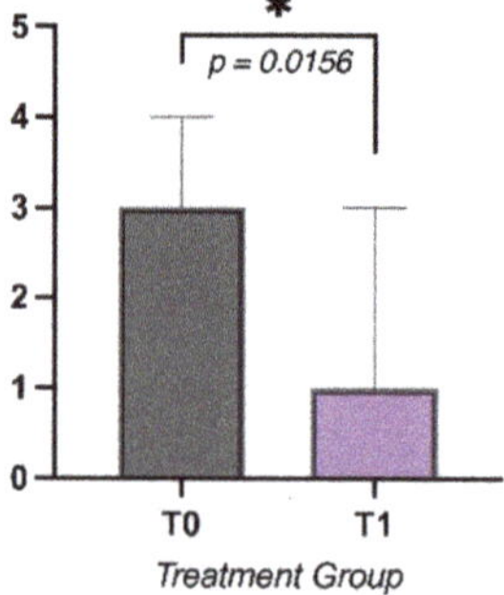

Figure 1. Clinical score at T0 and T1 in the Treatment group (* = $p<0.05$).

3.3. Endoscopic Score

The results of the endoscopic score at T0 and T1 in the Treatment and Control groups are reported in Figure 2. No significant differences were observed between the two groups at T0 for all the endoscopic scores. At T1, a significant difference ($p = 0.0156$) in the tracheal mucus score was detected in the Treatment group, while no other differences were present in either group for the other scores.

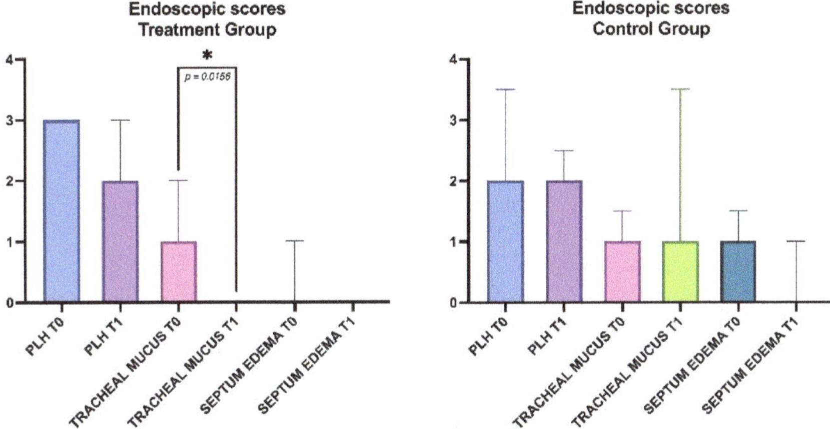

Figure 2. Results of the endoscopic scores at T0 and T1 in the two groups (* = $p<0.05$).

3.4. Cytological Examination of the BALf

At T0, the total nucleated cell (TNC) numbers were 489 ± 31 for the Treatment group and 340 ± 168 for the Control Group, while at T1, they were 487 ± 24 and 332 ± 166, respectively. A significant difference in TNC was present ($p = 0.028$) between the two groups at T0. No significant differences were present between T0 and T1 in either group.

The results of the differential count of the inflammatory cells of the BALf at T0 and T1 are reported in Figure 3. At T0, no significant differences were present between the two groups. At T1, no significant difference was found for all the cellular lines, either in the Treatment or in the Control group.

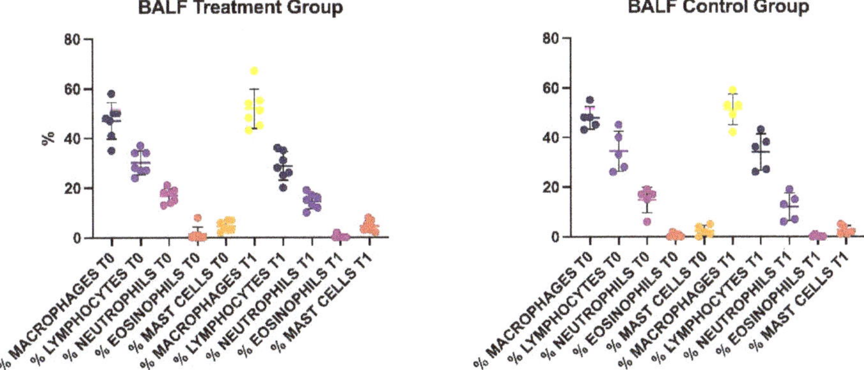

Figure 3. Results of the differential count of the inflammatory cells of the BALf in the two groups at T0 and T1.

4. Discussion

The present study aimed to evaluate the effect of the administration of a nutraceutical supplement on mild lower airway inflammation of racehorses. The study of nonpharmacological products that could have a beneficial action on the respiratory system of the horse may have a big impact on the equestrian world, as equine asthma is widely distributed in the horse population and the use of medications during competitions is almost always forbidden.

The age of the horses involved in this study was very young, and this is in accordance with the definition of MEA [2]; in fact, MEA can occur at any age, but is more commonly reported in young horses [2]. As the percentage of neutrophils has been described to increase [26] and that of eosinophils to decrease with age [22], it was important that the groups were age matched. Concerning the sex, on the other hand, no data are available regarding the prevalence of MEA in a particular gender. It must be noticed that most of the horses in this study were colts, as is normal in a population of young Thoroughbreds in training.

At the beginning of the study, the endoscopic scores and the cytological profile were analogous in the Treatment group when compared to the Control group; moreover, both groups underwent the same management modifications during the treatment period. This choice was important, as it allowed us to standardize the study and to eliminate any possible bias that could have interfered with the results. We decided to use this cut-off of the BALf inflammatory cells for the diagnosis of MEA [24] as it can be considered the most used and accurate in racehorses [4,5].

The clinical score at T0 was higher in the Treatment group than the Control group: this could be due to the fact that MEA is a paucisymptomatic syndrome [2], and some clinical symptoms such as cough or abnormal respiratory sound at rest might not have been present. Nevertheless, it should be noticed that a lower score in the Control group at T0 could have influenced the statistical results regarding the reduction of clinical signs.

Statistical analysis showed that, at T1, there was a significant reduction of the clinical score and of the tracheal mucus score in the Treatment group, but not in the Control group. These results suggest that the administration of the supplement was effective in the reduction of clinical signs such as cough, nasal discharge or abnormal respiratory sound at lung auscultation, probably due to a decreased mucus production that was confirmed by tracheal endoscopy.

The mechanism underlying the efficacy of this product was not investigated in this work, but some explanations may be hypothesized by investigating the properties of some of its components. In fact, some papers showed the efficacy of *Allium sativum* in the mitigation of inflammation and mucus production in a mouse model of human asthma [27], and in the reduction of tracheal mucus in equine asthmatic patients [28]. Furthermore, *Thymus vulgaris* showed effects in the modulation of mucus hypersecretion in vitro [29], while *Hedera helix* seems to play an important role in reducing cough [30] and to have a mucolytic activity similar to acetylcysteine [31] in human patients with bronchitis. Finally, *Glycyrrhiza glabra* represents one of the most used and effective compounds for asthma in Chinese traditional medicine, and it was effective in reducing ROS production, bronchial inflammation, and mucus production in a murine model of human asthma [32,33].

Another explanation could be considered in the antioxidant effect of the supplement. In fact, it contains, in association with antioxidant vitamins C and E, a high percentage of *Arthrospira platensis* (12%), a herbal extract that has been proved to have an important scavenging activity against ROS [34], and has shown its protective effect on human bronchial tissue [35]. It is recognized that neutrophilic accumulation in the airways leads to an accumulation of ROS, resulting in an increase in mucus production and airway hyperreactivity [14]. In contrast, controlling the level of reactive species in the airways could induce a reduction of mucus accumulation in the trachea and a mitigation of cough, as observed in our study. Further studies regarding the oxidative status of the lung tissue after the administration of this supplement are needed to confirm this mechanism. Moreover, recently it has been demonstrated that the presence of high oxidative stress in the airways of the horse is one of the cause of the neutrophil corticosteroid insensitivity [17]; it could be hypothesized that the administration of products with antioxidant properties in association with standard therapies may be useful in the treatment of equine asthma.

Conversely, in our study we did not find any significant differences for PLH after the treatment period, either in the Treatment or in the Control group; this could be explained by the age of the population, as in young horses the presence of PLH is very common and not associated with severity of lower airway inflammation [22,36]. In the same manner, the absence of significant differences in the tracheal bifurcation edema score between T0 and T1 in both groups can be explained by the fact that this finding is more common in older horses affected by the severe form of equine asthma [22,37].

Concerning the cytological differential count of the BALf, no differences were found in either group after the treatment period; this results means that the cytological characteristic of the lower airway inflammation remained the same. However, it must be noticed that also conventional bronchodilator and corticosteroid therapy, even if from a clinical and mechanical point of view is rapidly effective, it determines a reduction of BALf neutrophilia only after 8 weeks of treatment [38]. It has also been proven that the best management modification to reduce lung inflammation is pasture [9], which unfortunately was not possible in our study.

5. Conclusions

In conclusion, the administration of the feed supplement, which is the subject of the present study, for 21 days, in association with some environmental changes, was effective in reducing clinical score and tracheal mucus score in a population of 12 Thoroughbred racehorses affected by MEA, but did not have any effect on BALf cytology. As MEA has a negative impact on the athletic capacity of the horses, for the future, it would be interesting to evaluate the effect of the administration of this product on performance parameters. Moreover, as exercise-induced pulmonary hemorrhage is another respiratory condition that is highly frequent in the racehorse population, it could be useful to also evaluate the effect of this nutraceutical supplement for this disease.

Author Contributions: Conceptualization, L.S., C.M.L.F. and F.F.; methodology, L.S., C.M.L.F. and F.F.; formal analysis, L.S., C.M.L.F. and F.F.; investigation, L.S., C.M.L.F., G.S. and B.C.; writing—original draft preparation, L.S.; writing—review and editing, L.S., C.M.L.F., G.S., B.C. and F.F. All authors have read and agreed to the published version of the manuscript.

Funding: This study was funded by Acel Pharma s.r.l., which is the company that produces the supplement (CTE_NAZPR22FFERR_02).

Institutional Review Board Statement: The study was reviewed and approved by the Animal Welfare Organization of the University of Milan (Protocol Number OPBA 39 2021).

Informed Consent Statement: An informed owner consent was provided for the procedures and the use of data.

Data Availability Statement: The data presented in this study are available on request from the corresponding authors.

Conflicts of Interest: F.F. is responsible for the funding received by Acel Pharma s.r.l., for the execution of the study. L.S., C.M.L.F., B.C., G.S. declare no conflict of interest.

References

1. Couëtil, L.L.; Denicola, D.B. Blood gas, plasma lactate and bronchoalveolar lavage cytology analyses in racehorses with respiratory disease. *Equine Vet. J.* **1999**, *30*, 77–82. [CrossRef] [PubMed]
2. Couetil, L.L.; Cardwell, J.M.; Gerber, V.; Lavoie, J.P.; Leguillette, R.; Richard, E.A. Inflammatory airway disease of horses-revised consensus statement. *J. Vet. Intern. Med.* **2016**, *30*, 503–515. [CrossRef]
3. Stucchi, L.; Alberti, E.; Stancari, G.; Conturba, B.; Zucca, E.; Ferrucci, F. The relationship between lung inflammation and aerobic threshold in standardbred racehorses with mild-moderate equine asthma. *Animals* **2020**, *10*, 1278. [CrossRef] [PubMed]

4. Christley, R.M.; Hodgson, D.R.; Rose, R.J.; Hodgson, J.L.; Wood, J.L.N.; Reid, S.W.J. Coughing in thoroughbred racehorses: Riskfactors and tracheal endoscopic and cytological findings. *Vet. Rec.* **2001**, *148*, 99–104. [CrossRef]

5. Christley, R.M.; Hodgson, D.R.; Rose, R.J.; Wood, J.L.N.; Reid, S.W.J.; Whitear, K.G.; Hodgson, J.L. A case-control study of respiratory disease in Thoroughbred racehorses in Sydney, Australia. *Equine Vet. J.* **2001**, *33*, 256–264. [CrossRef]

6. Couetil, L.; Cardwell, J.M.; Leguillette, R.; Mazan, M.; Richard, E.; Bienzle, D.; Bullone, M.; Gerber, V.; Ivester, K.; Lavoie, J.P.; et al. Equine Asthma: Current Understanding and Future Directions. *Front. Vet. Sci* **2020**, *30*, 450. [CrossRef] [PubMed]

7. Kinnison, T.; McGilvray, T.A.; Couetil, L.L.; Smith, K.C.; Wylie, C.E.; Bacigalupo, S.A.; Gomez-Grau, E.; Cardwell, J.M. Mild-moderate equine asthma: A scoping review of evidence supporting the consensus definition. *Vet. J.* **2022**, *286*, 105865. [CrossRef]

8. Lange-Consiglio, A.; Stucchi, L.; Zucca, E.; Lavoie, J.P.; Cremonesi, F.; Ferrucci, F. Insights into animal models for cell-based therapies in translational studies of lung diseases: Is the horse with naturally occurring asthma the right choice? *Cytotherapy* **2019**, *21*, 525–534. [CrossRef]

9. Leclere, M.; Lavoie-Lamoureux, A.; Joubert, P.; Relave, F.; Lanctot Setlakwe, E.; Beauchamp, G.; Couture, C.; Martin, J.G.; Lavoie, J.P. Corticosteroids and Antigen Avoidance Decrease Airway Smooth Muscle Mass in an Equine Asthma Model. *Am. J. Respir Cell Mol. Biol* **2012**, *47*, 589–596. [CrossRef]

10. Calzetta, L.; Roncada, P.; di Cave, D.; Bonizzi, L.; Urbani, A.; Pistocchini, E.; Rogliani, P.; Matera, M.G. Pharmacological treatments in asthma-affected horses: A pair-wise and network meta-analysis. *Equine Vet. J.* **2017**, *49*, 710–717. [CrossRef]

11. Mönki, J.; Saastamoinen, M.; Karikoski, N.; Rajamäki, M.; Raekallio, M.; Junnila, J.; Särkijärvi, S.; Norring, M.; Valros, A.; Oranen Ben Fatma, S.; et al. Effects of Bedding Material on Equine Lower Airway Inflammation: A Crossover Study Comparing Peat and Wood Shavings. *Front. Vet. Sci.* **2021**, *8*, 656814. [CrossRef] [PubMed]

12. European Horseracing Scientific Liaison Committe website. Available online: https://www.ehslc.com/ (accessed on 4 August 2022).

13. Art, T.; Kirschvink, N.; Smith, N.; Lekeux, P. Indices of oxidative stress in blood and pulmonary epithelium lining fluid in horses suffering from recurrent airway obstruction. *Equine Vet. J.* **1999**, *31*, 397–401. [CrossRef] [PubMed]

14. Deaton, C.M. The role of oxidative stress in an equine model of human asthma. *Redox Rep.* **2006**, *11*, 46–52. [CrossRef] [PubMed]

15. Bullone, M.; Lavoie, J.P. The Contribution of Oxidative Stress and Inflamm-Aging in Human and Equine Asthma. *Int. J. Mol. Sci.* **2017**, *18*, 2612. [CrossRef] [PubMed]

16. Braga, P.C.; Dal Sasso, M.; Lattuada, N.; Greco, V.; Sibilia, V.; Zucca, E.; Stucchi, L.; Ferro, E.; Ferrucci, F. Antioxidant activity of hyaluronic acid investigated by means of chemiluminescence of equine neutrophil bursts and electron paramagnetic resonance spectroscopy. *J. Vet. Pharmacol. Therap.* **2013**, *38*, 48–54. [CrossRef]

17. Pourali Dogaheh, S.; Boivin, R.; Lavoie, J.P. Studies of molecular pathways associated with blood neutrophil corticosteroid insensitivity in equine asthma. *Vet. Immunol. Immunopat.* **2021**, *237*, 110265. [CrossRef] [PubMed]

18. Nogradi, N.; Couetil, L.L.; Messick, J.; Stochelski, M.A.; Burgess, J.A. Omega-3 fatty acid supplementation provides an additional benefit to a low-dust diet in the management of horses with chronic lower airway inflammatory disease. *J. Vet. Intern. Med.* **2015**, *29*, 299–306. [CrossRef] [PubMed]

19. Ferrucci, F.; Stucchi, L.; Salvadori, M.; Stancari, G.; Conturba, B.; Bronzo, V.; Ferro, E.; Zucca, E. Effetti dell'amikacina per via inalatoria in cavalli sportivi affetti da sindrome da calo di rendimento e confronto con la somministrazione endovenosa. *Ippologia* **2013**, *24*, 1–7.

20. Burrell, M.H. Endoscopic and virological observations on respiratory disease in a group of young Thoroughbred horses in training. *Equine Vet. J.* **1985**, *17*, 99–103. [CrossRef]

21. Gerber, V.; Straub, R.; Marti, E.; Hauptman, J.; Herholz, C.; King, M.; Imhof, A.; Tahon, L.; Robinson, N.E. Endoscopic scoring of mucus quantity and quality: Observer and horse variance and relationship to inflammation, mucus viscoelasticity and volume. *Equine Vet. J.* **2004**, *36*, 576–582. [CrossRef]

22. Lo Feudo, C.M.; Stucchi, L.; Alberti, E.; Stancari, G.; Conturba, B.; Zucca, E.; Ferrucci, F. The role of thoracic ultrasonography and airway endoscopy in the diagnosis of equine asthma and exercise-induced pulmonary hemorrhage. *Vet. Sci.* **2021**, *8*, 276. [CrossRef] [PubMed]

23. Lo Feudo, C.M.; Stucchi, L.; Stancari, G.; Alberti, E.; Conturba, B.; Zucca, E.; Ferrucci, F. Associations between Exercise-Induced Pulmonary Hemorrhage (EIPH) and Fitness Parameters Measured by Incremental Treadmill Test in Standardbred Racehorses. *Animals* **2022**, *12*, 449. [CrossRef] [PubMed]

24. Couetil, L.L. Inflammatory diseases of the lower airway of athletic horses. In *Equine Sports Medicine and Surgery*, 2nd ed.; Hinchcliff, K.W., Kaneps, A.J., Geor, R.J., Eds.; Elsevier Saunders: Philadelphia, PA, USA, 2014; pp. 605–632.

25. Leclere, M.; Lavoie-Lamoureux, A.; Gelinas-Lymburner, E.; David, F.; Martin, J.G.; Lavoie, J.P. Effect of Antigenic Exposure on Airway Smooth Muscle Remodeling in an Equine Model of Chronic Asthma. *Am. J. Respir. Cell Mol. Biol.* **2011**, *45*, 181–187. [CrossRef] [PubMed]

26. Widmer, A.; Doherr, M.G.; Tessier, C.; Koch, C.; Ramseyer, A.; Straub, R.; Gerber, V. Association of increased tracheal mucus accumulation with poor willingness to perform in show-jumpers and dressage horses. *Vet. J.* **2009**, *182*, 430–435. [CrossRef]

27. Shina, N.-R.; Kwonb, H.J.; Koa, J.W.; Kimc, J.S.; Leeb, I.C.; Kima, J.C.; Kimd, S.H.; Shina, I.S. S-Allyl cysteine reduces eosinophilic airway inflammation and mucus overproduction on ovalbumin-induced allergic asthma model. *Int. Immunopharmacol.* **2019**, *68*, 124–130. [CrossRef]

28. Saastamoinen, M.; Särkijärvi, S.; Hyyppä, S. Garlic (*Allium Sativum*) Supplementation Improves Respiratory Health but Has Increased Risk of Lower Hematologic Values in Horses. *Animals* **2019**, *9*, 13. [CrossRef]

29. Marinelli, O.; Iannarelli, R.; Morelli, M.B.; Valisi, M.; Nicotra, G.; Amantini, C.; Cardinali, C.; Santoni, G.; Maggi, F.; Nabissi, M. Evaluations of thyme extract effects in human normal bronchial and tracheal epithelial cell lines and in human lung cancer cell line. *Chem. -Biol. Interac.* **2016**, *256*, 125–133.

30. Sierocinski, E.; Holzinger, F.; Chenot, J.F. Ivy leaf (Hedera helix) for acute upper respiratory tract infections: An updated systematic review. *Eur. J. Clin. Pharmacol.* **2021**, *77*, 1113–1122. [CrossRef]

31. Kruttschnitt, E.; Wegener, T.; Zahner, C.; Henzen-Bucking, S. Assessment of the Efficacy and Safety of Ivy Leaf (Hedera helix) Cough Syrup Compared with Acetylcysteine in Adults and Children with Acute Bronchitis. *Evid. -Based Complementary Altern. Med.* **2020**, *2020*, 1910656. [CrossRef]

32. Xu, X.; Huang, M.; Duan, X.; Liu, H.; Zhang, W.; Li, D. Compound Glycyrrhiza Oral Solution alleviates oxidative stress and inflammation by regulating SRC/MAPK pathway in chronic obstructive pulmonary disease. *Immunopharmacol. Immunotoxicol.* **2022**, *11*, 1–12. [CrossRef]

33. Sun, X.; Nasab, E.M.; Athari, S.M.; Athari, S.S. Anti-inflammatory effect of herbal traditional medicine extract on molecular regulation in allergic asthma. *Allergol. Sel.* **2021**, *5*, 148–156. [CrossRef] [PubMed]

34. Jensen, G.S.; Attridge, V.L.; Beaman, J.L.; Guthrie, J.; Ehmann, A.; Benson, K.F. Antioxidant and Anti-Inflammatory Properties of an Aqueous Cyanophyta Extract Derived from Arthrospira Platensis: Contribution to Bioactivities by the Non-Phycocyanin Aqueous Fraction. *J. Med. Food* **2015**, *18*, 535–541. [CrossRef] [PubMed]

35. Tajvidi, E.; Nahavandizadeh, N.; Pournaderi, M.; Pourrashid, A.Z.; Bossaghzadeh, F.; Khoshnood, Z. Study the antioxidant effects of blue-green algae Spirulina extract on ROS and MDA production in human lung cancer cells. *Biochem. Biophys. Rep.* **2021**, *28*, 101139. [CrossRef]

36. Holcombe, S.J.; Jackson, C.; Gerber, V.; Jefcoat, A.; Berney, C.; Eberhardt, S.; Robinson, N.E. Stabling is associated with airway inflammation in young Arabian horses. *Equine Vet. J.* **2001**, *33*, 244–249. [CrossRef] [PubMed]

37. Koblinger, K.; Nicol, J.; McDonald, K.; Wasko, A.; Logie, N.; Weiss, M.; Leguillette, R. Endoscopic Assessment of Airway Inflammation in Horses. *J. Vet. Intern. Med.* **2011**, *25*, 1118–1126. [CrossRef] [PubMed]

38. Bullone, M.; Vargas, A.; Elce, Y.; Martin, J.G.; Lavoie, J.P. Fluticasone/salmeterol reduces remodelling and neutrophilic inflammation in severe equine asthma. *Sci. Rep.* **2017**, *7*, 8843. [CrossRef]

Article

Reference Values and Repeatability of Pulsed Wave Doppler Echocardiography Parameters in Normal Donkeys

Mohamed Marzok [1,2,*], Adel I. Almubarak [1], Zakriya Al Mohamad [1], Mohamed Salem [1], Alshimaa M. Farag [3], Hussam M. Ibrahim [3], Maged R. El-Ashker [3] and Sabry El-khodery [3]

[1] Department of Clinical Scienses, College of Veterinary Medicine, King Faisal University, Al-Ahsa 31982, Saudi Arabia

[2] Department of Surgery, Faculty of Veterinary Medicine, Kafrelsheikh University, Kafrelsheikh 33516, Egypt

[3] Department of Internal Medicine, Infectious and Fish Diseases, Faculty of Veterinary Medicine, Mansoura University, Manosura 35516, Egypt

* Correspondence: mmarzok@kfu.edu.sa

Citation: Marzok, M.; Almubarak, A.I.; Al Mohamad, Z.; Salem, M.; Farag, A.M.; Ibrahim, H.M.; El-Ashker, M.R.; El-khodery, S. Reference Values and Repeatability of Pulsed Wave Doppler Echocardiography Parameters in Normal Donkeys. *Animals* **2022**, *12*, 2296. https://doi.org/10.3390/ani12172296

Academic Editors: Francesco Ferrucci, Chiara Maria Lo Feudo and Luca Stucchi

Received: 22 July 2022
Accepted: 29 August 2022
Published: 5 September 2022

Publisher's Note: MDPI stays neutral with regard to jurisdictional claims in published maps and institutional affiliations.

Simple Summary: Cardiovascular disease is underreported in donkeys, possibly related to their limited athletic posture and frequent poor performance-related examinations. Reports on treatments for cardiovascular disease are anecdotal in donkey. Normal echocardiographic parameters have been reported in healthy donkeys. The aim of the present study was to establish the reference values and repeatability for Pulsed Wave Doppler echocardiographic variables of the mitral valve, aortic valve and myocardial performance. Two-dimensional Color flow mapping and spectral Doppler modes were performed. For the mitral valve, the mean velocity, pressure gradient and duration of E-wave were 57.7 ± 12.5 cm/s, 1.4 ± 0.7 mmHg and 0.4 ± 0.13 s, respectively. The results of the present study provide the reference values of PW echocardiographic parameters measurements in normal adult donkeys. Such reference values are helpful, especially when confronted with clinical cases with cardiovascular disorders.

Abstract: In the present study, thirty clinically healthy donkeys were used to establish the reference values and repeatability for Pulsed Wave Doppler echocardiographic variables of the mitral valve, aortic valve and myocardial performance. 2-dimensional Color flow mapping and spectral Doppler modes were performed. For the mitral valve, the mean velocity, pressure gradient and duration of E-wave were 57.7 ± 12.5 cm/s, 1.4 ± 0.7 mmHg and 0.4 ± 0.13 s, respectively. The velocity, pressure gradient and duration of the A-wave were 32.3 ± 9.1 cm/s, 0.3 ± 0.04 mmHg and 0.3 ± 0.1 s, respectively. The mitral valve area, pressure half time, pulsatility index (PI), resistance index (RI) and velocity time integral (VTI) were 1.8 ± 0.5 cm^2, 66 ± 17 ms, 2.8 ± 1.4, 0.9 ± 0.03 and 19.1 ± 5.7 cm, respectively. For the aortic valve, the mean velocity was 64.9 ± 10.4 cm/s, pressure gradient was 1.8 ± 0.4 mmHg, pulsatility index was 1.4 ± 0.3, resistance index was 0.9 ± 0.02, VTI was 25.02 ± 6.2 cm, systolic/diastolic was 19 ± 4.7 and heart rate was 95.7 ± 28.9 per minute. For Myocardial Performance Index (LV)–Tei Index, the mean ejection, isovolumic relaxation, isovolumic contraction time and myocardial performance index were 0.24 ± 0.01, 0.14 ± 0.01, 0.14 ± 0.02 and 1.2 ± 0.1 s, respectively. The results of the present study provide the reference values of PW echocardiographic parameter measurements in normal adult donkeys. Such reference values are helpful, especially when confronted with clinical cases with cardiovascular disorders.

Keywords: pulsed wave; echocardiography; donkeys; reference values

1. Introduction

Nowadays, donkeys are of high interest, both for clinicians and owners, due to their inner working abilities and as companion animals. Thus, in recent years, people's awareness of the well-being and care of these animals has continued to increase, and the demand for professional veterinary services has increased. According to reports, there are many differences between donkeys and horses. Therefore, clinical data, treatment and diagnostic protocols from horse to donkey may lead to misdiagnosis and unnecessary or inadequate treatment [1]. It was found that cardiovascular disease is underreported in donkeys, possibly related to their limited athletic posture and frequent poor performance-related examinations. Reports on treatments for cardiovascular disease are anecdotal in donkeys [2]. Normal echocardiographic parameters have also been reported in healthy donkeys [3]. Amory et al. [4] described normal values of echocardiographic dimensions and functional indexes, as well as quantitative reference values for Doppler flow in healthy donkeys. However, those references must be taken with restraint because marked variations can be seen depending on the breed, body size, age, growth rate and training of the animal.

Pulsed wave (PW) Doppler uses short ultrasound bursts, which are transmitted to a point (designated as "sample volume") distant from the transducer [5]. PW Doppler allows the calculation of blood flow velocity, direction and spectral characteristics from a specified point in the heart or blood vessel, but the measurement of the maximum velocity is limited as the pulse repetition frequency is limited [6]. Consequently, it is used for the evaluation of hemodynamic abnormalities in the heart, myocardial and pericardial disorders, transvalvular gradients, intracardiac pressures and shunts, diastolic and systolic cardiac performance and severity of valvular lesions. In humans, assessment of ventricular function can distinguish patients with normal LV function and LV dysfunction and provide knowledge about the hemodynamic alterations as a side effect of the use of therapeutic agents [7,8]. Blood flow velocity patterns are altered in human patients with cardiac dysfunction and valvular regurgitation [9]. Mitral and aortic valve regurgitation can be assessed by regurgitant volume or volumetric volume [10,11].

In humans and horses, the velocity time integral (VTI) is a hemodynamic echocardiographic parameter measured from the Doppler spectrum across the valves through the left ventricular outflow tract (LVOT) [12]. The area under the flow velocity curve represents the distance of the blood volume passing through the valve [13]. Doppler echocardiography has been used in the assessment of valvular regurgitation in horses with cardiac murmurs [14–16]. Moreover, peak blood flow velocity in the mitral valve has been assessed in healthy warm blood horses using PW Doppler echocardiography [17]. Furthermore, Blissitt and Bonagura [18] measured peak velocity and deceleration time of mitral inflow E-wave and peak velocity, peak acceleration, acceleration time and VTI of aortic outflow using Doppler echocardiography in thoroughbred and thoroughbred cross horses.

In horses, the pressure gradient, which was calculated using a simplified Bernoulli equation, is applied to insufficient valves, ventricular and atrial septal defects, and intracardiac pressure determination [19]. The pressure half time (PHT) is the time interval in milliseconds between the maximal mitral gradient in early diastole and the time point where the gradient is half the maximum initial value [20]. It is a simple Doppler method used for assessing the MVA, severity of aortic regurgitation and pressure deceleration [21].

Till now, echocardiographic parameters, including pulsatility index (PI), resistance index (RI) and myocardial performance index (MPI), which are used in human studies to evaluate the cardiac performance, was not recorded in horses with cardiac disorders [22,23]. PI and RI are useful in the measurement of blood flow resistance [24]. The myocardial performance index is an easily measured index used for the assessment of global heart function, combining both systolic and diastolic cardiac performance [22].

Data regarding reference values of Doppler echocardiographic parameters in healthy donkeys are scarce. Consequently, the current study was conducted to determine the reference values and repeatability for PW Doppler echocardiographic variables of the mitral valve, aortic valve and myocardial performance in normal donkeys.

2. Materials and Methods

2.1. Experimental Animals

Thirty clinically healthy donkeys (*Equus asinus*) were used for this study. The age of the donkeys was 5 to 9 years (7.2 ± 1.43), and their body weight was 100 to 220 Kg (172 ± 40.49). None of the donkeys had cardiovascular disorders nor any evidence of other systemic diseases based on clinical examination. All donkeys were free from significant valvular regurgitation using Doppler echocardiography. All donkeys under investigation were housed in straw-bedded boxes in the animal house of the veterinary teaching hospital, Faculty of Veterinary Medicine, Mansoura University. All donkeys were fed twice per day with 1.5 kg hay/100 kg B.W. and 1.5 kg concentrate with ad libitum water access at least two weeks before the trials. This study was carried out at the Department of Internal Medicine and Infectious Diseases, Faculty of Veterinary Medicine, Mansoura University, Mansoura, Egypt. The study was approved by the Animal welfare and Ethics Committee, Faculty of Veterinary Medicine, Mansoura University (approval no.R-136-2021).

2.2. Echocardiographic Examination

Transcutaneous echocardiographic examinations were performed according to the standard methods described by Youssef et al. [25]. All echocardiographic procedures and precautions were followed according to the recommendations of the American Society of Echocardiography. The 2-dimensional (2-D), Colour flow mapping and spectral Doppler modes were performed with a CHISON Digital Color Doppler Ultrasound System, iVis 60 EXPERT VET, (CHISON Medical Imaging Co., Ltd., Wuxi, China), using a 2–3.9 MHz phased array transducer, with a maximal depth of 24.1 cm. During spectral Doppler recording, the transducer was used in the high pulsed repetition frequency mode (HPRF) using a frequency of 6 MHz. The velocity scale was set at 150 cm/s so that only one sample volume was available for velocity recording. Flow velocities and time were displayed graphically, and accurate velocity recordings were obtained when the Doppler ultrasound beam was aligned parallel to the direction of flow according to [26]. Alignment with blood flow was initially assessed using a 2-D ultrasound image. A color flow study was used as a guide to determine the place of the sampling site in an area of maximal blood velocity. PW Doppler measurements were conducted according to guidelines previously described in [18,27]. For mitral inflow, a left parasternal long axis apical view of the left ventricular inlet was used [27]. For aortic outflow, a left parasternal long axis view (5-chambered) of the left ventricular outflow tract (LVOT) was used [27]. For a myocardial performance index of the left ventricle (MPILV), the apical five-chamber view was used [28].

2.3. PW Doppler Measurements

For mitral inflow, the sample volume was placed on the ventricular side of the mitral valve at the valve tips, with minor adjustments in transducer angles to obtain the flow velocity (Figure 1a,b). The velocity was measured during the rapid filling phase of the ventricle (E wave) and during the atrial contraction (A wave). Based on those, the duration of the E-wave and A-wave can be calculated. The pressure gradient was described according to the method described by Weyman [29]. Mitral PHT was calculated using the deceleration time, which is known as the time from the peak mitral velocity on the velocity decline extrapolated to the baseline [20]. The MVA was measured from PHT using an empirical formula as described in [30].

For aortic outflow, the sample volume was placed on the arterial side of the aortic valve [27] to obtain the velocity of blood flow. The pressure gradient was described according to Weyman 1994 [29].

PI and RI for mitral and aortic valves can be calculated according to standard methods [31,32]. The VTI under the velocity waveform was measured by tracing the modal velocity, which was represented by the bright line in the spectral Doppler waveform envelopes from the Doppler signal [33].

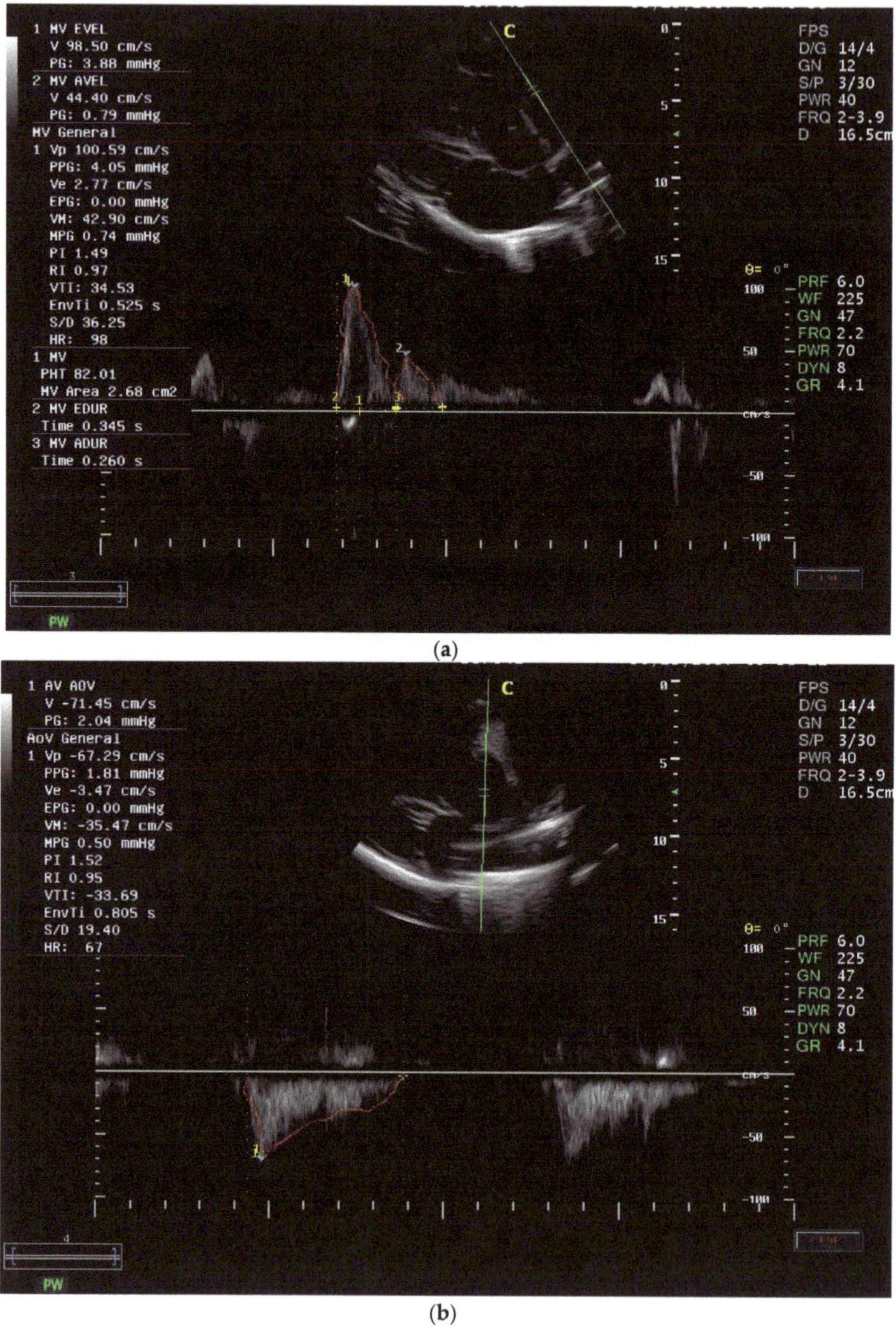

Figure 1. (**a**) Left parasternal long axis view of the left ventricular inlet of mitral valve view in one of the 30 normal adult donkeys. (**b**) Left parasternal long axis view (5-chambered) of left ventricular outflow tract (LVOT) with the sample volume on the arterial side of the aortic valve in 30 normal adult donkeys.

For the myocardial performance index of the left ventricle (MPILV), the sample volume was placed at the tips of the mitral valve leaflets. The isovolumic contraction time (ICT) interval was measured from the end to the onset of mitral inflow. Meanwhile, the isovolumic relaxation time (IRT) interval was measured at the time between the onset and the end of LV outflow when the sample volume was placed below the aortic valve. The myocardial performance index was later calculated from the equation (ICT + IRT)/ET [28].

2.4. Repeatability

Each donkey was examined via echocardiography three times at one-week intervals by the same observer. Each day, 3 PW echocardiographic measurements with a variance <5% were recorded on 3 non-consecutive cardiac cycles. Measurements with a variance >5% were discarded. Thus, by the end of the study, there were 9 PW echocardiographic measurements for each donkey.

2.5. Statistical Analysis

Statistical analysis followed the American Society for Veterinary Clinical Pathology (ASVCP) reference interval guidelines. Specifically, a histogram was used to illustrate the reference values. The data of PW echocardiography were checked for normality by the Kolmogorov–Smirnov normality test. The result revealed normally distributed data. The summary statistics and the frequency distribution (mean $\pm$ SD, 95% CI, median, range, and 10th, 25th, 75th, and 90th percentiles) of the PW echocardiographic measurements for the mitral and aortic valve in all donkeys were reported. The reproducibility of the PW echocardiographic measurements was assessed by calculation of the intra-assay and interassay coefficient of variation (CVs). For each of the 3 days that data were collected, the intra-assay CV was calculated by dividing the SD of the measurements for that day by the mean of the measurements for that day. For each donkey, the interassay CV was calculated for each of the 3 measurements (1, 2, and 3) of PW echocardiographic measurements obtained on each data collection day by dividing the SD for that particular measurement by the mean for that particular measurement. To assess the variation in the intra-assay CV among days and the interassay CV% among reads, repeated measures ANOVA was performed. Mauchly's sphericity test was used to detect the significant variations. When there was a significant result, one-way ANOVA with post-hoc Duncan multiple comparison tests were used to detect the specific variations. For all analyses, values of $p < 0.05$ were considered significant. All statistical procedures were performed with a commercially available software program (Graph Pad prism, San Diego, CA, USA).

3. Results

Frequency distribution for echocardiographic parameters of the mitral valve, aortic valve and myocardial Performance Index (LV)—Tei Index evaluated by PW Doppler echocardiography in healthy donkeys (*Equus asinus*) were summarized (Tables 1–3).

For echocardiographic parameters of the mitral valve, aortic valve and myocardial Performance Index (LV)—Tei Index evaluated by PW Doppler echocardiography in healthy donkeys (*Equus asinus*) were summarized (Tables 4–6).

The intra-assay and interassay coefficient of variation (CVs) of the mitral valve in healthy donkeys using PW Doppler echocardiography were summarized (Tables 7 and 8).

The intra-assay and interassay CVs of the aortic valve in healthy donkeys using PW Doppler echocardiography were summarized (Tables 9 and 10).

The intra-assay and interassay CVs of myocardial Performance Index (LV)–Tei Index in healthy donkeys using PW Doppler echocardiography were summarized (Tables 11 and 12).

Table 1. Frequency distribution of pulsed wave doppler echocardiographic measurements of mitral valve in normal donkeys.

Variables	Highest Frequency		Lowest Frequency	
	Value	No (270) (%)	Value	No (270) (%)
V E wave (cm/s)	≥12–14	66 (24.4%)	≥42–46	3 (1.1%)
PG E wave (mmHg)	≥15–25	63 (23.3%)	≥85–95	3 (1.1%)
Dur E wave (s)	≥15–25	62 (22.9%)	≥60–75	12 (4.4%)
V A wave (cm/s)	≥22–26	51 (18.9%)	≥40–46	3 (1.1%)
PG A wave (mmHg)	≥50–60	48 (17.8%)	≥80–90	3 (1.1%)
Dur A wave (s)	≥15–25	69 (25.6%)	≥60–70	9 (3.3%)
MV area (cm^2)	≥30–50	72 (26.7%)	≥100–120	3 (1.1%)
PHT (ms)	≥30–50	72 (26.7%)	≥100–120	9 (3.3%)
PI	≥15–25	75 (27.8%)	≥80–95	6 (2.2%)
RI	≥1.5–2.5	77 (28.5%)	≥3–4	7 (2.5%)
VTI (cm)	≥10–14	39 (14.4%)	≥34–38	9 (3.3%)
S/D	≥10–25	66 (24.4%)	≥60–70	3 (1.1%)

V E wave: Velocity E wave; PG E wave: pressure gradient E wave; Dur E wave: duration E wave; VA wave: Velocity A wave; PG A wave: pressure gradient A wave; Dur A wave: duration A wave; MV area: mitral valve area; PHT: pressure half time; PI: pulsatility index; RI: resistance index; VTI: velocity time integral; S/D: systolic/diastolic.

Table 2. Frequency distribution of PW doppler echocardiographic measurements of aortic valve in normal donkeys.

Variables	Highest Frequency		Lowest Frequency	
	Value	No (270) (%)	Value	No (270) (%)
V (cm/s)	≥12–16	60 (22.2%)	≥15–17	6 (2.2%)
PG (mmHg)	≥15–25	88 (32.5%)	≥60–65	7 (2.5%)
PI	≥15–25	72 (26.6%)	≥50–60	3 (1.1%)
RI	≥0.5–1.5	99 (36.6%)	≥5.5–6	3 (1.1%)
VTI (cm)	≥15–25	59 (21.8%)	≥55–65	3 (1.1%)
S/D	≥0–20	72 (26.6%)	≥90–100	3 (1.1%)

V: Velocity; PG: pressure gradient; PI: pulsatility index; RI: resistance index; VTI: velocity time integral; S/D: systolic/diastolic.

Table 3. Frequency distribution of PW doppler echocardiographic measurements of myocardial performance index (LV)—Tei Index valve in normal donkeys.

Variables	Highest Frequency		Lowest Frequency	
	Value	No (270) (%)	Value	No (270) (%)
ET (s)	≥15–25	75 (27.8%)	≥45–50	9 (3.3%)
IRT (s)	≥25–35	57 (21.1%)	≥70–85	15 (5.5%)
ICT (s)	≥30–40	78 (28.9%)	≥0–10	3 (1.1%)
MPI	≥6–10	63 (23.3%)	≥24–28	12 (4.4%)

ET: ejection time; IRT: isovolumic relaxation time; ICT: isovolumic contraction time; MPI: myocardial performance index.

Table 4. Summary statistics of PW doppler echocardiographic measurements of mitral valve in normal donkeys.

Variables	Mean ± SD	95% CI	Median (Range)	Percentile			
				10%	25%	75%	90%
V E wave (cm/s)	57.7 ± 12.4	56.2–59.2	53.4 (40.9–98.5)	45.1	50.6	62.4	76.7
PG E wave (mmHg)	1.4 ± 0.7	1.3–1.5	1.1 (0.7–3.9)	0.8	1.03	1.6	2.4
Dur E wave (s)	0.4 ± 0.14	0.3–0.4	0.4 (0.2–0.9)	0.2	0.3	0.4	0.5
V A wave (cm/s)	32.3 ± 9.1	31.2–33.4	31.6 (16.7–61.7)	22.2	24.9	37.5	42.3

Table 4. *Cont.*

Variables	Mean ± SD	95% CI	Median (Range)	Percentile			
				10%	25%	75%	90%
PG A wave (mmHg)	0.5 ± 0.3	0.4–0.5	0.4 (0.1–1.5)	0.2	0.3	0.6	0.7
Dur A wave (s)	0.3 ± 0.1	0.2–0.3	0.3 (0.1–0.6)	0.1	0.2	0.3	0.4
MV area (cm^2)	2.4 ± 1.5	2.2–2.6	1.9 (0.5–7.9)	0.7	1.4	2.9	4.3
PHT (ms)	91.3 ± 24.9	88.3–94.2	102.6 (27.6–125.3)	51.3	75.7	109.8	116.1
1.4 (0.01–2)	1.3–1.4	1.4 ± 0.4	PI		1.2	1.6	1.8
0.96 (0.8–1.1)	0.94–0.96	0.9 ± 0.03	RI		0.9	1	1.1
18.8 (9.9–36.5)	18.5–19.8	19.1 ± 5.7	VTI (cm)		15.1	22.4	26.5
22.5 (9.1–37.5)	21.9–23.7	22.8 ± 7.5	S/D		16	28.5	33.9

V E wave: Velocity E wave; PG E wave: pressure gradient E wave; Dur E wave: duration E wave; VA wave: Velocity A wave; PG A wave: pressure gradient A wave; Dur A wave: duration A wave; MV area: mitral valve area; PHT: pressure half time; PI: pulsatility index; RI: resistance index; VTI: velocity time integral; S/D: systolic/diastolic.

Table 5. Summary statistics of PW Doppler echocardiographic measurements of aortic valve in normal donkeys.

Variables	Mean ± SD	95% CI	Median (Range)	Percentile			
				10%	25%	75%	90%
V (cm/s)	64.9 ± 10.4	63.8–66.2	64.9 (42.3–86.7)	49.4	58.6	71.6	79.7
PG (mmHg)	1.7 ± 0.5	1.7–1.8	1.7 (0.7–3)	0.9	1.4	2.1	2.5
PI	1.4 ± 0.3	1.3–1.4	1.4 (0.5–1.9)	0.96	1.2	1.6	1.7
RI	0.9 ± 0.02	0.95–0.96	0.95 (0.9–1.02)	0.92	0.94	0.96	0.98
VTI (cm)	25.02 ± 6.2	24.3–25.8	26.3 (11.8–38.02)	15.7	20.3	29.9	32.7
S/D	23.6 ± 12.04	22.2–25.1	21.3 (6.7–86)	13.2	17.3	25.5	36

V: Velocity; PG: pressure gradient; PI: pulsatility index; RI: resistance index; VTI: velocity time integral; S/D: systolic/diastolic.

Table 6. Summary statistics of PW Doppler echocardiographic measurements of myocardial performance index (LV)—Tei Index in normal donkeys.

Variables	Mean ± SD	95% CI	Median (Range)	Percentile			
				10%	25%	75%	90%
ET (s)	0.4 ± 0.1	0.4 ± 0.1	0.4 (0.2–0.6)	0.2	0.3	0.4	0.5
IRT (s)	0.3 ± 0.2	0.3 ± 0.2	0.3 (0.1–0.8)	0.1	0.2	0.4	0.5
ICT (s)	0.3 ± 0.1	0.3 ± 0.1	0.3 (0.1–0.7)	0.1	0.2	0.4	0.5
MPI	1.7 ± 0.7	1.7 ± 0.7	1.7 (0.6–3.3)	0.9	1.3	2.1	2.9

ET: ejection time; IRT: isovolumic relaxation time; ICT: isovolumic contraction time; MPI: myocardial performance index.

Table 7. Intra-assay CVs of pulsed wave doppler echocardiographic measurements of mitral valve in normal donkeys.

Variables	Mean ± SD	95% CI	Median (Range)	Percentile			
				10%	25%	75%	90%
V E wave (cm/s)	13.01 ± 9.5	11.1–14.9	12.6 (0.8–45.8)	2.8	4.9	16.4	22.9
PG E wave (mmHg)	25.9 ± 18.9	21.9–29.8	24.3 (1.9–90.7)	5.6	9.9	32.2	47.2
Dur E wave (s)	24.8 ± 17.2	21.1–28.4	20.02 (0.8–71.4)	3.1	13.6	31.4	50.8
V A wave (cm/s)	20.9 ± 8.3	19.2–22.7	21.2 (5.1–35.2)	7.7	14.9	25.7	31.1
PG A wave (mmHg)	40.6 ± 15.9	37.3–43.9	41.9 (9.5–69.1)	14.6	28.4	52.8	62.5
Dur A wave (s)	27.8 ± 15.5	24.6–31.1	23.8 (3.7–69.6)	10.3	16.5	40.7	44.8
MV area (cm^2)	46.3 ± 25.3	41.1–51.6	43.8 (9.4–119.6)	17.7	26.4	57.8	87.9
PHT (ms)	49.1 ± 24.2	43.9–54.1	47.1 (9.9–91.4)	18.1	25.7	68.3	84.5
PI	23.1 ± 19.1	19.1–27.1	15.3 (3.2–85.6)	6.7	12.1	28.1	50.2
RI	1.8 ± 1.5	1.4–2.1	1.5 (0.1- 6.6)	0.6	0.61	2.2	5.3

Table 7. *Cont.*

Variables	Mean ± SD	95% CI	Median (Range)	Percentile			
				10%	25%	75%	90%
VTI (cm)	21.2 ± 11.2	18.9–23.6	19.6 (5.7–45.4)	7.2	11.6	25.3	42.5
S/D	32.8 ± 19.5	28.8–36.9	28.6 (2.2–89.8)	14.9	20.9	37.3	62.1

V E wave: Velocity E wave; PG E wave: pressure gradient E wave; Dur E wave: duration E wave; VA wave: Velocity A wave; PG A wave: pressure gradient A wave; Dur A wave: duration A wave; MV area: mitral valve area; PHT: pressure half time; PI: pulsatility index; RI: resistance index; VTI: velocity time integral; S/D: systolic/diastolic.

Table 8. Interassay CVs of PW doppler echocardiographic measurements of mitral valve in normal donkeys.

Variables	Mean ± SD	95% CI	Median (Range)	Percentile			
				10%	25%	75%	90%
V E wave (cm/s)	13.5 ± 7.1	11.9–14.9	13.1 (1.5–30.1)	3.2	7.9	16.8	22.7
PG E wave (mmHg)	26.5 ± 13.5	23.7–29.3	26.7 (3.2–53.9)	6.2	15.8	34.2	46.1
Dur E wave (s)	27.6 ± 17.6	23.9–31.3	24.6 (5.7–67.1)	6.7	12.9	37.7	57.3
V A wave (cm/s)	19.9 ± 9.6	17.9–21.9	20.9 (4.5–44.8)	5.8	13.2	26.4	31.2
PG A wave (mmHg)	38.7 ± 18.5	34.8–42.5	42.3 (9.2–87.9)	11.5	26.2	48.8	60.9
Dur A wave (s)	28.3 ± 15.3	25.1–31.5	21.8 (11.4–65.2)	12.9	18.3	37.6	56.4
MV area (cm^2)	48.1 ± 26.4	42.5–53.6	39.4 (6.9–93.5)	15.1	26.4	68.5	91.02
PHT (ms)	51.4 ± 30.8	44.9–57.8	44.8 (6.9–110.7)	13.9	27.6	79.9	97.8
PI	22.7 ± 17.8	18.9–26.4	21.2 (4.7–92.6)	6.5	11.1	27.4	33.8
RI	2.3 ± 1.8	1.8–2.1	1.7 (0.6–6.9)	0.6	1.1	2.6	6.5
VTI (cm)	21.9 ± 8.5	20.2–23.7	23.8 (6.9–34.9)	11.3	15.3	28.5	31.7
S/D	36.1 ± 20.6	31.7–40.4	34.1 (5.8–91.9)	11.1	23.9	41.1	76.4

V E wave: Velocity E wave; PG E wave: pressure gradient E wave; Dur E wave: duration E wave; VA wave: Velocity A wave; PG A wave: pressure gradient A wave; Dur A wave: duration A wave; MV area: mitral valve area; PHT: pressure half time; PI: pulsatility index; RI: resistance index; VTI: velocity time integral; S/D: systolic/diastolic.

Table 9. Intra-assay CVs of PW doppler echocardiographic measurements of aortic valve in normal donkeys.

Variables	Mean ± SD	95% CI	Median (Range)	Percentile			
				10%	25%	75%	90%
V (cm/s)	11.6 ± 6.9	10.2–13.1	10.4 (1.3–29.7)	2.6	5.6	15.7	20.9
PG (mmHg)	23.04 ± 13.6	20.6–25.5	20.9 (2.7–59.3)	5.3	13.7	30.2	39.9
PI	17.1 ± 11.5	14.7–19.5	14.9 (3.4–45.2)	3.9	7.2	25.2	37.2
RI	1.9 ± 1.3	1.7–2.3	1.6 (0.6–5.4)	0.6	1.1	3.1	3.9
VTI (cm)	19.2 ± 13.1	16.4–21.9	15.1 (1.2–60.4)	4.7	10.2	30.5	34.1
S/D	32.6 ± 21.6	28.1–37.1	31.2 (5.5–81.8)	7.4	15.2	40.3	66.3

V: velocity; PG: pressure gradient; PI: pulsatility index; RI: resistance index; VTI: velocity time integral; S/D: systolic/diastolic.

Table 10. Interassay CVs of PW doppler echocardiographic measurements of aortic valve in normal donkeys.

Variables	Mean ± SD	95% CI	Median (Range)	Percentile			
				10%	25%	75%	90%
V (cm/s)	13.6 ± 6.9	12.2–15.1	13.7 (0.9–31.4)	4.9	8.3	17.7	24.1
PG (mmHg)	26.9 ± 13.3	24.2–29.8	26.7 (2.04–63.9)	9.7	17.4	33.5	46.9
PI	20.6 ± 11.6	18.3–23.1	18.5 (3.6–59.7)	6.1	13.8	28.6	35.3
RI	2.03 ± 1.3	1.7–2.3	1.6 (0.6–6.5)	0.6	1.1	2.5	3.9
VTI (cm)	22.9 ± 13.1	20.2–25.7	18.6 (6.9–47.9)	7.6	13.3	32.6	43.9
S/D	35.6 ± 22.1	31.1–40.2	31.2 (7.6–93.1)	12.1	19.4	51.5	72.2

V: velocity; PG: pressure gradient; PI: pulsatility index; RI: resistance index; VTI: velocity time integral; S/D: systolic/diastolic.

Table 11. Intra-assay CVs of PW doppler echocardiographic measurements of myocardial performance index (LV)—Tei Index in normal donkeys.

Variables	Mean ± SD	95% CI	Median (Range)	Percentile			
				10%	25%	75%	90%
ET (s)	19.5 ± 8.4	17.7–21.2	19.4 (4.8–34.5)	9.3	10.8	23.8	33.6
IRT (s)	44.3 ± 19.5	40.3–48.4	40.4 (24.5–80.9)	24.8	27.8	56.8	80.1
ICT (s)	38.7 ± 11.3	36.3–40.9	37.7 (25.01–63.9)	25.9	29.1	43.8	56.7
MPI	30.2 ± 14.7	33.2–37.6	33.2 (5.92–54.4)	13.6	18.6	40.8	53.9

ET: ejection time; IRT: isovolumic relaxation time; ICT: isovolumic contraction time; MPI: myocardial performance index.

Table 12. Interassay CVs of PW doppler echocardiographic measurements of myocardial performance index (LV)–Tei Index in normal donkeys.

Variables	Mean ± SD	95% CI	Median (Range)	Percentile			
				10%	25%	75%	90%
ET (s)	28.03 ± 12.8	25.4–30.7	28.2 (5.3–49.8)	8.2	21.1	39.4	42.9
IRT (s)	38.7 ± 17.6	35.01–42.4	38.3 (10.9–66.8)	14.9	20.3	53.8	63.4
ICT (s)	37.4 ± 17.3	33.9–42.9	36.7 (3.5–71.1)	14.5	22.1	47.8	65.5
MPI	36.4± 15.6	35.6–46.2	38.7 (9.2–66.01)	12.3	24.5	48.6	57.03

ET: ejection time; IRT: isovolumic relaxation time; ICT: isovolumic contraction time; MPI: myocardial performance index.

4. Discussion

PW Doppler is used in combination with the 2D image to assess flow velocities within discrete regions of the heart and great vessels, which are used to evaluate the cardiac performance [21]. In the current study, E-wave was higher than A-wave for the mitral inflow during filling of the left ventricle in healthy donkeys. The same results were recorded in normal thoroughbred and thoroughbred cross horses [18], normal dogs [34] and normal human subjects [35].

PW Doppler evaluates the mitral velocity, which provides intuition into the dynamics of LV filling and helps to evaluate the diastolic function [36]. Furthermore, the evaluation of transmitral velocity, together with tricuspid and hepatic vein velocities, is useful when evaluating cardiac tamponade and constrictive pericarditis [37].

The ventricular filling results from isovolumic relaxation, ventricular compliance, filling pressures from the left atrium to left ventricle, pericardial restraint, ventricular interaction and atrial function, and it may be influenced by afterload and contractility [7]. Consequently, changes in the ventricular relaxation affect early filling of the left ventricular chamber, while changes in ventricular compliance affect late diastolic filling of the ventricle with a resultant increase of E-wave [38]. Moreover, E-wave velocity is increased with increased left atrial pressure, decreased left ventricular pressure associating the increased rate of ventricular relaxation, and decreased MVA [39]. Meanwhile, the early ventricular filling is decreased in cases of decreased atrial pressure, decreased rate of ventricular relaxation, increased ventricular compliance and increased MVA with a subsequent decrease of transmitral E-wave amplitude, which resulted in increased A-wave velocity as late diastole contributes more to total left ventricular filling [39]. Thus, changes in the velocity occur with alterations in the left atrial and left ventricular diastolic pressures [40].

The pressure gradient of the E-wave is higher than the A-wave of the mitral inflow during filling of the left ventricle in healthy donkeys. In equine, the pressure gradient is a reflection of normal intra-cardiac pressure and pathological increased pressure [41]. The pressure gradient quantifies the severity of stenotic lesions and can differentiate the unknown pressure from the known pressure. The pressure gradient will be increased in the presence of conduct-type lesions as tunnel sub-valvular stenosis and in cases of decreased blood viscosity. In contrast, the increased blood viscosity may underestimate the pressure gradient [42].

In the current study, the velocity of the aortic outflow is 64.9 ± 10.4 during the first third of systole in healthy donkeys. The aortic flow pattern is found in normal thoroughbred and thoroughbred cross horses (0.937 ± 0.094) [18], in normal adult Turkmen horses (101.948 ± 15.341) [43], in clinically normal dogs (106.0 ± 21.0) [34] and in human (92 ± 11) [44]. This is probably due to differences in alignment with aortic flow in donkeys. However, this may represent a species variation in actual flow velocities.

The flow velocity of aortic flow is affected by heart rate. A faster heart rate will increase peak and mean velocity [42]. The velocity of blood flow depends on the blood volume moving through the vessels or orifice. When there is a high blood flow [18], severe valvular insufficiency or stenosis, a coexisting shunt, anemia and/or sepsis [45], the pressure gradient will be inaccurate.

In the current study, the MVA equals 2.4 ± 1.5 cm^2, and the PHT equals 135 ± 93.9 ms in healthy donkeys. However, in humans, the MVA is found to be 4.0–6.0, and the PHT is 40–70 ms [20].

The Doppler study can be used to calculate pressure half-time (PHT), which is defined as the time required for the pressure gradient across an obstruction to decrease to half of its maximal value. Thus, PHT increases as the severity of stenosis increases.

Overestimation of the MVA occurs when PHT is shortened by concomitant significant aortic insufficiency, decreased ventricular compliance and atrial septal defect [21]. PHT is useful in the detection of mitral stenosis with coexistent mitral regurgitation [21].

In patients with aortic regurgitation, the Doppler velocity becomes significantly shorter <250 ms because of the rapid increase in left ventricular diastolic pressure and decrease in aortic pressure. Furthermore, it may be affected by severe diastolic dysfunction with marked elevation of left ventricular diastolic pressure without severe aortic regurgitation [30].

In the current study, PI was 1.4 ± 0.4, RI was 0.9 ± 0.03 for the mitral valve, the PI was 1.4 ± 0.3 and RI was 0.9 ± 0.02 for the aortic valve in healthy donkeys. Normal values for the PI and RI are 1.36–1.56 and 0.6–0.8, respectively.

PI is equal to the difference between the peak systolic velocity and the minimum diastolic velocity divided by the mean velocity during the cardiac cycle. The value of PI decreases with distance from the heart [46]. The two indices, pulsatility index and resistance index, measure the resistance of blood flow. Furthermore, they are affected by input pressure waveform pulsatility, impedance and resistance changes [24].

In the current study, the VTI is 19.1 ± 5.7 cm for mitral inflow and equals 25.02 ± 6.2 for aortic outflow in healthy donkeys. The VTI is found to be 25.369 ± 3.209 cm in normal horses for aortic flow [18], 0.146 ± 0.029 cm in dog for aortic outflow [34] and 25.1 ± 3.4 cm in humans for aortic flow [47]. Left ventricular outflow tract velocity time integral (LVOT VTI) is a measure of cardiac systolic function and cardiac output. Heart failure patients with low cardiac output are known to have poor cardiovascular outcomes. Thus, extremely low LVOT VTI may predict heart failure patients at highest risk for mortality [12].

In the present study, Myocardial Performance Index (LV)–Tei Index was 1.7 ± 0.7, isovolumic contraction time was 0.3 ± 0.1, isovolumic relaxation time was 0.3 ± 0.1, and ejection time was 0.4 ± 0.1. The MPI was found to be 0.52 ± 0.12 in dog [48] and in human (0.39 ± 0.05) [22]. Systolic dysfunction prolongs pre-ejection (ICT) and a shortening of the ejection time (ET). Both systolic and diastolic dysfunction results in abnormality in myocardial relaxation, which prolongs the relaxation period (IRT) [22].

The Myocardial Performance Index (LV)–Tei Index is independent of heart rate, arterial blood pressure, ventricular geometry, atrioventricular valve regurgitation, loading condition as afterload and preload, and can be used to evaluate the function of both the LV and the RV [49]. The Myocardial Performance Index (LV)–Tei Index have strong prognostic value in severe cardiac diseases such as cardiac amyloidosis (0.54 ± 0.16), dilated cardiomyopathy (0.59 ± 0.10), ischemic heart diseases (0.85 ± 0.32), pulmonary hypertension (0.93 ± 0.34), congestive heart failure (0.37 ± 0.05), valvular diseases (0.6 ± 0.2), myocardial infarction, congenital heart diseases (0.35 ± 0.03) and cardiotoxicity (0.45 ± 0.06) [50].

5. Conclusions

Results of the present study provided an initial reference for PW (PW) Doppler echocardiographic variables of the mitral valve, aortic valve and myocardial performance in donkeys, which will be beneficial for clinicians who perform cardiac examinations in these animals. The intra-assay and interassay CVs for the mitral valve, aortic valve measurements and myocardial performance indicated that the technique is a feasible and precise method for determining cardiac measurements and functions in donkeys.

Author Contributions: Conceptualization, M.M., S.E.-k., H.M.I., A.M.F., M.R.E.-A. and A.I.A.; methodology, M.S.; software, Z.A.M. and A.I.A.; validation, M.M., M.R.E.-A. and A.M.F.; formal analysis, S.E.-k.; investigation, M.M.; resources, H.M.I., M.R.E.-A. and A.M.F.; data curation, H.M.I., A.M.F. and M.S.; writing—original draft preparation, M.M.; writing—review and editing, S.E.-k., A.M.F. and M.R.E.-A.; visualization, A.I.A. and M.S.; supervision, M.M.; project administration, M.M.; funding acquisition, M.M. All authors have read and agreed to the published version of the manuscript.

Funding: This work was supported through the Annual Funding track by the Deanship of Scientific Research, Vice Presidency for Graduate Studies and Scientific Research, King Faisal University, Saudi Arabia [Project No. AN000430].

Institutional Review Board Statement: The animal study was reviewed and approved by The Ethics Committee of Mansoura University.

Informed Consent Statement: Not applicable.

Data Availability Statement: All data sets collected and analyzed during the current study are available from the corresponding author on fair request.

Acknowledgments: The authors acknowledge the Deanship of Scientific Research at King Faisal University for the financial support under the Annual track (Grant No. 430).

Conflicts of Interest: The authors declare no conflict of interest.

References

1. Perez-Ecija, A.; Mendoza, F.J. Characterisation of clotting factors, anticoagulant protein activities and viscoelastic analysis in healthy donkeys. *Equine Vet. J.* **2017**, *49*, 734–738. [CrossRef] [PubMed]
2. Mendoza, F.J.; Toribio, R.E.; Perez-Ecija, A. Donkey Internal Medicine—Part II: Cardiovascular, Respiratory, Neurologic, Urinary, Ophthalmic, Dermatology, and Musculoskeletal Disorders. *J. Equine Vet. Sci.* **2018**, *65*, 86–97. [CrossRef]
3. Roberts, S.L.; Dukes-McEwan, J. Echocardiographic reference ranges for sedentary donkeys in the UK. *Vet. Rec.* **2016**, *179*, 332. [CrossRef] [PubMed]
4. Amory, H.; Bertrand, P.; Delvaux, V.; Sandersen, C. Doppler Echocardiographic Reference Values in Healthy Donkeys. In *Veterinary Care of Donkeys*; International Veterinary Information Service: Ithaca, NY, USA, 2004.
5. Quinones, M.A.; Otto, C.M.; Stoddard, M.; Waggoner, A.; Zoghbi, W.A. Recommendations for quantification of Doppler echocardiography: A report from the Doppler Quantification Task Force of the Nomenclature and Standards Committee of the American Society of Echocardiography. *J. Am. Soc. Echocardiogr.* **2002**, *15*, 167–184. [CrossRef]
6. McConachie, E.; Barton, M.; Rapoport, G.; Giguère, S. Doppler and Volumetric Echocardiographic Methods for Cardiac Output Measurement in Standing Adult Horses. *J. Vet. Intern. Med.* **2013**, *27*, 324–330. [CrossRef]
7. Edler, I.; Lindström, K. The history of echocardiography. *Ultrasound Med. Biol.* **2004**, *30*, 1565–1644. [CrossRef]
8. Oh, J.K.; Seward, J.B.; Tajik, A.J. (Eds.) *The Echo Manual*, 3rd ed.; Lippincott Williams & Wilkins: Philadelphia, PA, USA, 2006.
9. Marcus, R.H.; Neumann, A.; Borow, K.M.; Lang, R.M. Transmitral flow velocity in symptomatic severe aortic regurgitation: Utility of Doppler for determination of preclosure of the mitral valve. *Am. Heart J.* **1990**, *120*, 449–451. [CrossRef]
10. Currie, P.J.; Hagler, D.J.; Seward, J.B.; Reeder, G.S.; Fyfe, D.A.; Bove, A.A.; Taji, A.J. Instantaneous pressure gradient: A simultaneous doppler and dual catheter correlative study. *J. Am. Coll. Cardiol.* **1986**, *7*, 800–806. [CrossRef]
11. Goldberg, S.J.; Allen, H.D.; Marx, G.R.; Donnerstein, R.L. Flow computation. In *Doppler Echocardiography*, 2nd ed.; Lea and Febiger: Philadelphia, PA, USA, 1988.
12. Tan, C.; Rubenson, D.; Srivastava, A.; Mohan, R.; Smith, M.R.; Billick, K.; Bardarian, S.; Heywood, J.T. Left ventricular outflow tract velocity time integral outperforms ejection fraction and Doppler-derived cardiac output for predicting outcomes in a select advanced heart failure cohort. *Cardiovasc. Ultrasound* **2017**, *15*, 18. [CrossRef]
13. Roger, V.L.; Go, A.S.; Lloyd-Jones, D.M.; Benjamin, E.J.; Berry, J.D.; Borden, W.B.; Bravata, D.M.; Dai, S.; Ford, E.S.; Writing Group Members; et al. Heart disease and stroke statistics—2012 update: A report from the American Heart Association. *Circulation* **2012**, *125*, e2–e220.

14. Reef, V.B. Pulsed and continuous wave Doppler echocardiographic evaluation of mitral and tricuspid insufficiency in horses. In Proceedings of the 6th American College Veterinary Internal Medicine Forum Madison, Washington, DC, USA, 28–29 May 1988.
15. Marr, C.M.; Love, S.; Pirie, H.M.; Northridge, D.B. Confirmation by Doppler echocardiography of valvular regurgitation in a horse with a ruptured chorda tendinea of the mitral valve. *Vet. Rec.* **1990**, *127*, 376–379. [PubMed]
16. Reimer, J.M.; Reef, V.B.; Sommer, M. Echocardiographic detection of pulmonic valve rupture in a horse with right-sided heart failure. *J. Am. Vet. Med. Assoc.* **1991**, *198*, 880–882. [PubMed]
17. Stadler, P.; Weinberger, T.; Deegen, E. Pulsed Doppler echocardiography measurement in healthy warmblood horses. *Zentralblatt Vet. Reihe A* **1993**, *40*, 757–778. [CrossRef]
18. Blissitt, K.J.; Bonagura, J.D. Pulsed wave Doppler echocardiography in normal horses. *Equine Vet. J.* **2010**, *27* (Suppl. S19), 38–46. [CrossRef]
19. Reef, V.B. Evaluation of ventricular septal defects in horses using two-dimensional and Doppler echocardiography. *Equine Vet. J.* **2010**, *27*, 86–95. [CrossRef] [PubMed]
20. Baumgartner, H.; Hung, J.; Bermejo, J.; Chambers, J.B.; Evangelista, A.; Griffin, B.P.; Iung, B.; Otto, C.M.; Pellikka, P.A.; Quiñones, M. Echocardiographic Assessment of Valve Stenosis: EAE/ASE Recommendations for Clinical Practice. *J. Am. Soc. Echocardiogr.* **2009**, *22*, 1–23. [CrossRef]
21. Harris, P.; Kuppurao, L. Quantitative Doppler echocardiography. *BJA Educ.* **2016**, *16*, 46–52. [CrossRef]
22. Tei, C.; Ling, L.H.; Hodge, D.O.; Bailey, K.R.; Oh, J.K.; Rodeheffer, R.J.; Tajik, A.J.; Seward, J.B. New index of combined systolic and diastolic myocardial performance: A simple and reproducible measure of cardiac function—A study in normals and dilated cardiomyopathy. *J. Cardiol.* **1995**, *26*, 357–366.
23. Shabanian, R.; Mirzaaghayan, M.R.; Dadkhah, M.; Hosseini, M.; Rahimzadeh, M.; Akbari Asbagh, P.; Navabi, M.A. Echocardiographic Assessment of Pulmonary Arteries Pulsatility Index in Fontan Circulation. *J. Cardiovasc. Ultrasound* **2015**, *23*, 228–232. [CrossRef]
24. Saunders, H.M.; Burns, P.N.; Needleman, L.; Liu, J.B.; Boston, R.; Wortman, J.A.; Chan, L. Hemodynamic factors affecting uterine artery Doppler waveform pulsatility in sheep. *J. Ultrasound Med.* **1998**, *17*, 357–368. [CrossRef]
25. Youssef, M.A.; Ibrahim, H.M.; Farag, E.-S.M.; El-Khodery, S.A. Effects of Tilmicosin Phosphate Administration on Echocardiographic Parameters in Healthy Donkeys (*Equus asinus*): An Experimental Study. *J. Equine Vet. Sci.* **2016**, *38*, 24–29. [CrossRef]
26. Hatle, L.; Angelsen, B. *Doppler Ultrasound in Cardiology*, 2nd ed.; Lea and Febiger: Philadelphia, PA, USA, 1985.
27. Long, K.J.; Bonagura, J.D.; Darke, P.G. Standardised imaging technique for guided M-mode and Doppler echocardiography in the horse. *Equine Vet. J.* **1992**, *24*, 226–235. [CrossRef] [PubMed]
28. Lakoumentas, J.A.; Panou, F.K.; Kotseroglou, V.K.; Aggeli, K.I.; Harbis, P.K. The Tei index of myocardial performance: Applications in cardiology. *Hell. J. Cardiol.* **2005**, *46*, 52–58.
29. Weyman, A.E. (Ed.) Cross-sectional scanning: Technical principles and instrumentation. In *Principles and Practice of Echocardiography*; Lea and Fehiger: Philadelphia, PA, USA, 1994; pp. 29–56.
30. Anavekar, N.S.; Oh, J.K. Doppler echocardiography: A contemporary review. *J. Cardiol.* **2009**, *54*, 347–358. [CrossRef]
31. Bude, R.O.; Rubin, J.M. Relationship between the Resistive Index and Vascular Compliance and Resistance. *Radiology* **1999**, *211*, 411–417. [CrossRef]
32. Garcia, S.; Calvo, D.; Spitznagel, M.B.; Sweet, L.; Josephson, R.; Hughes, J.; Gunstad, J. Dairy intake is associated with memory and pulsatility index in heart failure. *Int. J. Neurosci.* **2014**, *125*, 247–252. [CrossRef]
33. Go, A.S.; Mozaffarian, D.; Roger, V.L.; Benjamin, E.J.; Berry, J.D.; Borden, W.B.; Bravata, D.M.; Dai, S.; Ford, E.S.; Fox, C.S.; et al. Heart disease and stroke statistics—2013 update: A report from the American Heart Association. *Circulation* **2013**, *127*, e6–e245. [CrossRef]
34. Brown, D.J.; Knight, D.H.; King, R.R. Use of pulsed-wave Doppler echocardiography to determine aortic and pulmonary velocity and flow variables in clinically normal dogs. *Am. J. Vet. Res.* **1991**, *52*, 543–550.
35. Pye, M.P.; Pringle, S.D.; Cobbe, S.M. Reference values and reproducibility of Doppler echocardiography in the assessment of the tricuspid valve and right ventricular diastolic function in normal subjects. *Am. J. Cardiol.* **1991**, *67*, 269–273. [CrossRef]
36. Cohen, G.I.; Pietrolungo, J.F.; Thomas, J.D.; Klein, A.L. A practical guide to assessment of ventricular diastolic function using doppler echocardiography. *J. Am. Coll. Cardiol.* **1996**, *27*, 1753–1760. [CrossRef]
37. Oh, J.K.; Hatle, L.K.; Seward, J.B.; Danielson, G.K.; Schaff, H.V.; Reeder, G.S.; Tajik, A. Diagnostic role of Doppler echocardiography in constrictive pericarditis. *J. Am. Coll. Cardiol.* **1994**, *23*, 154–162. [CrossRef]
38. Reef, V.B.; Lalezari, K.; De Boo, J.; Van Der Belt, A.J.; Spencer, P.A.; Dik, K.J. Pulsed-wave Doppler evaluation of intracardiac blood flow in 30 clinically normal Standardbred horses. *Am. J. Vet. Res.* **1989**, *50*, 75–83.
39. Long, K.J. Doppler echocardiography in the horse. *Equine Vet. Educ.* **1990**, *2*, 15–17. [CrossRef]
40. Appleton, C.P.; Galloway, J.M.; Gonzalez, M.S.; Gaballa, M.; Basnight, M.A. Estimation of left ventricular filling pressures using two-dimensional and Doppler echocardiography in adult patients with cardiac disease: Additional value of analyzing left atrial size, left atrial ejection fraction and the difference in duration of pulmonary venous and mitral flow velocity at atrial contraction. *J. Am. Coll. Cardiol.* **1993**, *22*, 1972–1982. [CrossRef] [PubMed]
41. Reef, V.B.; Bain, F.T.; Spencer, P.A. Severe mitral regurgitation in horses: Clinical, echocardiographic and pathological findings. *Equine Vet. J.* **1998**, *30*, 18–27. [CrossRef] [PubMed]
42. Darke, P.G.G. Doppler echocardiography. *J. Small Anim. Pract.* **1992**, *33*, 104–112. [CrossRef]

43. Vajhi, A.R.; Mirshahi, A.; MOKHBER, D.M.; Masoudifard, M.; Veshkini, A.; Sorouri, S.; Torki, E.; Azizzadeh, M.; Ghiadi, A.J. Normal pulsed wave Doppler echocardiographic parameters of Turkmen horses of Iran. *Iran. J. Vet. Res.* **2013**, *14*, 42–49. [CrossRef]
44. Gardin, J.M.; Burn, C.S.; Childs, W.J.; Henry, W.L. Evaluation of blood flow velocity in the ascending aorta and main pulmonary artery of normal subjects by Doppler echocardiography. *Am. Heart J.* **1984**, *107*, 310–319. [CrossRef]
45. Paul, J.J.; Tani, L.Y.; Williams, R.V.; Lambert, L.M.; Hawkins, J.A.; Minich, L. Relation of the discrete subaortic stenosis position to mitral valve function. *Am. J. Cardiol.* **2002**, *90*, 1414–1416. [CrossRef]
46. Tortora, G.J.; Derrickson, B. The cardiovascular system: Blood vessels and hemodynamics. In *Principles of Anatomy and Physiology*, 13th ed.; John Wiley & Sons: Hoboken, NJ, USA, 2012; pp. 610–635.
47. Pees, C.; Glagau, E.; Hauser, J.; Michel-Behnke, I. Reference Values of Aortic Flow Velocity Integral in 1193 Healthy Infants, Children, and Adolescents to Quickly Estimate Cardiac Stroke Volume. *Pediatr. Cardiol.* **2013**, *34*, 1194–1200. [CrossRef]
48. Hori, Y.; Kunihiro, S.I.; Hoshi, F.; Higuchi, S.I. Comparison of the myocardial performance index derived by use of pulsed Doppler echocardiography and tissue Doppler imaging in dogs with volume overload. *Am. J. Vet. Res.* **2007**, *68*, 1177–1182. [CrossRef] [PubMed]
49. Vonk, M.C.; Sander, M.H.; Van Den Hoogen, F.H.; Van Riel, P.L.; Verheugt, F.W.; Van Dijk, A.P. Right ventricle Tei-index: A tool to increase the accuracy of non-invasive detection of pulmonary arterial hypertension in connective tissue diseases. *Eur. J. Echocardiogr.* **2007**, *8*, 317–321. [CrossRef] [PubMed]
50. Goroshi, M.; Chand, D. Myocardial Performance Index (Tei Index): A simple tool to identify cardiac dysfunction in patients with diabetes mellitus. *Indian Heart J.* **2016**, *68*, 83–87. [CrossRef] [PubMed]

Review

Inhalative Nanoparticulate CpG Immunotherapy in Severe Equine Asthma: An Innovative Therapeutic Concept and Potential Animal Model for Human Asthma Treatment

John Klier [1], Sebastian Fuchs [2], Gerhard Winter [2] and Heidrun Gehlen [3,*]

[1] Equine Clinic, Centre for Clinical Veterinary Medicine, Ludwig-Maximilians-University, 85764 Oberschleißheim, Germany
[2] Pharmaceutical Technology and Biopharmaceutics, Faculty of Chemistry and Pharmacy, Ludwig-Maximilians-University, 81377 Munich, Germany
[3] Equine Clinic, Surgery and Radiology, Department of Veterinary Medicine, Free University of Berlin, 14163 Berlin, Germany
* Correspondence: heidrun.gehlen@fu-berlin.de; Tel.: +49-30-838-62299; Fax: +49-30-838-4-62529

Simple Summary: Severe equine asthma is the most common globally widespread non-infectious equine respiratory disease (together with its mild and moderate form), which is associated with exposure to hay dust and mold spores, has certain similarities to human asthma, and continues to represent a therapeutic problem. Immunomodulatory DNA sequences (CpG) bound to nanoparticles were successfully administered by inhalation to severe asthmatic horses in several studies. It was possible to demonstrate a significant, sustained, one-to-eight-week improvement in important clinical parameters: partial oxygen pressure in the blood, quantity and viscosity of tracheal mucus secretion in the airways, and the amount of inflammatory cells in the respiratory tracts of severe asthmatic horses. The immunotherapy with CpG is performed independent of specific allergens. At an immunological level, the treatment leads to decreases in allergic and inflammatory parameters. This innovative therapeutic concept thus opens new perspectives in severe equine asthma treatment and possibly also in human asthma treatment.

Citation: Klier, J.; Fuchs, S.; Winter, G.; Gehlen, H. Inhalative Nanoparticulate CpG Immunotherapy in Severe Equine Asthma: An Innovative Therapeutic Concept and Potential Animal Model for Human Asthma Treatment. *Animals* **2022**, *12*, 2087. https://doi.org/10.3390/ani12162087

Academic Editors: Francesco Ferrucci, Chiara Maria Lo Feudo and Luca Stucchi

Received: 16 May 2022
Accepted: 9 August 2022
Published: 16 August 2022

Publisher's Note: MDPI stays neutral with regard to jurisdictional claims in published maps and institutional affiliations.

Abstract: Severe equine asthma is the most common globally widespread non-infectious equine respiratory disease (together with its mild and moderate form), which is associated with exposure to hay dust and mold spores, has certain similarities to human asthma, and continues to represent a therapeutic problem. Immunomodulatory CpG-ODN, bound to gelatin nanoparticles as a drug delivery system, were successfully administered by inhalation to severe equine asthmatic patients in several studies. It was possible to demonstrate a significant, sustained, and allergen-independent one-to-eight-week improvement in key clinical parameters: the arterial partial pressure of oxygen, the quantity and viscosity of tracheal mucus, and neutrophilic inflammatory cells in the respiratory tracts of the severe equine asthmatic subjects. At the immunological level, an upregulation of the regulatory antiallergic and anti-inflammatory cytokine IL-10 as well as a downregulation of the proallergic IL-4 and proinflammatory IFN-γ in the respiratory tracts of the severe equine asthmatic patients were identified in the treatment groups. CD4$^+$ T lymphocytes in the respiratory tracts of the asthmatic horses were demonstrated to downregulate the mRNA expression of Tbet and IL-8. Concentrations of matrix metalloproteinase-2 and -9 and tissue inhibitors of metalloproteinase-2 were significantly decreased directly after the treatment as well as six weeks post-treatment. This innovative therapeutic concept thus opens new perspectives in the treatment of severe equine asthma and possibly also that of human asthma.

Keywords: asthma; immunotherapy; allergy; nanoparticle; CpG

1. Introduction

The global prevalence and morbidity of human asthma has increased significantly in the last four decades [1]. The prevalence increases by 50% every decade [1]. This has been evident in humans [1]. Severe equine asthma has become the most common non-infectious, chronic, obstructive, inflammatory airway disease in industrialized countries in the northern hemisphere, together with its mild and moderate form, formerly known as inflammatory airway disease (IAD). Severe equine asthma has also been known as recurrent airway obstruction (RAO) (until 2016 when "equine asthma" became the officially accepted term), heaves, chronic obstructive bronchitis/bronchiolitis (COB), and chronic obstructive pulmonary disease (COPD) of the horse (the term is now considered to be obsolete due to pathogenetic differences to human COPD, such as exposure to smoke and minimal reversibility) [2–4]. Early descriptions of the clinical presentation of severe equine asthma with the characteristic "heave line" in horses with obstructive pulmonary issues appear in ancient Greek texts by Aristotle from 333 BC [2,5]. In 1656, Markham described heaves in association with housing horses in stables [5]. The first scientific representation in modern times of heaves as an asthma-like syndrome in horses took place in 1964 [2,6].

This overreaction of the airways, which, depending on the reference, has an occurrence in local latitudes between 10 and 20% [7] or maybe even higher [8,9] of adult horses, starting at about 7 years old (reviewed by [2–4]) or maybe even earlier [9], with no breed or sex predisposition [2], and possesses certain similarities to human asthma [2,4,10,11]. There are many different human asthma phenotypes, and not all share similarities with severe equine asthma [11]. The most common human asthma phenotypes are allergic (extrinsic) and non-allergic asthma, late-onset asthma, asthma with fixed airflow limitation, and asthma in obese patients [11]. The first three show a good accordance with severe equine asthma [11]. Severe equine asthma has therefore been proposed as an ideal model for the human severe allergic, non-allergic, and late-onset asthma phenotypes [11,12]. Severe equine and human asthma are both characterized by reversible airway obstruction with bronchoconstriction, increased mucus production, airway hyperresponsiveness, and pulmonary remodeling [12]. The pulmonary remodeling features in severe equine asthma closely resemble those of human asthma and make this naturally occurring animal model unique [12]. According to Bond et al. [11], the *allergic phenotype* is characterized by an allergenic trigger (e.g., molds) associated with clinical signs and pathology (increased neutrophil % in BAL and increased respiratory effort at rest), multiple hypersensitivities in some families of horses (insect bite hypersensitivity, urticaria, and increased parasite resistance), and a good response to inhaled corticosteroids. An association exists between IL-4Rα and severe equine asthma, where IL-4Rα upregulates IL-4 expression during disease exacerbation, which promotes isotype switching from IgM to IgE and results in increased IgE in BAL from horses with severe equine asthma. The *non-allergic phenotype* is characterized by a neutrophilic or paucigranulocytic BAL (in severe cases where BAL return is low), chronic innate immune activation, and the chronic activation of peripheral neutrophils, and it often responds less well to inhaled corticosteroids. The *late-onset phenotype* is characterized by decreased baseline pulmonary function during disease exacerbation, occurs in mature/older animals, and can require higher doses of corticosteroids for control [11]. Hulliger et al. [13] revealed a strong similarity on the transcriptomic level between severe equine asthma and severe neutrophilic asthma in humans, potentially through affecting Th17 cell differentiation. This study also showed that several dysregulated miRNAs and mRNAs are involved in airway remodeling [13].

1.1. Pathophysiology of Severe Equine Asthma

The pathophysiology of severe equine asthma involves recurring, reversible, cholinergic bronchospasms with air trapping, hyperresponsiveness, or hyperreactivity of the airways, hypercrinia and dyscrinia (dysfunctional sol and gel layer of the physiological airway epithelial secretion) with dysfunctional mucociliary clearance, a decrease in the club/Clara cells (physiological secretory cells), goblet cell metaplasia and hyperplasia, and mucosal edema [2]. A significant luminal migration of primarily neutrophilic granulocytes is observed in the airways as well as submucosal and peribronchial lymphocytic infiltration, partially with mast cells, plasma cells, and eosinophilic granulocytes [2]. This can lead to airway remodeling with hyperplasia and hypertrophy of the smooth muscles and subepithelial and interstitial fibrosis [2]. Clinical signs include cough, exercise intolerance, recurrent seromucosal up to mucopurulent nasal discharge, and depending on the exacerbation phase, severe respiratory distress and significant dyspnea at rest (elevated resting respiratory rate, increased abdominal lift, nasal flaring, intercostal breathing, biphasic exhalation, and protrusion of the anus during exhalation) [2,9–11]. On the immunological level, there are conflicting study findings and opinions, so some authors consider the reaction a type I IgE-mediated immediate reaction [14–16], while others deem this unlikely since the bronchospasm following allergen exposure is delayed, IgE cannot be regularly demonstrated [17,18], and mast cells usually play a subordinate role [17]. The delayed neutrophilic inflammatory reaction could possibly be indicative of a type III Arthus immune complex reaction [2,19,20]. A cell-mediated delayed type IV immune reaction is also plausible [19], which these authors consider the most likely explanation. According to Fey [9], not further specified nonspecific immune/defense reactions could also play a role. The existence of different phenotypes and endotypes within severe equine asthma, similar to human asthma, is probable [10,11].

Numerous studies on the role of cytokines indicate an excessive expression of proallergic Th2 cytokines in the lungs of asthmatic horses [20–22]. Others have additionally demonstrated increased Th1 cytokines in asthmatic horses [23,24], which points to the involvement of the proinflammatory track in the pathogenesis of severe equine asthma and contradicts the possibility of a sole Th2 overreaction. Despite this, it is assumed that severe equine asthma involves a dysfunctional Th1/Th2 balance, with a shift to an excessive Th2 response [25]. Other studies have additionally demonstrated the involvement of the proinflammatory and chemotactic IL-17 from Th17 cells in the pathogenic mechanism of severe equine asthma [26,27]. Transcriptomic data derived from bronchial epithelium (in vivo) stimulated with hay dust extract to identify differentially expressed genes and pathways in severe equine asthma indicate that the most upregulated genes are those involved in immune cell trafficking, neutrophil chemotaxis, immune and inflammatory responses, cell cycle regulation, and apoptosis (reviewed by [11]). The most upregulated chemokine was CXCL13, a B-cell chemoattractant predominantly produced by Th17 (reviewed by [11]). CXCL13 has been shown to be upregulated 8-fold in BALF from human asthmatics compared to controls (reviewed by [11]). The treatment of a sensitized murine asthma model with an anti-CXCL13 antibody reduces inflammatory cell recruitment, bronchial-associated lymphoid tissue formation, and airway inflammation [11].

There is a genetic association between severe equine asthma and microsatellite markers with the IL-4 receptor α-chain (IL-4Rα) gene on equine chromosome 13 [11]. The IL-4Rα gene is associated with the development of asthma, skin allergies, and parasite defense in humans [11]. Severe equine asthma is associated with multiple hypersensitivities, including insect bite hypersensitivity, urticaria, and increased parasite resistance in one high-incidence family [11]. Besides a genetic predisposition with higher familial incidence [28–31], the high prevalence of severe equine asthma is primarily attributed to the widespread stabling of horses, with permanent exposure to antigens [2,3]. The disease is chronic and is currently not curable (with the exception of the mild and moderate form of equine asthma, previously termed "inflammatory airway disease" (IAD)) [2]. The consistent avoidance of antigens is the foundation for the management of the disease and is essential for sustained therapeutic success in the sense of achieving clinical remission [2,9]. A large number of potentially proinflammatory particles have been identified in stall dust, such as bacterial endotoxins, over 50 different species of mold, peptidoglycans, proteases, microbial toxins, storage mites, organic plant particles, and inorganic dust as well as harmful gases and ammonia that contribute to the detrimental effect [32,33]. A subtype of severe equine asthma, summer pasture-associated severe equine asthma—severe equine pasture asthma (EPA) or its former term summer pasture-associated obstructive pulmonary disease (SPAOPD)—is clinically identical but occurs during the summer months (sometimes also in spring and autumn), as it is caused by seasonal pollen in pastures [3].

1.2. Diagnosis of Severe Equine Asthma

The diagnosis of severe equine asthma is made based on anamnesis, thorough clinical examination, and additional tests such as bronchoscopy, bronchoalveolar lavage fluid cytology, and additional tests for the evaluation of lung function and arterial blood gas parameters, if available (reviewed by [4,34]). A reliable evaluation and grading of the disease status of severe equine asthma, including coughing, nasal discharge, respiratory rate at rest and during exercise, and the performance and willingness of the horses to work, can be achieved via the standardized HOARSI questionnaire (Horse Owner Assessed Respiratory Signs Index) with four grades [35]. It is important to exclude other clinically similar diseases of the airways with mainly infectious and rarely vascular, neoplastic, toxic, or metabolic backgrounds. In addition, it is necessary to distinguish the severe form from mild to moderate equine asthma, which can be challenging if severe equine asthma is in partial clinical remission [4]. Mild to moderate equine asthma (formerly termed inflammatory airway disease, IAD) can occur at any age (usually in young and middle-aged horses) [4]. Affected horses have no increased respiratory effort at rest; clinical signs persist for more than 3 weeks, including poor performance and occasional coughing, and often improve spontaneously or with treatment; and recurrence is rare [4]. The diagnosis is confirmed via endoscopy, with excess mucus evident in the tracheobronchial tree (score $\geq$ 2 with larger but non-confluent blobs for racehorses and $\geq$3 with confluent or stream-forming mucus for sport/pleasure horses). Regardless of the technique used concerning the instilled volume of fluid, site of sampling, selection of aliquot, and/or sample preparation, BALF cytology values of >10% neutrophils, >5% mast cells, and >5% eosinophils are consistent with mild to moderate equine asthma (references for healthy controls are: neutrophils $\leq$ 5%, eosinophils $\leq$ 1%, and metachromatic cells $\leq$ 2% with 250 mL instilled volume) [4]. Neutrophils above 25% and a mucus score of >2/5 within the trachea are consistent with severe equine asthma [4]. In mild to moderate equine asthma, there is no evidence of airflow limitation based on the esophageal balloon catheter technique (DPmax < 10 cm H_2O), but with more sensitive methods it is possible to detect airflow limitation and even airway hyperresponsiveness [4]. A moldy hay provocation test can be used in a research setting to discriminate between mild/moderate and severe equine asthma (in remission) based on the development of respiratory effort at rest, but it is not recommended for diagnosis in clinical practice [4]. The analysis of metabolomics in the exhaled breath condensate, such as methanol and ethanol, could offer new diagnostic perspectives in the future for severe equine asthma comparable to those in human medicine (reviewed by [34]).

1.3. Treatment Options for Severe Equine Asthma

The cornerstone of the treatment of severe equine asthma is the imperative improvement in stabling conditions with steamed or soaked hay (avoiding dry hay), packaged shavings as bedding (avoiding straw), or permanent outdoor living and the avoidance of potentially triggering antigens, which is, in many cases, very difficult. Additional medical treatment concentrates on improving the clinical symptoms of airway inflammation, bronchoconstriction, and mucus accumulation. The route of administration can be systemic and/or inhalative. Inhalative corticosteroids (ciclesonide, fluticasone, budesonide, and beclomethasone) or systemic corticosteroids (prednisolone and dexamethasone) can be administered to improve clinical signs. In a large, prospective, multicenter, placebo-controlled, double-blinded, clinical trial with 224 horses with severe equine asthma, using a soft mist inhaler and a ten-day (5 d twice daily and 5 d once daily) inhalation treatment, Pirie et al. [36] demonstrated that ciclesonide is efficacious in the treatment of severe equine asthma, with at least a 30% or greater reduction in the weighted clinical score of 73% of the horses in the ciclesonide group and 43% of the horses in the placebo group. The reduction in the mean weighted clinical score (severe: 15–23 points; moderate: 11–15 points; mild: 5–10 points) after 10 days of treatment was 7.2 ± 4.8 in the ciclesonide-treated group, compared to 3.8 ± 4.4 in the placebo group. The reduction in the weighted clinical score (82% sensitivity and 70% specificity [37]) after ciclesonide administration was greater in horses with severe clinical signs compared with horses with moderate clinical signs [36]. Owners recognized an improved quality of life after 10 days of treatment in 69% of ciclesonide-treated horses, compared to 43% of placebo-treated horses [36]. Few systemic and local adverse events of ciclesonide were observed [36]. No lung function testing or lower airway cytology of BALF or mucus scoring via endoscopy were performed in this study [36]. Lavoie et al. [37] showed that dexamethasone per os (0.066 mg/kg once daily for 14 d) and inhalative ciclesonide (twice daily for 14 d) significantly improved lung function, and dexamethasone was superior to ciclesonide. The effect was lost 7 days post-treatment for both [37]. Usually, clinical signs reappear quickly after treatment cessation if the environment is not improved concurrently [38]. Beclomethasone [39–41] and fluticasone [42–45] have been shown to be efficacious in controlling airway obstruction in severe equine asthma, although the magnitude of the response was higher with systemically administered dexamethasone than with inhaled beclomethasone [46]. While improving clinical signs and lung function, corticosteroids have generally been found to be only mildly to not effective in controlling the neutrophilic inflammation in equine asthma (reviewed by [38,40,47,48]), unless combined with antigen avoidance [24,45]. The poor response of neutrophilic inflammation to corticosteroids is not limited to horses, as similar findings have also been reported in human patients with neutrophilic asthma [49]. The improvement in lung function via ciclesonide was lost one week after the discontinuation of the therapy, but the weighted clinical score improvement remained significant up to one week post-treatment [36].

Bronchoconstriction can be improved medically via systemic bronchodilators (clenbuterol and butylscopolamine) or inhalative bronchodilators (salbutamol, salmeterol, and ipratropium bromide) [2–4,50]. Finally, expectorants (dembrexine and acetylcysteine) or inhaled saline (isotonic or hypertonic) can be administered to improve the liquefaction and transport of mucus, and mild exercise can also be beneficial. Expectorants are certainly less important than the aforementioned strategies in controlling inflammation and bronchoconstriction. Allergen-specific immunotherapy could potentially become an interesting and promising therapy approach in the future for some cases of allergic severe equine asthma. However, to the best of the authors' knowledge, the current scientific literature scarcely includes any reports of allergen-specific immunotherapy as a successful treatment option for severe equine asthma, likely due to the multifactorial nature, the different phenotypes of the disease, and the diagnostic limitations of the currently commercially available allergy tests. The question arises whether it is possible to affect the pathophysiology on the immunological level and prevent or reduce the occurrence of a hypersensitivity reaction.

1.4. Th1/Th2 Balance

The Th1/Th2 balance in the body is vital to the homeostasis of the immune response [51], and a shift towards an excessive Th2 response can result in allergic disease [51,52]. Physiologically, Th2 cells play an essential role in fighting extracellular parasites such as intestinal helminths [53]. This path is oriented towards the mucosa of the gastrointestinal and respiratory tracts, where contact with the external world occurs [53]. These surfaces are predisposed to the parasitic invasions against which IgE defends [53]. Antigen-presenting cells (APCs) at these barriers to the external world are programmed to the Th2 response [53]. Natural infections by bacteria and viruses result in a cell-mediated proinflammatory Th1 immune response by the immune system [51]. In contrast, an allergy-favoring humoral Th2 response is dominant in the neonatal immune system [51]. It is suspected that the decline of infectious diseases during the immune system's early phase of development could be a cause of the increased incidence of allergies [54–56]. Contact with microbial pathogens during this developmental phase appears to have a significant effect on the imprinting of the immune system [54,56]. Certain studies have shown that children who grow up on farms and have regular contact with animals develop fewer allergies than children in cities [54,56]. The interaction between the environment, the microbiome, and the immune system is probably far more complicated. The number of recent studies analyzing the microbiome of the gut and lung of healthy and severe asthma-affected horses is increasing, offering new insights and perspectives in the treatment and understanding of the pathophysiology of severe equine asthma [57,58]. The immunological interaction between the gut and lungs, termed the gut–lung axis, has been intensely studied in humans and provides new insights into how metabolites produced in the gut influence the immune system in the lungs (reviewed by [34]).

1.5. Toll-Like Receptors

The intracellular recognition of the CpG motif occurs via toll-like receptor 9 (TLR9), one of the most important pathogen recognition receptors of the innate immune system [59] and an evolutionarily highly conserved type I transmembrane protein [60]. To date, ten different TLR classes are known in humans, and thirteen are known in vertebrates [61]. Due to individual ligand specificity, each receptor class only recognizes certain binding partners. TLRs can differentiate between "endogenous" and "foreign" and only react to a pathogen-associated molecular pattern (PAMP) during an infection [62,63]. The majority of TLRs (TLR1, TLR2, TLR4, TLR5, TLR6, and TLR11) are located on the cell surface [64], while TLR3, TLR7, TLR8, and TLR9 are localized intracellularly within endosomes [62]. The ligand binding here occurs in the acidic environment of the endosomes, which is a prerequisite for the cellular activity, the dimerization of the TLRs, and their stabilization [63]. TLR9 is specialized to single-stranded CpG DNA [63]. In equine lungs, Schneberger et al. [65] could identify TLR9 in intravascular macrophages, alveolar macrophages, bronchial epithelial cells, capillary endothelial cells of the lungs, type II epithelial cells of the alveolar septa, and neutrophilic granulocytes. Via lipopolysaccharide (LPS) treatment, the level of TLR9 expression as well as the number of TLR9-positive cells could be significantly increased [65]. Cytosine methylation, as it most commonly exists in the mammal genome, or an inversion of the cytosine–guanine dinucleotides (GC) inhibit TLR9 activation [59]. The differentiation between endogenous and foreign DNA is dependent both on the low CpG content and the high rate of cytosine methylation [66] as well as on the protection of the endosomal localization of TLR9 against constant activation by endogenous DNA [67,68].

The nucleotide sugar backbone is of particular importance for TLR9 activation [63]. The synthetic ODN possesses a modified sugar backbone (phosphorothioate), in contrast to the naturally occurring ODN (phosphodiester) [63], and therefore possesses a receptor affinity 100 times stronger than the natural phosphodiester [63].

1.6. CpG-ODN

Unmethylated cytosine–phosphate–guanine oligodeoxynucleotides (CpG-ODN) occur primarily in prokaryotic DNA and stimulate the eukaryotic immune system [51,64,69,70]. In mammalian DNA these motifs are quite rare (suppressed) and they are usually methylated (shut down) [70]. These CpG-ODN motifs are able to downregulate excessive allergic immune reactions [51,64,69–71]. The use of synthetic CpG-ODN mimics the effect of a bacterial or viral infection and makes use of the body's own immune system. Thus, CpG-ODN represents a large therapeutic potential for allergic diseases [64]. Gelatin nanoparticles (GNP), as a molecular transport system, protect CpG-ODN from premature degradation by ubiquitous nucleases and simultaneously improve the cellular absorption of the DNA molecules in the target cells of the immune system [72]. First, a CpG motif effective for use in equine bronchoalveolar cells (BAL cells) was identified in an in vitro study on the basis of existing sequence-dependent species specificity [73,74]. Second, the extent to which a specific immunomodulatory effect of the administered CpG-ODN can be demonstrated in equine BAL cells was determined [73,74]. The ability of GNP to serve as an effective molecular transport system for CpG-ODN in equine BAL cells was confirmed [74]. The previously identified CpG-ODN sequence was administered in subsequent in vivo studies to horses with severe equine asthma and was thereby examined for its local and systemic tolerability and its therapeutic effect on clinical and immunological parameters [75–79].

1.7. Immunology of CpG-ODN

Natural bacterial and viral infections train the immune system toward a cell-mediated proinflammatory Th1 immune response [51]. The contact with microbial pathogens, or synthetic CpG-ODN, leads to the differentiation of naive CD4$^+$ T-helper cells into a specialized subclass of Th1 lymphocytes (Figure 1) [70]. These, in turn, release the proinflammatory cytokines IL-12 and IL-18, which mediate an increase in IFN-γ [70]. This central Th1 cytokine has an inhibitory effect on allergy-mediating Th2 cytokines such as IL-4, IL-5, and IL-13 (Figure 1) [64]. This differentiation in favor of Th1 cells (Th1 shift) is largely dependent on the synthesis of IL-12 [70]. Of particular importance in this immunologic event is the activation of regulatory T cells (Treg) by CpG-ODN [80,81]. This leads to the production of IL-10, which causes a peripheral T-cell tolerance, possesses anti-inflammatory as well as antiallergic effects [82], and inhibits Th1 cytokines (IFN-γ) as well as Th2 cytokines (IL-4) [83]. It is therefore of particular interest in the study of excessive allergic and inflammatory diseases such as severe equine asthma. The CpG sequences activate the transcription of the cells within 15 min [84], inducing the production of T-helper 1 (Th1) cytokines by antigen-presenting cells (APC) (Figure 1) [84]. In addition, this causes an increased expression of the major histocompatibility complex (MHC I and II) and costimulatory molecules by APC [69]. T cells can only recognize antigens in their processed form in the presence of endogenous molecules (MHC restriction). If, upon contact with an antigen, CD4$^+$ T-helper cells are subjected to a cytokine milieu of IL-4 by APC, the Th2 lineage is stimulated, which as a result, leads to IgE production by the activated B lymphocytes [85]. Furthermore, B-cell activation, independent of T cells, occurs through toll-like receptor 9 (TLR9) [64]. The B lymphocytes, as part of the APC, differentiate themselves in short-lived plasma and memory cells. The memory cells, upon new contact with antigens, increase the production of antibodies by eight to ten times [53].

1.8. Molecular Signal Transduction of the CpG-ODN

CpG-ODN are taken up via endocytosis and transported to the endosomal compartments of the cells through acidic vesicles (Figure 2) [64,71]. The intracellular signal transduction of the CpG-ODN occurs through a dimerization of the receptor molecules [71] with an allosteric conformational change in the cytoplasmic domain that leads to the recruitment of signal adaptor molecules (MyD88) and signal transduction molecules such as the IL-1 receptor-associated kinase (IRAK) and the mitogen-activated protein kinase (MAPK) as well as IFN regulatory factors (Figure 2) [71]. This signal cascade leads to the activation of the transcription factor nuclear factor-κB (NF-κB), with subsequent cytokine production (Figure 2) [71]. One of the central transcription factors of the CpG effect is Tbet (Th1-specific T-box transcription factor) [71]. This inhibits the Th2-associated antibody isotypes and increases the Th1-associated antibody isotypes [71]. The regulatory cytokine IL-10, in a negative feedback loop through the upregulation of CpG activity, leads to a downregulation of the CpG effects [71].

1.9. Nanoparticulate Transport Systems of CpG-ODN

CpG-ODN can be successfully transferred to target cells with the aid of various delivery systems [72,86]. This improves the cellular uptake of the CpG-ODN, while steric shielding protects the CpG-ODN molecules from premature degradation by endogenous nucleases and results in a sustained enhancement of the immunostimulatory effect of the CpG-ODN [72]. However, free CpG-ODN display problems in their cellular uptake, stability, and specificity for target cells [72]. When packaged in nanoparticulate delivery systems, this leads to a local lymphatic uptake without systemic circulation [72,87,88] and increases the local rate of phagocytosis, through which CpG-ODN reaches TLR9 in the endosomes directly [87]. The CpG-ODN A class consists of a central palindromic sequence and poly guanine tails on both ends as well as a mixed phosphorothioate and phosphodiester sugar backbone, contrasting with the CpG-ODN B class's pure phosphorothioate backbone and the C class's mixture of both classes with an effect on IFN-α, IL-6, and IgM production [89]. The CpG-ODN A class is quickly degraded in vivo by ubiquitous DNases but can be protected against degradation by steric shielding on gelatin nanoparticles or by packaging it in virus-like particles (VLP) [86]. Unpackaged CpG-ODN, on the other hand, spread throughout the organism, and under certain circumstances can lead to shock reactions or splenomegaly [90,91]. Lipid nanoparticles have been proven to be an effective delivery system for CpG-ODN [86]. The packaging of CpG-ODN in nanoparticles leads to enhanced effectiveness, a longer life span, a passive accumulation at the site of the disease incidence, and a reservoir effect of the CpG-ODN with delayed release [86]. Colloid particle formulations move in the same proportions as microorganisms and are therefore better phagocytosed through corresponding defense cells [72].

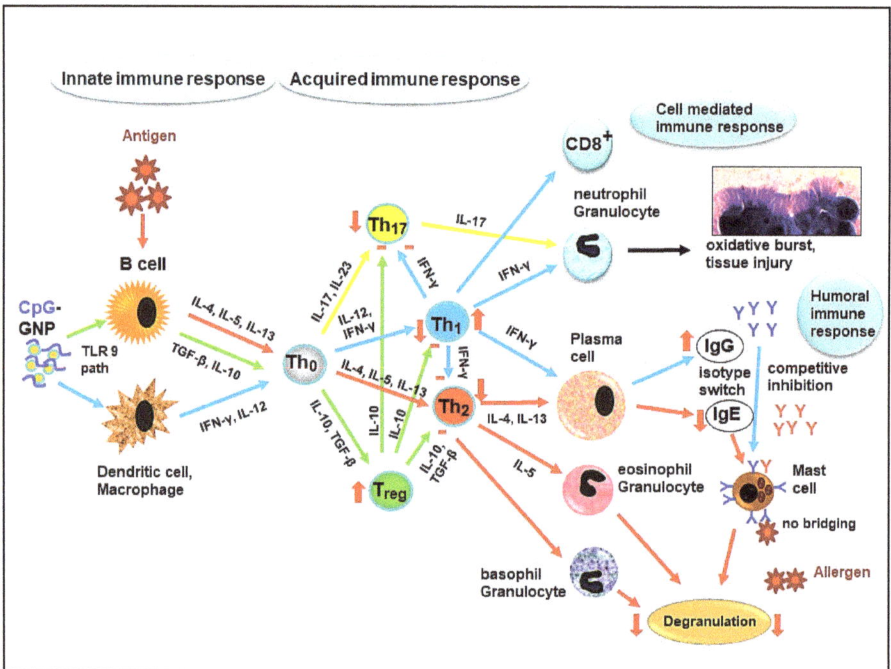

Figure 1. Schematic drawing of the CpG-induced immunological mechanism. The influence of CpG-GNP on the innate and acquired immune system is depicted with particular consideration of the T-helper-cell subsets. After recognition by dendritic cells and macrophages or B lymphocytes upon antigen contact, naive Th0 cells differentiate into Th1, Th2, Th17, or Treg in the appropriate cytokine patterns (Th1 lineage: IL-12 and IFN-γ; Th2 lineage: IL-4, IL-5, and IL-13; Treg lineage: IL-10 and TGF-β; Th17 lineage: IL-17 and IL-23). Depending on the path taken, stimulation (arrow up) or inhibition (arrow down or minus sign) results. This influences immunoglobulin class switching (to IgG or IgE) by plasma cells, the competitive inhibition of IgE by IgG on mast cells, the prevention of cross-linking upon recurrent allergen contact, and no degranulation of the effector cells (mast cells, eosinophils, and basophils). The regulating influence on the oxidative burst and tissue injury in the respiratory tract is caused by migrating neutrophils through the activation of Treg and the regulation of excessive Th1 and Th17 immune responses. (Figure modified according to [73]).

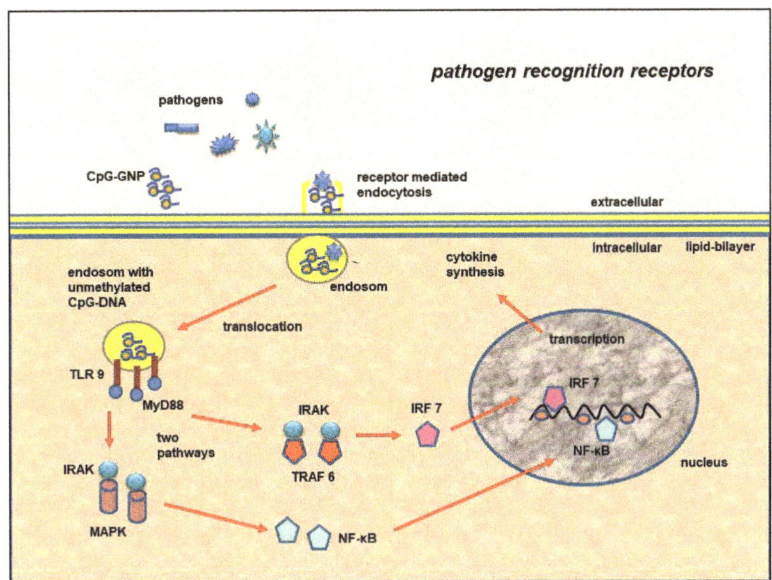

Figure 2. Schematic drawing of TLR9-mediated signal transduction and the intracellular uptake of the pathogens or CpG-GNP over receptor-mediated endocytosis and translocation in endosomes. Recognition by pathogen recognition receptor TLR9 (toll-like receptor 9) triggers a signal cascade over MyD88 (myeloid differentiation primary response gene 88), IRAK (IL-1-receptor-associate kinase), TRAF 6 (TNF receptor-associated factor 6), and IRF 7 (interferon-regulatory factor 7) or IRAK and MAPK (mitogen-activated protein kinase), with the resulting activation of transcription factor NF-κB (nuclear factor-κB). The induction of the transcription of genes encoding specific cytokines leads to the subsequent protein biosynthesis of the cytokines. (Figure modified according to [73]).

1.10. Gelatin Nanoparticles

Zwiorek et al. [72] established a new delivery system with gelatin nanoparticles (GNP) out of type A porcine gelatin (175 Bloom) combined with cholamine with a pH-independent cationic surface via a quaternary amino group. The loading of the cationic GNP with negatively charged CpG-ODN occurs through electrostatic attraction. The optimal load of GNP is a ratio of 1:20 (5%). At higher loads, the GNP may lose their stability and display a tendency to aggregate, most likely due to their neutral surfaces [72]. The gelatin nanoparticles themselves are immunologically inert, which is a condition for repeated application in the organism and conjugation with immunomodulatory substances [72]. GNP polymers display excellent biodegradability and very good biological compatibility [72,88,92]. The functional groups on the nanoparticles can be conjugated with ligands, making them suitable as a delivery system for pharmacological substances [72].

The aerodynamic stability of GNP after nebulization has been proven [92,93]. Fuchs et al. [94] extensively investigated various nebulizer systems, the appropriate measurements of the particle sizes after nebulization, and other studies on the alveolar mobility of the nebulized particles and the in vitro cytokine expression of nebulized particles in cell cultures of equine BAL cells. With an active vibrating mesh nebulizer (AeroNeb Go micropump nebulizer, Aerogen, Galway, Ireland) the second-stage particle deposition in an impinger was up to $65.68 \pm 11.2\%$ of the nebulized dose [94]. The higher the viscosity of the dispersion and the lower the surface tension of the particle-comprising droplets to be nebulized, the smaller the nebulized particles become [40,94]. This is of great significance for alveolar mobility (1–5 µm particle size window). The ideal load has been determined to be 30 µg CpG-ODN per 600 µg GNP (5%) [72]. Phagocytosis has been identified as the main uptake mechanism in the target cells. Zwiorek et al. [72] could demonstrate in vitro that positively charged particles, in contrast to neutral and negatively charged formulations, are better phagocytosed through DC and macrophages. GNP-bound CpG-ODN stimulate a cytokine response that is two to three times greater than the response to unbound CpG-ODN [72]. CpG-GNP leads to an upregulation of the expression of MCH II and CD86 surface molecules [72]. CpG-GNP increases the immunogenicity of additionally transported antigens and enhances the activation and imprinting of T cells by DC [72]. This is significant for a combined application with potential allergens in specific immunotherapy (SIT). The phagocytosis of larger particles (>200 nm) results in a longer retention in endosomal vesicles and positively influences the induction of cytokine synthesis [72]. Unloaded GNP displayed no induction of cytokine expression, meaning that they were immunologically inert. Particles with greater diameters (300 nm) were superior to the smaller particles (150 nm) in reference to cytokine induction (IL-12 production was three times greater) [72]. At no point in vitro or in vivo were any toxic effects observed. None of the mice treated with CpG-ODN/GNP displayed detectable levels of antibodies against GNP after three weeks. Zwiorek et al. [72] therefore concluded that no direct immune response to the protein matrix of the delivery system occurs.

1.11. CpG-ODN as an Adjuvant

Many commercial vaccines have used aluminum hydroxide (alum) as an adjuvant for decades. It, however, blocks the activation of $CD8^+$ cytotoxic T lymphocytes [84]. Most adjuvants only enhance the Th2 response without stimulating the cellular immune response, with the exception of the *Bacillus Calmette-Guérin* (BCG) adjuvant, which also activates the Th1 lineage [84]. CpG-ODN activate both the humoral as well as the cell-mediated immune responses [95]. In contrast to other adjuvants in vaccines, such as Freund's adjuvant, CpG-ODN display no sterile abscess formation, regardless of the administration route [84]. Ziegler and colleagues [96] compared two TLR agonists, monophosphoryl lipid A (MPLA) and a C-class CpG-ODN, in vitro with equine PBMCs from healthy and insect bite hypersensitivity (IBH)-affected horses. MPLA induced IL-10 secretion in all horses, with and without *Culicoides* allergens, while suppressing the antigen-induced production of IFN-γ, IL-4, and IL-17. CpG-ODN significantly increased IFN-α, IFN-γ, and IL-4 production. MPLA was seen as a promising adjuvant candidate for allergen-specific immunotherapy (ASIT) in horses, while C-class CpG-ODN was considered to be an unsuitable adjuvant for ASIT because of the induction of IFN-γ and IFN-α and thus may be a useful adjuvant in combination with vaccines for equine infectious or neoplastic diseases [96]. Ziegler et al. [97] showed in vitro that the addition of an adjuvant (MPLA or CpG-ODN) to equine PBMCs obtained from healthy and IBH-affected horses further enhanced the effect of dendritic cell-binding peptides by significantly increasing the production of IFN-γ, IL-4, IL-10, and IFN-α (CpG-ODN) and IL-10 (MPLA) while simultaneously suppressing IFN-γ, IL-4, and IL-17 production (MPLA). The combination with MPLA seems to be a promising option for improving ASIT efficacy in horses, while the combination with CpG-ODN increases the effector immune response to recombinant antigens [97].

Nearly all vaccine studies on CpG-ODN to date have used B classes as adjuvants due to their specific stimulatory effect on B lymphocytes [71]. The synergistic effects between TLR9 and B-cell receptors result in an activation of the humoral immune response, which leads to the stimulation of antigen-specific B cells, inhibits B-cell apoptosis, and increases the survival rate and IgG class switching [71]. No other vaccine adjuvant effects a Th1 immune response as strong as that of CpG-ODN [71]. The mucosal administration activates local and systemic humoral and cellular immune responses and thus imparts superior protection against infection [71]. Vaccine studies have demonstrated that the low protective antibody concentrations after vaccination in HIV infections can be significantly increased through the combination with CpG-B (CpG 7909) [71]. Some RNA viruses display CpG suppression in their genomes and thus evade recognition by the immune system [98]. This is the case with HIV, which displays clear CpG suppression [98]. The introduction of CpG motifs in the HIV genome demonstrated a clear inhibition of the gene expression of the human immunodeficiency virus [98].

In the vaccination against hepatitis B, the seroprotective antibody level with the combined CpG-ODN B administration increased dramatically (100%) and more quickly and was also sustained for a longer period of time (over 3.5 years) (reviewed by [71]). In an anthrax vaccine study, CpG-ODN reached the toxin neutralization level at half the number of days (22 days) in comparison to the control group (including an antibody titer with an $8\times$ increase) (reviewed by [71]). In mice, vaccine doses could be decreased by up to 99% in combination with a CpG-ODN adjuvant [71]. A reduction in the vaccine dose is especially relevant to influenza vaccines in order to enable adequate production of the vaccines [71]. A tenth of the normal influenza vaccine dose, in combination with CpG-ODN, is sufficient to obtain an equal antigen-specific IFN-γ level (reviewed by [71]). The prophylactic administration of CpG-ODN in mice led to transient protection against a wide range of viral, bacterial, and parasitic pathogens such as *Bacillus anthracis*, *Listeria monocytogenes*, the Ebola virus, and the vaccinia virus (reviewed by [71]). Depending on the route of administration (oral, inhalative, or by injection), this protection upon one-time administration of CpG-ODN ranges from one day to two weeks [71].

The direct combination of CpG-ODN with certain allergens leads to an antigen-specific Th1 immune response and simultaneously to the suppression of Th2-mediated allergic asthma [89]. The combination with allergens as a specific immunotherapy is used to combat the major human allergen *Ambrosia artemisiifolia* [61]. A conjugation of the CpG-ODN to specific antigens enhances the uptake of antigens and reduces the necessary antigen quantity [71].

In the horse, CpG-ODN has been administered as a vaccine adjuvant for equine influenza and *Rhodococcus* vaccines [99,100]. Furthermore, CpG-ODN has been successfully administered in vivo in a pilot study on canine atopic dermatitis in combination with liposomes and specific allergens, resulting in a significant improvement in the pruritus [101] and in a study on canine atopic dermatitis without the addition of allergens with gelatin nanoparticles as the drug delivery system [102]. In both studies, the CpG-GNP treatment resulted in a significant decrease in IL-4 mRNA expression in the blood [101,102]. The clinical improvement was determined to be similar to that of specific immunotherapy [102]. An in vivo study on allergic rhinitis and asthma (phase I and IIa) in humans demonstrated the tolerability and clinical effectiveness of CpG-ODN administered in combination with allergens [89]. CpG-ODN have also been used in human tumor therapy. Several clinical studies reviewed by Adamus and Kortylewski [103] with TLR9 agonists as a monotherapy have demonstrated the very good tolerability and security of CpG-ODN.

1.12. Inhalative Nanoparticulate CpG Immunotherapy of Horses with Severe Equine Asthma

Because no other comparable studies on this specific subject have been performed by other groups, the authors must note here that independent replication was not possible, and the risk of author bias in the available data cannot be excluded.

1.12.1. In Vitro Trials

The goal of an in vitro study on inhalative nanoparticulate CpG immunotherapy was to identify the optimal stimulatory CpG sequence for horses in consideration of the species specificity of the CpG motifs in equine bronchoalveolar lavage (BAL) cells with regard to an immunomodulatory effect (Th2/Th1 shift) [74]. Gelatin nanoparticles (GNP) were used as the drug delivery system. BAL cells were obtained from horses with severe equine asthma and healthy horses and subsequently incubated with five different CpG-ODN sequences (classes A, B, and C) and an ODN sequence without a CpG motif. The cytokine release of IL-4, IL-10, and IFN-γ was then determined via quantitative equine capture ELISA (R&D systems, Minneapolis, MN, USA) in order to detect an allergy-mediated Th2 immune response (IL-4) and/or a proinflammatory Th1 response (IFN-γ). Due to its specific anti-inflammatory and antiallergic effects, IL-10 was considered a positive regulatory cytokine in the pathophysiology of severe equine asthma. The results in the asthmatic horses revealed both a significant upregulation of IL-10 and IFN-γ as well as a downregulation of IL-4 [74]. The cell cultures of the healthy horses had a significantly greater cytokine release in response to the administered stimuli in contrast to the cell cultures of the horses with severe equine asthma. In the comparison of all five CpG sequences, the A class 2216 displayed the strongest immunomodulatory effect on equine BALF cells [74], and for that reason, it was selected for the follow-up clinical studies.

1.12.2. In Vivo Trials

In the following in vivo trial, the previously identified sequence was administered via inhalation to healthy and asthmatic horses [73,75]. Without altering external environmental factors, a significant improvement in the examined parameters (neutrophil percentage in tracheobronchial secretion (TBS), arterial oxygen pressure, tracheal mucus, and respiratory rate) was evident after three and five inhalations. However, the percentage of neutrophils within the trachea does not correlate well in all cases with the percentage of neutrophils within the lungs gained via BAL. Consequently, the reduction in neutrophils within the trachea does not necessarily allow a direct conclusion regarding the neutrophils within the lungs. The inhalations were administered every second day (once daily for ca. 10 min) (Figure 3).

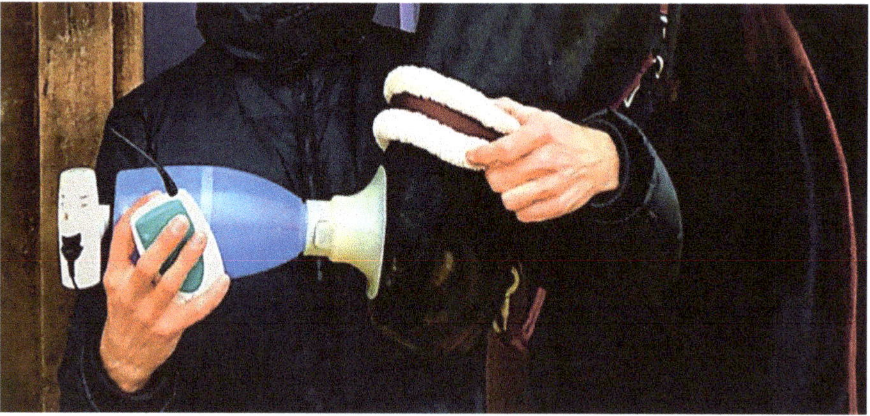

Figure 3. Inhalation device. Combination of "AeroNeb Go micropump nebulizer" (Aerogen, Galway, Ireland) and "equine haler" (Equine HealthCare Aps, Hoersholm, Denmark) for in vivo inhalation studies on horses.

In addition, the BAL cells obtained after inhalation displayed a significantly higher stimulation potential (increased IL-10 release and decreased IL-4 release in vitro in contrast to before the inhalation) [73,75]. Consequently, it was concluded that the inhalation treatment leads to an alteration of the cell population in vivo and results in the upregulation of Treg cells (increased IL-10 production). The upregulation of IL-10 and IFN-γ in the BAL could also be determined in vivo after the inhalation treatment [75].

The clinical and immunological parameters of the GNP-bound CpG-ODN formulations were examined further in the subsequent double-blind, placebo-controlled, and prospective randomized clinical phase I and IIa field study [76]. In the study, 24 horses with severe equine asthma received inhalative treatment (verum group n = 16; placebo group n = 8) five times with an interval of two days and were examined before and immediately after treatment cessation as well as four weeks after the treatment's conclusion. The CpG-GNP treatment achieved a 4-week persistent and significant improvement in 70% of the examined parameters, including breathing type, auscultation, alveolar–arterial oxygen gradient (AaDO$_2$) (calculated according to the current atmospheric pressure and the measured blood gas values PaO$_2$ and PaCO$_2$ (AaDO$_2$ = (atmospheric pressure $-$ 47 mmHg)

$\times$ 0.2095 $-$ PaCO$_2$ $-$ PaO$_2$)) [78], the neutrophil percentage in the tracheobronchial secretion, and the amount and viscosity of tracheal mucus as well as the nasal discharge of the horses in their accustomed environmental conditions with sustained exposure to allergens [76]. The positive results of this exploratory field study revealed new possibilities beyond the conventional symptomatic treatments and could therefore also serve as a potential therapy model for human asthma.

A further study investigated whether two additional specific allergens in the sense of an allergen-specific immunotherapy (ASIT) (according to the results of a functional in vitro test on each horse) could enhance the immunomodulatory capacity of the CpG-GNP formulation in horses with severe equine asthma both in regards to a strengthened immunological response as well as a longer sustained effect in contrast to monotherapy with CpG-ODN [78]. Furthermore, the study investigated whether a longer inhalation therapy (seven inhalations in comparison to five) could achieve a superior therapy effect that was sustained longer or was stronger in comparison to the earlier study [76]. Twenty horses with severe equine asthma were divided into two groups. Based on a functional in vitro test, eleven horses were administered two specific allergens (Artu Biologicals, Europe B.V., Lelystad, the Netherlands) in increasing concentrations (beginning with 0.6 mL and increasing to 1.2 mL) every second day for a total of seven administrations in addition to CpG-GNP. The treatment with solely CpG-GNP (n = 9 severe equine asthma horses), as well as in combination with relevant allergens (n = 11), resulted in no significant improvement in the parameters of respiratory rate, breathing type, nasal discharge, and the quantity and viscosity of tracheal mucus six weeks after the inhalation treatment [78]. There were no significant differences between the two treatment groups in either the clinical parameters or the cytokine profiles in tracheal wash sampling (IL-10, IFN-γ, and IL-17). The IL-4 concentrations decreased significantly in both groups. An allergen-independent CpG-GNP immunotherapy therefore has great potential as a treatment for equine and possibly also human asthma. An additional allergen component did not achieve any significant advantage.

In severe equine asthma, increased matrix metalloproteinase (MMP) expression contributes to pathological pulmonary tissue damage, while tissue inhibitors of metalloproteinases (TIMPs) combat MMP overexpression and pulmonary fibrosis (reviewed by [79]). Barton et al. [79] showed that CpG-GNP inhalation presents a possible effective therapy that can work against the imbalance in MMP and TIMP expression in pulmonary tissue in severe equine asthma. Matrix metalloproteinases (MMP-2/MMP-9) and tissue inhibitors of metalloproteinase (TIMP-1/TIMP-2) concentrations were determined in tracheal wash sampling by equine ELISAs before and two and six weeks after CpG-GNP inhalation [79]. MMP-2, MMP-9, and TIMP-2 concentrations were significantly decreased directly after the treatment as well as six weeks post-treatment [79]. The imbalance in the elastolytic activity appears to be improved by the CpG-GNP inhalation for at least six weeks after treatment, which could possibly reduce the remodeling of the extracellular matrix [79]. The CpG-GNP inhalation could therefore represent an effective therapy for the prevention of pulmonary fibrosis in severe equine asthma [79].

Finally, a prospective, randomized, double-blind clinical field study examined 29 horses with severe equine asthma to explore the dose-dependent effect (the single dose of 187 μg CpG from the previous studies and the double dose of 374 μg CpG) of the inhalative immunotherapy with CpG-GNP, a repeated inhalation treatment in the sense of a booster effect (10 inhalations every second day in comparison to 5 and 7 in the former studies), and a later follow-up examination after a period of 8 weeks (in comparison to 4 and 6 weeks) without further treatment or changes in environmental factors [77]. Here, the therapy concept was compared with a traditional inhalative treatment with beclomethasone (once-daily inhalation with 1600 μg beclomethasone over 10 days). No significant difference could be found between the two CpG-GNP doses (single and double doses) [77]. With regard to a sustained effect over 8 weeks, the CpG-GNP treatment proved advantageous compared to the beclomethasone inhalation in the parameters of respiratory rate, the quantity and viscosity of tracheal secretion, and neutrophils in the BAL [77]. The single-dose CpG-GNP resulted in a significant improvement in 82% of the examined parameters, while the double-dose CpG-GNP resulted in a significant improvement in 72% of the parameters examined directly after the inhalation regimen [77]. With regard to a persistent effect over 8 weeks, the single-dose CpG-GNP showed a significant improvement in 100% of the parameters in comparison to the initial values, and the double-dose CpG-GNP showed a significant improvement in 67% of the parameters [77]. Regarding the immunological parameters in the bronchoalveolar lavage, significant decreases in IL-4 and IFN-γ were evident with the single-dose CpG-GNP treatment [77]. CD4$^+$ T lymphocytes gained via BAL were demonstrated to significantly downregulate mRNA expression (realtime PCR) of Tbet and IL-8 8 weeks after CpG-GNP administration in comparison to baseline values [77]. The double dose did not present any advantage in comparison to the original single dose. On the immunological level, an anti-inflammatory and immunomodulatory effect away from a Th2-dominated immune response was detected.

The referenced studies investigated the first inhalative nanoparticulate equine immunotherapy (Figure 4) [73,75–79]. Beyond the conventional therapy approaches through symptomatic treatment and allergen avoidance, this treatment method opens a new innovative therapy strategy on the immunomodulatory level for severe equine asthma. Especially noteworthy is the comparatively low number of inhalations (10 × q48h) and the sustained significant improvement over at least eight weeks for all examined clinical, endoscopic, and cytological parameters [77]. The clinical effectiveness could thus be proven in several clinical field studies. The proof of concept (IIa) was verified [76], and the dose (IIb) was determined [77].

This nanoparticulate immunotherapy is independent of specific allergens [78]. In this way, the difficulty of identifying clinically relevant allergens in cases of severe equine asthma can be avoided. This difficulty lies in the multifactorial genesis of severe equine asthma, the existence of different phenotypes, and the diagnostic limitations of commercially available allergy tests [104]. However, White et al. [105] recently developed a microarray platform to detect allergen-specific equine IgE in the serum of severe equine asthma-affected horses against a wide range of presumed allergenic proteins. The microarray revealed an abundance of novel pollen, bacteria, mold, and arthropod proteins, which could play a role in the etiology of severe equine asthma [105]. Furthermore, an IgE latex protein antibody was identified, which showed an association with severe equine asthma-affected horses [105], as this protein is ubiquitous to the horses' environments in riding surfaces and race tracks [105]. This could potentially open new perspectives in future diagnosis and therapy. As anticipated and in agreement with previous studies [72], the pure GNP application (as a placebo control) showed no clinical or immunological effect in the examined parameters [75,76]. All of the patients involved in the field studies remained in their customary housing conditions (home stables) and were examined without any change in their customary bedding, feed (hay), and exercise in order to accurately test the inhalative CpG-GNP treatment under natural environmental conditions.

The inhalation interval of two days was selected due to the known half-life of CpG-ODN in vivo being up to 48 h [106] and therefore has an advantage over the otherwise necessary once to twice daily administration of inhalative corticosteroids and bronchodilators [2,4]. In order to ensure a safe application of this formulation, possible local and systemic inflammatory reactions or side effects were documented with the help of a modified scoring system according to the Veterinary Cooperative Oncology Group-Common Terminology Criteria of Adverse Events (VCOG-CTCAE) V1.0 [73,75,76]. None of the studies revealed any local (e.g., irritation, redness, follicular hyperplasia, increased mucus production, and cough) or systemic signs of inflammation (e.g., fever, including fibrogen tests and differential blood counts) or any other side effects as a result of the CpG-GNP inhalation treatment. The inhalation enables noninvasive administration with the lowest possible stress and physical impact on the animal and, in comparison to other systemic administration routes (e.g., intralymphatic and subcutaneous [89,102]), has a direct effect on the local pulmonary immune system as well as fewer systemic interactions and thus proves itself to be an ongoing, target-organ-specific, and promising therapeutic approach.

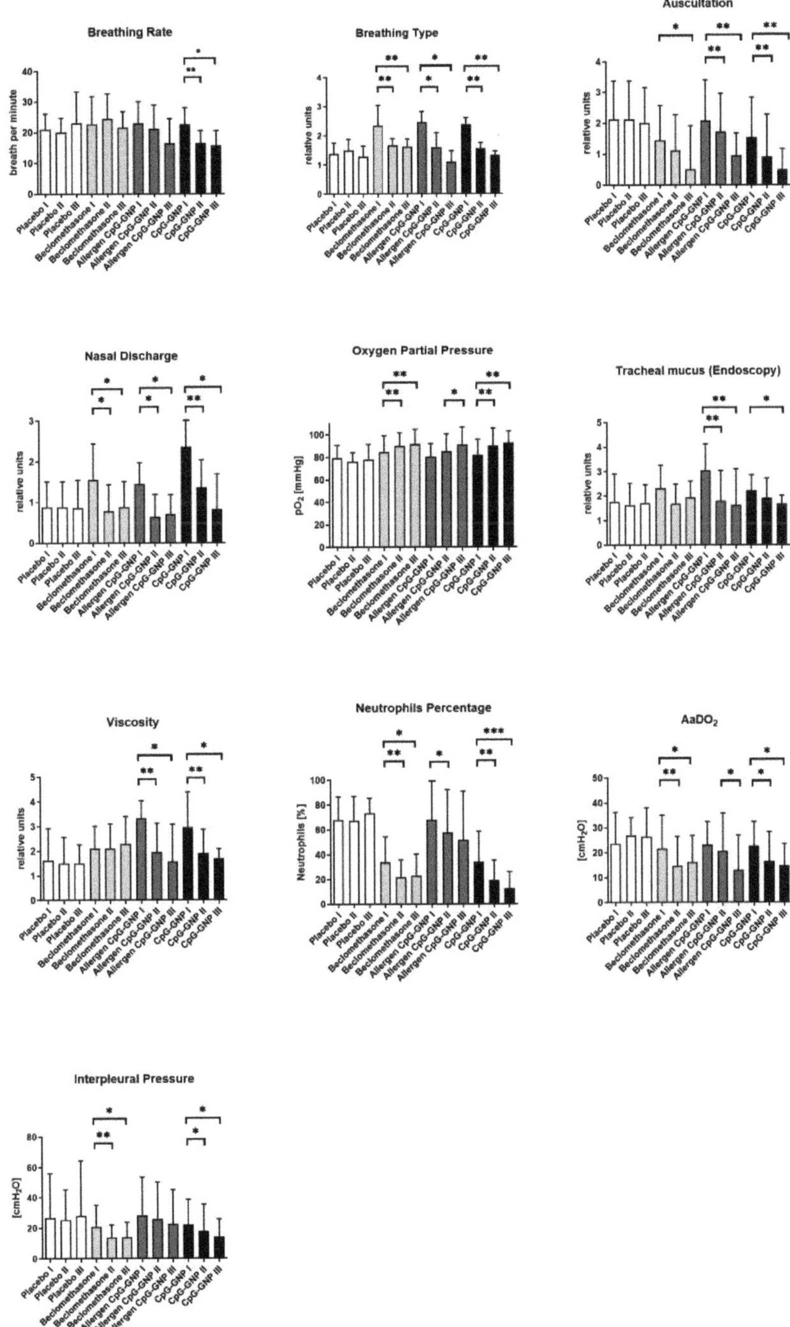

Figure 4. Comparison of the results of the three studies with placebo, beclomethasone, allergen and CpG-GNP, and CpG-GNP inhalation in reference to clinical, endoscopic, cytological, and laboratory parameters: Calculated values (means ± SDs) of $p < 0.05$ represented as *; $p < 0.001$ as **; $p < 0.0001$ as ***. Increase or

decrease in respiratory rate, breathing type, nasal flaring, auscultatory findings of the lungs, indirect measurement of pulmonary pressure, partial oxygen pressure, AaDO$_2$, neutrophilic granulocytes in the BAL, nasal discharge, quantity and viscosity of secretion in the endoscopy between first (I), second (II), and third (III) examination with placebo (white bar, n = 8 severe equine asthma-affected horses; GNP and highly purified water; five inhalations, 48 h dosing interval), beclomethasone (light gray bar, n = 9 severe equine asthma-affected horses, seven inhalations, 48 h dosing interval, CpG-GNP with two specific allergens chosen according to an allergy test), and CpG-GNP (dark gray bar, n = 11 severe equine asthma-affected horses, 10 inhalations, 48 h dosing interval, normal single dose of 187 ug CpG). The time of the third examination (follow-up without further treatment or improvements in stabling during this time) was dependent on the respective study: with placebo, after 4 weeks [76]; allergen and CpG-GNP after 6 weeks [78]; CpG-GNP after 8 weeks [77].

The partially varying initial values of the groups occurred despite random assignment [76,78] and stratified randomization [77]. Due to the heterogenous groups of horses (different stables, stages, and durations of the disease), the character of the study as a field study and the comparatively low number of horses in the individual groups, especially in the placebo group (this, due to the animal welfare aspect, was unavoidable), the starting points of the horses could not be identical. In the placebo group, the differences are especially apparent in the initial values of nasal discharge, breathing type, and the quantity and viscosity of tracheal mucus (four of the ten parameters examined), and in the allergen group, differences are most apparent in neutrophils and viscosity, which were lower than in the other groups, and made the comparison between the groups difficult. The tracheobronchial neutrophil percentage does not correlate well in all cases with the BAL neutrophil percentage, which makes the comparison of some of the results with other studies difficult and does not necessarily allow a direct conclusion about the neutrophil percentage in the lungs. Recently, however, a study comparing tracheal wash and BAL samples in 145 horses, together with endoscopy and mucus findings, found that only 17.5% of horses would have been classified differently if the other method would have been used (reviewed by [10,107]).

The direct comparison of the inhalative beclomethasone application in the study [77] is also problematic, since although the total number of inhalations ($10\times$) is identical to the CpG-GNP application, the interval is different (once daily for the beclomethasone group vs. every second day for the CpG-GNP group) due to their different effect durations. This creates a differing duration of administration (10 days vs. 20 days) and allows no direct comparison between inhalative beclomethasone and CpG-GNP. In addition, cortisone inhalation is typically recommended to be administered twice daily due to the short effect duration, but in practice, this is often not feasible, including in the context of this study due to organizational difficulties (horses were located in their home stables and inhalations were performed by the veterinarians).

In order to observe the influence of the different numbers of inhalations (5, 7, or $10\times$) with CpG-GNP and to assess an ongoing effect, the effect sizes (Cohen's d) of all three studies [76–78] were calculated and compared for the individual parameters (Table 1). The comparison was made between the initial values before treatment and after four weeks with five inhalation treatments [76], after six weeks with seven inhalations [78], and after eight weeks with ten inhalations [77]. With the exception of the parameters of auscultation and tracheal mucus, the effect sizes for all examined parameters were higher after ten inhalation treatments in comparison to seven or five inhalations. All examined parameters, except for interpleural pressure and clinical scoring, showed a significant clinical effect, $d > 0.8$ (large clinical effect), while seven parameters were $d > 1$ (Table 1).

Limiting factors of the inhalative nanoparticulate CpG studies in severe equine asthma-affected horses include the comparatively low numbers of horses per group, as mentioned above. Other factors are the heterogeneity of the groups already referenced (different ages, breeds, uses, seasonal variance in symptom occurrence, genetics, duration of the disease, etc.) and various external influencing factors and housing conditions (influence of various allergen factors) that could not be avoided. With the exception of one study [76], all the others mentioned were performed without a placebo control, which reduces their reliability, and improvements could potentially be attributed to other factors. However, the absence of a placebo control was due to animal welfare concerns and was in accordance with the strict regulations for animal studies. In addition, the subjects used were client-owned horses. The other studies [77,78] compared different therapeutic concepts. Despite randomization, the groups are not exactly homogenous due to the small number of subjects and individual differences, leading to partially varying starting points of the horses within and between groups. However, according to the authors' opinion, the inhalative CpG immunotherapy shows, at this point, its potential to effect clinically significant improvements, even under these diverse influencing factors in the field. Finally, as there have been no comparable studies on this specific subject by other groups, it has not been possible to provide independent replication, so a risk of author bias in the available data cannot be excluded.

Table 1. Comparison of effect sizes of different parameters after 5, 7, and 10 inhalations of CpG-GNP. In order to evaluate the impact of the number of inhalations with CpG-GNP and to determine the long-term effect after a period of no treatment, the effect sizes (Cohen's d) of the three different studies were calculated for each individual parameter to facilitate comparisons between different numbers of inhalations and different periods of re-evaluation. The comparisons were performed between the starting point before treatment and after four weeks with five inhalation treatments [76], after six weeks with seven treatments [78], and after eight weeks with ten treatments [77]. With the exception of auscultation and tracheal mucus, the effect sizes for all examined parameters were higher with ten inhalation treatments in comparison to seven or five treatments (Cohen's *d*: *d* > 0.8: large clinical effect; 0.5–0.8: medium effect; 0.2–0.5: small effect).

Parameters	Effect Size 5 Inhalations	Effect Size 7 Inhalations	Effect Size 10 Inhalations
HOARSI	-	-	2.135
Nasal discharge	1.268	0.764	1.787
Clinical scoring	-	-	0.757
Breathing type	1.333	1.183	1.670
Nasal flaring	-	-	1.538
Respiratory rate	0.341	1.025	1.334
Viscosity	1.177	0.615	1.230
Neutrophils	0.408	0.119	1.064
Auscultation	1.436	1.233	0.973
Tracheal mucus	1.456	0.720	0.964
PaO_2	0.224	0.739	0.856
$AaDO_2$	0.389	0.588	0.829
Interpleural pressure	0.291	0.337	0.551

The use of additional allergens after an allergy test as a hyposensitization did not show any significant improvement in comparison to the monotherapy [78]. Because of the difficulty in determining clinically relevant allergens via allergy tests in horses with severe equine asthma due to the disease's multifactorial nature [104], the authors consider an allergen-independent immunotherapy with CpG-GNP to be much more promising. Human medicine has also seen successful approaches to allergen-independent immunotherapy with CpG and VLPs [108].

The decrease in the quantity and viscosity of secretion in the respiratory tract was one of the most noticeable effects of the CpG-GNP treatment [75,76]. The reduction in neutrophils in the airways, as one of the most important pathomechanisms of the disease, is also a decisive therapeutic effect of the immunotherapy [75–77]. That these changes occurred in many horses despite sustained allergen contact is of particular significance, especially considering that corticosteroid treatment often has only mild to no benefit for controlling the neutrophilic inflammation in equine asthma (reviewed by [38,40,47,48]). Interestingly, a recent randomized, double-blind, controlled pilot clinical trial by Mahalingam-Dhingra et al. [109] with lidocaine inhalation treatment (1 mg/kg q12h, 14 d) in severe equine asthma-affected horses showed a significant decrease in bronchoalveolar lavage neutrophil percentage and tracheal mucus score. Both lidocaine and budesonide groups (positive control) had significant decreases in clinical scores [109]. Lidocaine may therefore be an effective and safe treatment for severe equine asthma in horses that cannot tolerate treatment with corticosteroids [109].

The improvements in the partial oxygen pressure at rest and the alveolar–arterial oxygen gradient as well as the interpleural pressure indicate improved ventilation of the lungs [75,76]. This is also reflected in the significantly improved breathing type and the reduction in active expiratory effort [75,76]. This is likely due to a reduction in the secretion in the lumen as well as a decrease in cholinergic bronchospasm [75,76].

As the authors already demonstrated in vivo in equine BAL cells, the CpG-GNP treatment stimulates the upregulation of IL-10 [72–74]. This cytokine, produced by Treg cells and others, has a regulatory effect on excessive proinflammatory Th1 and proallergic Th2 immune responses as they occur in the lungs of horses with severe equine asthma. We hypothesize that the modulatory effect of this immunotherapy on inflammatory and allergic reactions mediated by the activation of TLR9 receptors in the lungs will stimulate Treg cells to regulate the dysfunctional Th1/Th2 balance. A unilateral overstimulation of Th1 or Th2 would lead to an enhanced inflammatory or allergic reaction, which explains the importance of this balance and regulation.

In humans, an increase in IL-10 resulting from the activation of Tregs could be attributed to an antiallergic therapy (e.g., the administration of glucocorticoids or allergen-specific immunotherapy) [84]. Asthma patients display significantly lower numbers of Tregs in the BAL of the affected airways in comparison to similar healthy individuals, which correlates with a loss of peripheral allergen tolerance in asthma patients [110]. Jarnicki et al. [81] showed that the application of CpG-ODN effects a release of IL-12 and IL-10 by dendritic cells, which leads to an IL-12-mediated Th1 induction and an IL-10-mediated Treg induction. It can therefore be concluded for the present studies that CpG-ODN effected an IL-10 release via Tregs and thus possibly led to a peripheral tolerance, which would represent an innovative treatment form for allergic equine diseases [52,64,80,85]. Studies have shown that an increase of IFN-γ in conjunction with IL-10 is effective in the inhibition of asthmatic and allergic human diseases (reviewed by [111]).

The significance of IL-10 lies in its inhibitory effect on proallergic IL-4 and IL-5 as well as proinflammatory IFN-γ [82,83]. CpG-ODN also activate the release of IL-10 by B lymphocytes, which regulates and limits CpG-mediated proinflammatory reactions [112]. In humans, CpG-ODN activate plasmacytoid dendritic cells [80]. It has been demonstrated that plasmacytoid dendritic cells (pDCs) form a close-knit network within the human airway epithelia of the upper and lower respiratory tract [113]. Assuming similar conditions in equine lungs, pDCs could also function as the first line of defense and a mediator between the external world and the innate immune system in horses. Moseman et al. [80] demonstrated that CpG-ODN A 2216, via the TLR9 pathway, activates human pDCs to induce $CD4^{+}CD25^{-}$ cells in the direction of IL-10-producing $CD4^{+}CD25^{+}$ cells in direct cell-to-cell contact. This could also occur in horses in a similar manner, indicating a CpG-ODN-induced Treg activation with IL-10 production. Fewer Tregs are evident in the BAL of asthmatic children than in healthy children [110]. Tregs in the BAL make up 25% of the T-helper cells (in contrast to 5% in PBMCs) [110]. If these results from humans are similarly present in horses, they would explain the elevated IL-10 values in healthy horses after CpG-ODN application via TLR9 activation [74].

The significant downregulation of mRNA expression of CD4$^+$ T lymphocytes gained via BAL of Tbet (Th1) and IL-8 (chemotactic to neutrophils) 8 weeks after CpG-GNP administration in comparison to baseline values demonstrates the immunomodulatory effect of CpG-GNP on CD4$^+$ T lymphocytes within the airways of severe equine asthma-affected horses [77].

In contrast to other studies [89,99,100,114] in which CpG have been used, this was the first to apply CpG as a monotherapy, via inhalation, and/or bound to gelatin nanoparticles. The advantage of the gelatin nanoparticles as a drug delivery system as opposed to other systems (e.g., virus-like particles) [89] is their good biological tolerability, biodegradability, and aerosol stability as well as their immunologically inert structure [72,88,92–94,115]. The mucosal administration activates the local and systemic humoral and cellular immune responses [71].

According to Montamat et al. [116], CpG-ODN have been recently rediscovered for their immune-tolerance-promoting properties, bringing them back into a prominent position as an immune modulator for the treatment of allergic diseases. It has been shown that the appropriate dosage is essential in promoting immune regulation via the recruitment of pDCs. High doses of CpG-ODN trigger an immune tolerance response that can reverse an established allergic milieu. CpG-ODN have been demonstrated to stimulate IL-10-producing B cells and have shown the capacity to prevent and reverse allergic immune reactions in several animal models, indicating their potential as a preventive and active treatment for allergic disease.

2. Conclusions

Immunotherapy with CpG-GNP, independent of the causal antigens that may vary geographically and with different endo- and phenotypes, is an effective, allergen-independent, inhalative immunomodulatory therapy with a demonstrated clinical and immunological ongoing effect over at least eight weeks in horses that have suffered for years from severe equine asthma and are often otherwise resistant to conventional therapy. The presently proposed immunotherapy thus opens new perspectives beyond the conventional therapy consisting of symptomatic treatment with corticosteroids and bronchodilators and the often difficult-to-achieve avoidance of antigens. This could also be of interest for human asthma treatment.

3. Patents

The authors are co-proprietors of patents EP2399608B1 and US9504760B2.

Author Contributions: Each author has made substantial contributions to the conception of the work and the acquisition, analysis, and interpretation of data. Conceptualization: J.K., S.F., H.G. and G.W.; Methodology: J.K., S.F., G.W. and H.G.; Software: J.K. and S.F.; Validation: J.K. and S.F.; Formal analysis: J.K. and S.F.; Investigation: J.K. and S.F.; Resources: G.W. and H.G.; Data curation: J.K., S.F., G.W. and H.G.; Writing—original draft preparation, J.K.; Writing—review and editing, J.K., S.F. Visualization: J.K. Supervision: J.K., S.F., G.W. and H.G.; Project administration: J.K., S.F., G.W. and H.G.; Funding acquisition: H.G. and G.W. All authors have read and agreed to the published version of the manuscript.

Funding: This review received no external funding.

Institutional Review Board Statement: For this review, no ethical approval was required.

Informed Consent Statement: For this review, no informed consent statement was required.

Data Availability Statement: The data that support the findings of this review are available from the corresponding author upon reasonable request.

Acknowledgments: The authors thank Elena Serkin and Ann-Kristin Barton for the critical revision of the paper and all our colleagues who participated in the equine asthma CpG studies. The publication of this article was funded by Freie Universität Berlin.

Conflicts of Interest: The authors declare no competing financial interest.

Abbreviations

AaDO$_2$, alveolar–arterial oxygen gradient/arterio-alveolar oxygen pressure difference; APC, antigen-presenting cells; ASIT, allergen-specific immunotherapy; bpm, breaths per minute; BAL, bronchoalveolar lavage; COB, chronic obstructive bronchitis/bronchiolitis; COPD, chronic obstructive pulmonary disease; CpG-GNP, cytosine–phosphate–guanosine gelatin nanoparticles; CpG-ODN, cytosine–phosphate–guanosine oligodeoxynucleotides; DC, dendritic cells; GNP, gelatin nanoparticle; HOARSI, Horse Owner Assigned Respiratory Sign Index; RAO, recurrent airway obstruction; TBS, tracheobronchial secretion; IAD, inflammatory airway disease; IBH, insect bite hypersensitivity; IL, interleukin; IRAK, IL-1 receptor-associated kinase; IFN, interferon; MAPK, mitogen-activated protein kinase; MMP, matrix metalloproteinase; MPLA, monophosphoryl lipid A; MyD88; myeloid differentiation primary response gene 88; NF-κB, nuclear factor-κB; ODN, oligodeoxynucleotide; PAMP, pathogen-associated molecular pattern; PaO$_2$ and PaCO$_2$, partial oxygen pressure and partial carbon dioxide pressure; PBMCs, peripheral blood mononuclear cells; pDCs, plasmacytoid dendritic cells; Th, T-helper cell; TIMPs, tissue inhibitors of metalloproteinases; TLR-9, toll-like receptor 9; TRAF 6, TNF receptor-associated factor 6; Tregs, T-regulatory cells.

References

1. Braman, S.S. The global burden of asthma. *Chest* **2006**, *130*, 4S–12S. [CrossRef] [PubMed]
2. Leclere, M.; Lavoie-Lamoureux, A.; Lavoie, J.P. Heaves, an asthma-like disease of horses. *Respirology* **2011**, *16*, 1027–1046. [CrossRef] [PubMed]
3. Robinson, N.E. Recurrent airway obstruction (Heaves). In *Equine Respiratory Diseases*; Lekeux, P., Ed.; International Veterinary Information Service: Ithaca, NY, USA, 2001.
4. Couetil, L.; Cardwell, J.M.; Gerber, V.; Lavoie, J.P.; Leguillette, R.; Richard, E.A. Inflammatory Airway Disease of Horses—Revised Consensus Statement. ACVIM Consensus Statement. *J. Vet. Intern. Med.* **2016**, *30*, 503–515. [CrossRef] [PubMed]
5. Moran, G.; Folch, H. Recurrent airway obstruction in horses—An allergic inflammation: A review. *Vet Med.* **2011**, *56*, 1–13. [CrossRef]
6. Lowell, F.C. Observations on heaves. An asthma-like syndrome in the horse. *J. Allergy Clin. Immunol.* **1964**, *35*, 322–330. [CrossRef]
7. Hotchkiss, J.W.; Reid, S.W.; Christley, R.M. A survey of horse owners in Great Britain regarding horses in their care. Part 2: Risk factors for recurrent airway obstruction. *Equine Vet. J.* **2007**, *39*, 301–308. [CrossRef]
8. Bracher, V.; von Fellenberg, R.; Winder, C.N.; Gruenig, G.; Hermann, M.; Kraehenmann, A. An investigation of the incidence of chronic obstructive pulmonary disease (COPD) in random populations of Swiss horses. *Equine Vet. J.* **1991**, *23*, 136–141. [CrossRef]
9. Fey, K. Chronisch obstruktive bronchi(oli)tis. In *Handbuch Pferdepraxis*, 4th ed.; Brehm, W., Gehlen, H., Ohnesorge, B., Wehrend, A., Eds.; Enke Verlag: Stuttgart, Germany, 2017; pp. 368–377.
10. Couetil, L.; Cardwell, J.M.; Leguillette, R.; Mazan, M.; Richard, E.; Bienzle, D.; Bullone, M.; Gerber, V.; Ivester, K.; Lavoie, J.-P.; et al. Equine Asthma: Current Understanding and Future Directions. *Front. Vet. Sci.* **2020**, *7*, 450. [CrossRef]
11. Bond, S.; Léguillette, R.; Richard, E.A.; Couetil, L.; Lavoie, J.-P.; Martin, J.G.; Pirie, R.S. Equine asthma: Integrative biologic relevance of a recently proposed nomenclature. *J. Vet. Intern. Med.* **2018**, *32*, 2088–2098. [CrossRef]
12. Bullone, M.; Lavoie, J.P. Asthma "of horses and men"—How can equine heaves help us better understand human asthma immunopathology and its functional consequences? *Mol. Immunol.* **2015**, *66*, 97–105. [CrossRef]
13. Hulliger, M.F.; Pacholewska, A.; Vargas, A.; Lavoie, J.-P.; Leeb, T.; Vincent Gerber, V.; Jagannathan, V. An Integrative miRNA-mRNA Expression Analysis Reveals Striking Transcriptomic Similarities between Severe Equine Asthma and Specific Asthma Endotypes in Humans. *Genes* **2020**, *28*, 1143. [CrossRef]
14. Halliwell, R.E.W.; McGorum, B.C.; Irving, P.; Dixon, P.M. Local and systemic antibody production in horses affected with chronic obstructive pulmonary disease. *Vet. Immunol. Immunopathol.* **1993**, *38*, 201–215. [CrossRef]
15. Schmallenbach, K.H.; Rahman, I.; Sasse, H.H.L.; Dixon, P.M.; Halliwell, R.E.; McGorum, B.C.; Crameri, R.; Miller, H.R. Studies on pulmonary and systemic *Aspergillus fumigatus*-specific IgE and IgG antibodies in horses affected with chronic obstructive pulmonary disease (COPD). *Vet. Immunol. Immunopathol.* **1998**, *66*, 245–256. [CrossRef]
16. Künzle, F.; Gerber, V.; Van Der Haegen, A.; Wampfler, B.; Straub, R.; Marti, E. IgE-bearing cells in bronchoalveolar lavage fluid and allergen-specific IgE levels in sera from RAO-affected horses. *J. Vet. Med. A Physiol. Pathol. Clin. Med.* **2007**, *54*, 40–47. [CrossRef] [PubMed]
17. Marti, E. Role of IgE and perspectives on clinical allergy testing. In *4th World Equine Airways Symposium (WEAS)*; Tessier, C., Gerber, V., Eds.; Pabst Science Publishers: Bern, Switzerland, 2009; pp. 109–111.
18. Wagner, B. IgE in horses: Occurrence in health and disease. *Vet. Immunol. Immunopathol.* **2009**, *132*, 21–30. [CrossRef]

19. Franchini, M.; Gill, U.; Von Fellenberg, R.; Bracher, V.V. Interleukin-8 concentration and neutrophil chemotactic activity in bronchoalveolar lavage fluid of horses with chronic obstructive pulmonary disease following exposure to hay. *Am. J. Vet. Res.* **2000**, *11*, 1369–1374. [CrossRef]

20. Lavoie, J.P.; Maghni, K.; Desnoyers, M.; Taha, R.; Martin, J.G.; Hamid, Q.A. Neutrophilic airway inflammation in horses with heaves is characterized by a Th2-type cytokine profile. *Am. J. Respir. Crit. Care Med.* **2001**, *164*, 1410–1413. [CrossRef]

21. Cordeau, M.E.; Joubert, P.; Dewachi, O.; Hamid, Q.; Lavoie, J.P. IL-4, IL-5 and IFN-gamma mRNA-expression in pulmonary lymphocytes in equine heaves. *Vet. Immunol. Immunopathol.* **2004**, *97*, 87–96. [CrossRef]

22. Horohov, D.W.; Beadle, R.E.; Mouch, S.; Pourciau, S.S. Temporal regulation of cytokine mRNA expression in equine recurrent airway obstruction. *Vet. Immunol. Immunopathol.* **2005**, *108*, 237–245. [CrossRef]

23. Giguere, S.; Viel, L.; Lee, E.; MacKay, R.J.; Hernandez, J.; Franchini, M. Cytokine induction in pulmonary airways of horses with heaves and effect of therapy with inhaled fluticasone propionate. *Vet. Immunol. Immunopathol.* **2002**, *85*, 147–158. [CrossRef]

24. Ainsworth, D.M.; Grünig, G.; Matychak, M.B.; Young, J.; Wagner, B.; Erb, H.N.; Antczak, D.F. Recurrent airway obstruction (RAO) in horses is characterized by IFN-γ and IL-8 production in bronchoalveolar lavage cells. *Vet. Immunol. Immunopathol.* **2003**, *96*, 83–91. [CrossRef]

25. Horohov, D.W.; Mills, W.R.; Gluck, M. Specific and innate immunity in the lung as it relates to equine RAO. In *4th World Equine Airways Symposium (WEAS)*; Tessier, C., Gerber, V., Eds.; Pabst Science Publishers: Bern, Switzerland, 2009; pp. 106–108.

26. Debrue, M.; Hamilton, E.; Joubert, P.; Lajoie-Kadoch, S.; Lavoie, J.P. Chronic exacerbation of equine heaves is associated with an increased expression of interleukin-17 mRNA in bronchoalveolar lavage cells. *Vet. Immunol. Immunopathol.* **2005**, *105*, 25–31. [CrossRef] [PubMed]

27. Ainsworth, D.M. Just how important is IL-17 in horses with RAO? In *4th World Equine Airways Symposium (WEAS)*; Tessier, C., Gerber, V., Eds.; Pabst Science Publishers: Bern, Switzerland, 2009; pp. 100–102.

28. Gerber, V.; Ramseyer, A.; Laumen, E.; Nussbaumer, P.; Klukowska-Rötzler, J.; Swineburne, J.E.; Mart, E.; Leeb, T.; Dolf, G. Genetics of equine RAO. In *4th World Equine Airways Symposium (WEAS)*; Tessier, C., Gerber, V., Eds.; Pabst Science Publishers: Bern, Switzerland, 2009; pp. 34–36.

29. Gerber, V.; Baleri, D.; Klukowska-Rötzler, J.; Swinburne, J.E.; Dolf, G. Mixed inheritance of equine recurrent airway obstruction. *J. Vet. Intern. Med.* **2009**, *23*, 626–630. [CrossRef] [PubMed]

30. Gerber, V.; Tessier, C.; Marti, E. Genetics of upper and lower airway diseases in the horse. *Equine Vet. J.* **2015**, *47*, 390–397. [CrossRef]

31. Gerber, V. Genetics of Equine Respiratory Disease. *Vet. Clin. N. Am. Equine Pract.* **2020**, *36*, 243–253. [CrossRef]

32. Pirie, R.S. Recurrent airway obstruction: A review. *Equine Vet. J.* **2014**, *46*, 276–288. [CrossRef]

33. McGorum, B.C.; Pirie, R.S. A review of recurrent airway obstruction and summer pasture associated obstructive pulmonary disease. *Ippologia* **2008**, *19*, 11–19.

34. Simões, J.; Batista, M.; Tilley, P. The Immune Mechanisms of Severe Equine Asthma-Current Understanding and What Is Missing. *Animals* **2022**, *12*, 744. [CrossRef]

35. Laumen, E.; Doherr, M.G.; Gerber, V. Relationship of horse owner assessed respiratory signs index to characteristics of recurrent airway obstruction in two Warmblood families. *Equine Vet. J.* **2010**, *42*, 142–148. [CrossRef]

36. Pirie, R.S.; Mueller, H.-W.; Engel, O.; Albrecht, B.; von Salis-Soglio, M. Inhaled ciclesonide is efficacious and well tolerated in the treatment of severe equine asthma in a large prospective European clinical trial. *Equine Vet. J.* **2021**, *53*, 1094–1104. [CrossRef]

37. Lavoie, J.-P.; Bullone, M.; Rodrigues, N.; Germim, P.; Albrecht, B.; von Salis-Soglio, M. Effect of different doses of inhaled ciclesonide on lung function, clinical signs related to airflow limitation and serum cortisol levels in horses with experimentally induced mild to severe airway obstruction. *Equine Vet. J.* **2019**, *51*, 779–786. [CrossRef] [PubMed]

38. Lavoie, J.-P.; Leclere, M.; Rodrigues, N.; Lemos, K.R.; Bourzac, C.; Lefebvre-Lavoie, J.; Beauchamp, G.; Albrecht, B. Efficacy of inhaled budesonide for the treatment of severe equine asthma. *Equine Vet. J.* **2019**, *51*, 401–407. [CrossRef] [PubMed]

39. Ammann, V.J.; Vrins, A.A.; Lavoie, J.P. Effects of inhaled beclomethasone dipropionate on respiratory function in horses with chronic obstructive pulmonary disease (COPD). *Equine Vet. J.* **1998**, *30*, 152–157. [CrossRef] [PubMed]

40. Rush, B.R.; Flaminio, M.J.; Matson, C.J.; Hakala, J.E.; Shuman, W. Cytologic evaluation of bronchoalveolar lavage fluid from horses with recurrent airway obstruction after aerosol and parenteral administration of beclomethasone dipropionate and dexamethasone, respectively. *Am. J. Vet. Res.* **1998**, *59*, 1033–1038.

41. Couetil, L.L.; Art, T.; de Moffarts, B.; Becker, M.; Melotte, D.; Jaspar, F.; Bureau, F.; Lekeux, P. Effect of beclomethasone dipropionate and dexamethasone isonicotinate on lung function, bronchoalveolar lavage fluid cytology, and transcription factor expression in airways of horses with recurrent airway obstruction. *J. Vet. Intern. Med.* **2006**, *20*, 399–406. [CrossRef] [PubMed]

42. Viel, L. Therapeutic efficacy of inhaled fluticasone propionate in horses with chronic obstructive pulmonary disease. *Proc. Am. Ass. Equine Practnrs.* **1999**, *45*, 306–307.

43. Couetil, L.L.; Chilcoat, C.D.; DeNicola, D.B.; Clark, S.P.; Glickman, N.W.; Glickman, L.T. Randomized, controlled study of inhaled fluticasone propionate, oral administration of prednisone, and environmental management of horses with recurrent airway obstruction. *Am. J. Vet. Res.* **2005**, *66*, 1665–1674. [CrossRef]

44. Robinson, N.E.; Berney, C.; Behan, A.; Derksen, F.J. Fluticasone propi- onate aerosol is more effective for prevention than treatment of recurrent airway obstruction. *J. Vet. Intern. Med.* **2009**, *23*, 1247–1253. [CrossRef]

45. Leclere, M.; Lavoie-Lamoureux, A.; Joubert, P.; Relave, F.; Setlakwe, E.L.; Beauchamp, G.; Couture, C.; Martin, J.G.; Lavoie, J.P. Corticosteroids and antigen avoidance decrease airway smooth muscle mass in an equine asthma model. *Am. J. Respir. Cell Mol. Biol.* **2012**, *47*, 589–596. [CrossRef]
46. Rush, B.R.; Raub, E.S.; Rhoads, W.S.; Flaminio, M.J.; Matson, C.J.; Hakala, J.E.; Gillespie, J.R. Pulmonary function in horses with recurrent airway obstruction after aerosol and parenteral administration of beclomethasone dipropionate and dexamethasone, respectively. *Am. J. Vet. Res.* **1998**, *59*, 1039–1043.
47. Lavoie, J.P.; Leguillette, R.; Pasloske, K.; Charette, L.; Sawyer, N.; Guay, D.; Murphy, T.; Hickey, G.J. Comparison of effects of dexamethasone and the leukotriene D4 receptor antagonist L-708,738 on lung function and airway cytologic findings in horses with recurrent airway obstruction. *Am. J. Vet. Res.* **2002**, *63*, 579–585. [CrossRef] [PubMed]
48. Robinson, N.E.; Jackson, C.; Jefcoat, A.; Berney, C.; Peroni, D.; Derksen, F.J. Efficacy of three corticosteroids for the treatment of heaves. *Equine Vet. J.* **2002**, *34*, 17–22. [CrossRef] [PubMed]
49. Mazan, M.R.; Lascola, K.; Bruns, S.J.; Hoffman, A.M. Use of a novel one-nostril mask-spacer device to evaluate airway hyperresponsiveness (AHR) in horses after chronic administration of albuterol. *Can. J. Vet. Res.* **2014**, *78*, 214–220.
50. Niedermaier, G.; Gehlen, H. Möglichkeiten der Inhalationstherapie zur Behandlung der chronisch obstruktiven Bronchitis des Pferdes. *Pferdeheilkunde* **2009**, *25*, 327–332. [CrossRef]
51. Kline, J.N. Immunotherapy of asthma using CpG Oligodeoxynucleotides. *Immunol. Res.* **2007**, *39*, 279–286. [CrossRef] [PubMed]
52. Umetsu, D.T.; DeKruyff, R.H. The regulation of allergy and asthma. *Immunol. Rev.* **2006**, *212*, 238–255. [CrossRef] [PubMed]
53. Lunn, P.; Horohov, D.W. The equine immune system. In *Equine Internal Medicine*, 3rd ed.; Reed, S.M., Bayly, W.M., Sellon, D.C., Eds.; Saunders Elsevier: St. Louis, MO, USA, 2010; pp. 2–56.
54. Braun-Fahrländer, C. Environmental exposure to endotoxin and other microbial products and the decreased risk of childhood atopy: Evaluating developments since April 2002. *Curr. Opin. Allergy Clin. Immunol.* **2003**, *3*, 325–329. [CrossRef]
55. Waser, M.; von Mutius, E.; Riedler, J.; Nowak, D.; Maisch, S.; Carr, D.; Eder, W.; Tebow, G.; Schierl, R.; Schreuer, M.; et al. Exposure to pets, and the association with hay fever, asthma, and atopic sensitization in rural children. *Allergy* **2005**, *60*, 177–184. [CrossRef]
56. Braun-Fahrländer, C. Role of environment in asthma, endotoxin and other factors. In *4th World Equine Airways Symposium (WEAS)*; Tessier, C., Gerber, V., Eds.; Pabst Science Publishers: Bern, Switzerland, 2009; pp. 44–46.
57. Kaiser-Thoma, S.; Hiltyb, M.; Gerbera, V. Effects of hypersensitivity disorders and environmental factors on the equine intestinal microbiota. *Vet. Q.* **2020**, *40*, 97–107. [CrossRef]
58. Fillion-Bertrand, G.; Dickson, R.P.; Boivin, R.; Lavoie, J.-P.; Gary, B.; Huffnagle, G.B.; Leclere, M. Lung microbiome is influenced by the environment and asthmatic status in an equine model of asthma. *Am. J. Resp. Cell Mol. Biol.* **2019**, *60*, 189–197. [CrossRef]
59. Krieg, A.M. CpG motifs in bacterial DNA and their immune effects. *Annu. Rev. Immunol.* **2002**, *20*, 709–760. [CrossRef] [PubMed]
60. Anderson, K.V. Toll signaling pathways in the innate immune response. *Curr. Opin. Immunol.* **2000**, *12*, 13–19. [CrossRef]
61. Richter, C. Aufklärung der Struktur-Wirkungsbeziehungen von CpG-A und CpG-C-Oligodesoxynukleotiden als Grundlage für die Entwicklung immunstimulatorischer Nanopartikel. Ph.D. Thesis, Medizinische Fakultät der Ludwig-Maximilians-Universität, München, Germany, 2006.
62. Akira, S.; Uematsu, S.; Takeuchi, O. Pathogen recognition and innate immunity. *Cell* **2006**, *124*, 783–801. [CrossRef] [PubMed]
63. Haas, T.; Metzger, J.; Schmitz, F.; Heit, A.; Müller, T.; Latz, E.; Wagner, H. The DNA sugar backbone 2′deoxyribose determines toll-like receptor 9 activation. *Immunity* **2008**, *28*, 315–323. [CrossRef] [PubMed]
64. Fonseca, D.; Kline, J. Use of CpG oligonucleotides in treatment of asthma and allergic disease. *Adv. Drug Deliv. Rev.* **2009**, *61*, 256–262. [CrossRef]
65. Schneberger, D.; Caldwell, S.; Singh Suri, S.; Singh, B. Expression of Toll-Like Receptor 9 in Horse Lungs. *Anat. Rec.* **2009**, *292*, 1068–1077. [CrossRef]
66. Bird, A.P. CpG-rich islands and the function of DNA methylation. *Nature* **1986**, *321*, 209–213. [CrossRef]
67. Latz, E.; Schoenemeyer, A.; Visintin, A.; Fitzgerald, K.A.; Monks, B.G.; Knetter, C.F.; Lien, E.; Nilsen, N.J.; Espevik, T.; Golenbock, D.T. TLR9 signals after translocating from the ER to CpG DNA in the lysosome. *Nat. Immunol.* **2004**, *5*, 190–198. [CrossRef]
68. Yasuda, K.; Rutz, M.; Schlatter, B.; Metzger, J.; Luppa, P.B.; Schmitz, F.; Haas, T.; Heit, A.; Bauer, S.; Wagner, H. CpG motif-independent activation of TLR9 upon endosomal translocation of "natural" phosphodiester DNA. *Eur. J. Immunol.* **2006**, *36*, 431–436. [CrossRef]
69. Krieg, A.M.; Yi, A.-K.; Matson, S.; Waldschmidt, T.J.; Bishop, G.A.; Teasdale, R.; Koretzky, G.A.; Klinman, D.M. CpG motifs in bacterial DNA trigger direct B-cell activation. *Nature* **1995**, *374*, 546–549. [CrossRef]
70. Bohle, B. CpG motifs as possible adjuvants for the treatment of allergic diseases. *Int. Arch. Allergy Immunol.* **2002**, *129*, 198–203. [CrossRef] [PubMed]
71. Vollmer, J.; Krieg, A.M. Immunotherapeutic applications of CpG oligodeoxynucleotide TLR9 agonists. *Adv. Drug Deliv. Rev.* **2009**, *61*, 195–204. [CrossRef] [PubMed]
72. Zwiorek, K.; Bourquin, C.; Battiany, J.; Winter, G.; Endres, S.; Hartmann, G.; Coester, C. Delivery by cationic gelatin nanoparticles strongly increases the immunostimulatory effects of CpG oligonucleotides. *Pharm. Res.* **2008**, *25*, 551–562. [CrossRef]
73. Klier, J. Neuer Therapieansatz zur Behandlung der COB des Pferdes durch Immunstimulation von BAL-Zellen mit verschiedenen CpG-. Klassen. Thesis, Tierärztliche Fakultät der Ludwig-Maximilians-Universität, München, Germany, 2011.

74. Klier, J.; May, A.; Fuchs, S.; Schillinger, U.; Plank, C.; Winter, G.; Gehlen, H.; Coester, C. Immunostimulation of bronchoalveolar Lavage cells from recurrent airway obstruction-affected horses by different CpG-classes bound to gelatin nanoparticles. *Vet. Immunol. Immunopathol.* **2011**, *144*, 79–87. [CrossRef] [PubMed]

75. Klier, J.; Fuchs, S.; May, A.; Schillinger, U.; Plank, C.; Winter, G.; Coester, C.; Gehlen, H. A nebulized gelatin nanoparticle-based CpG formulation is effective in Immunotherapy of allergic horses. *Pharm. Res.* **2012**, *29*, 1650–1657. [CrossRef]

76. Klier, J.; Lehmann, B.; Fuchs, S.; Reese, S.; Hirschmann, A.; Coester, C.; Winter, G.; Gehlen, H. Nanoparticulate CpG immunotherapy in RAO-affected horses: Phase I and IIa study. *J. Vet. Intern. Med.* **2015**, *29*, 286–293. [CrossRef] [PubMed]

77. Klier, J.; Bartl, C.; Geuder, S.; Geh, K.J.; Reese, S.; Goehring, L.S.; Winter, G.; Gehlen, H. Immunomodulatory asthma therapy in the equine animal model: A dose-response study and evaluation of a long-term effect. *Immun. Inflamm. Dis.* **2019**, *7*, 130–149. [CrossRef]

78. Klier, J.; Geis, S.; Steuer, J.; Geh, K.; Reese, S.; Fuchs, S.; Mueller, R.S.; Winter, G.; Gehlen, H. A comparison of nanoparticulate CpG immunotherapy with and without allergens in spontaneously equine asthma-affected horses, an animal model. *Immun. Inflamm. Dis.* **2018**, *6*, 81–96. [CrossRef]

79. Barton, A.K.; Shety, T.; Klier, J.; Geis, S.; Einspanier, R.; Gehlen, H. Metalloproteinases and their Inhibitors under the Course of Immunostimulation by CPG-ODN and Specific Antigen Inhalation in Equine Asthma. *Mediators. Inflamm.* **2019**, *2019*, 7845623. [CrossRef]

80. Moseman, E.A.; Liang, X.; Dawson, A.J.; Panoskaltsis-Mortari, A.; Krieg, A.M.; Liu, Y.-J.; Blazar, B.R.; Chen, W. Human plasmacytoid dendritic cells activated by CpG oligodeoxynucleotides induce the generation of CD4+ CD 25+ regulatory T cells. *J. Immunol.* **2004**, *173*, 4433–4442. [CrossRef]

81. Jarnicki, A.G.; Conroy, H.; Brereton, C.; Donnelly, G.; Toomey, D.; Walsh, K.; Sweeney, C.; Leavy, O.; Fletcher, J.; Lavelle, E.C.; et al. Attenuating regulatory T cell induction by TLR agonists through inhibition of p38 MAPK signalling in dendritic cells enhances their efficacy as vaccine adjuvants and cancer immunotherapeutics. *J. Immunol.* **2008**, *180*, 256–262. [CrossRef] [PubMed]

82. Taylor, A.; Verhagen, J.; Blaser, K.; Akdis, M.; Akdis, C.A. Mechanisms of immune suppression by interleukin-10 and transforming growth factor-beta: The role of Tregulatory cells. *Immunology* **2006**, *117*, 433–442. [CrossRef] [PubMed]

83. Van Scott, M.R.; Justice, J.P.; Bradfield, J.F.; Enright, E.; Sigounas, A.; Sur, S. IL-10 reduces Th2 cytokine production and eosinophilia but augments airway reactivity in allergic mice. *Am. J. Physiol. Lung Cell Mol. Physiol.* **2000**, *278*, 667–674. [CrossRef] [PubMed]

84. Weiner, G.J. The immunobiology and clinical potential of immunostimulatory CpG oligodeoxynucleotides. *J. Leukoc. Biol.* **2000**, *68*, 455–463.

85. Ryanna, K.; Stratigou, V.; Safinia, N.; Hawrylowicz, C. Regulatory T cells in in bronchial asthma. *Allergy* **2009**, *64*, 335–347. [CrossRef]

86. Wilson, K.D.; de Jong, S.D.; Tam, Y.K. Lipid-based delivery of CpG oligonucleotides enhances immunotherapeutic efficacy. *Adv. Drug Deliv. Rev.* **2009**, *61*, 233–242. [CrossRef]

87. Storni, T.; Ruedl, C.; Schwarz, K.; Schwendener, R.A.; Renner, W.A.; Bachmann, M.F. Nonmethylated CG motifs packaged into virus-like particles induce protective cytotoxic T cell responses in the absence of systemic side effects. *J. Immunol.* **2004**, *172*, 1777–1785. [CrossRef]

88. Bourquin, C.; Anz, D.; Zwiorek, K.; Lanz, A.-L.; Fuchs, S.; Weigel, S.; Wurzenberger, C.; von der Borch, P.; Golic, M.; Moder, S.; et al. Targeting CpG oligodeoxynucleotides to the lymph node by nanoparticles elicits efficient antitumoral immunity. *J. Immunol.* **2008**, *181*, 2990–2998. [CrossRef]

89. Senti, G.; Johansen, P.; Haug, S.; Bull, C.; Gottschaller, C.; Müller, P.; Pfister, T.; Maurer, P.; Bachmann, M.F.; Graf, N.; et al. Use of A-type CpG oligodeoxynucleotides as an adjuvant in allergen-specific immunotherapy in humans: A phase I/IIa clinical trial. *Clin. Exp. Allergy* **2009**, *39*, 562–570. [CrossRef]

90. Sparwasser, T.; Hültner, L.; Koch, E.S.; Luz, A.; Lipford, G.B.; Wagner, H. Immunostimulatory CpG-Oligodeoxynucleotides cause extramedullary murine hemopoiesis. *J. Immunol.* **1997**, *162*, 2368–2374.

91. Von Beust, B.R.; Johansen, P.; Smith, K.A.; Bot, A.; Storni, T.; Kündig, T.M. Improving the therapeutic index of CpG oligodeoxynucleotides by intralymphatic administration. *Eur. J. Immunol.* **2005**, *35*, 1869–1876. [CrossRef] [PubMed]

92. Tseng, C.-L.; Yueh-Hsiu Wu, S.; Wang, W.-H.; Peng, C.-L.; Lin, F.-H.; Lin, C.-C.; Young, T.-H.; Shieh, M.-J. Targeting efficiency and biodistribution of biotinylated-EGF-conjugated gelatine nanoparticles administered via aerosol delivery in nude mice with lung cancer. *Biomaterials* **2008**, *29*, 3014–3022. [CrossRef] [PubMed]

93. Tseng, C.-L.; Su, W.-Y.; Yen, K.-C.; Yang, K.-C.; Lin, F.-H. The use of biotinylated-EGF-modified gelatine nanoparticle carrier to enhance cisplatin accumulation in cancerous lungs via inhalation. *Biomaterials* **2009**, *30*, 3476–3485. [CrossRef] [PubMed]

94. Fuchs, S.; Klier, J.; May, A.; Winter, G.; Coester, C.; Gehlen, H. Towards an inhalative in vivo application of immunomodulating gelatin nanoparticles in horse-related preformulation studies. *J. Microencapsul.* **2012**, *29*, 615–625. [CrossRef]

95. Flaminio, M.J.; Borges, A.S.; Nydam, D.V.; Horohov, D.W.; Hecker, R.; Matychak, M.B. The effect of CpG-ODN on antigen presenting cells of the foal. *J. Immune Based Ther. Vaccines* **2007**, *25*, 1. [CrossRef]

96. Ziegler, A.; Gerber, V.; Marti, E. In vitro effects of the toll-like receptor agonists monophosphoryl lipid A and CpG-rich oligonucleotides on cytokine production by equine cells. *Vet. J.* **2017**, *219*, 6–11. [CrossRef]

97. Ziegler, A.; Olzhausen, J.; Hamza, E.; Stojiljkovic, A.; Stoffel, M.H.; Garbani, M.; Rhyner, C.; Marti, E. An allergen-fused dendritic cell-binding peptide enhances in vitro proliferation of equine T-cells and cytokine production. *Vet. Immunol. Immunopathol.* **2022**, *243*, 110351. [CrossRef]

98. Ficarelli, M.; Antzin-Anduetza, I.; Hugh-White, R.; Firth, A.E.; Sertkaya, H.; Wilson, H.; Neil, S.J.D.; Schulz, R.; Swanson, C.M. CpG Dinucleotides Inhibit HIV-1 Replication through Zinc Finger Antiviral Protein (ZAP)-Dependent and -Independent Mechanisms. *J. Virol.* **2020**, *94*, e01337-19. [CrossRef]

99. Lopez, A.M.; Hecker, R.; Mutwiri, G.; van Drunen Little-van den Hurk, S.; Babiuk, L.A.; Townsend, H.G.G. Formulation with CpG-ODN enhances antibody responses to an equine influenza virus vaccine. *Vet. Immunol. Immunopathol.* **2006**, *114*, 103–110. [CrossRef]

100. Hullmann, A.G. Prophylaxe der Rhodococcus equi-Pneumonie bei Fohlen durch Vakzination mit Rhodococcus equi-Impfstoff und Adjuvans CpG. XXXX. Thesis, School of Veterinary Medicine Hannover, Hannover, Germany, 2006.

101. Mueller, R.S.; Veir, J.; Fieseler, K.V.; Dow, S.W. Use of immunostimulatory liposome-nucleic acid complexes in allergen-specific immunotherapy of dogs with refractory atopic dermatitis—A pilot study. *Vet. Dermatol.* **2005**, *16*, 61–68. [CrossRef]

102. Wagner, I.; Geh, K.J.; Hubert, M.; Winter, G.; Weber, K.; Classen, J.; Klinger, C.; Mueller, R.S. Preliminary evaluation of cytosine-phosphate-guanine oligodeoxynucleotides bound to gelatine nanoparticles as immunotherapy for canine atopic dermatitis. *Vet. Rec.* **2017**, *29*, 118. [CrossRef] [PubMed]

103. Adamus, T.; Kortylewski, M. The revival of CpG oligonucleotide-based cancer immunotherapies. *Contemp. Oncol.* **2018**, *22*, 56–60. [CrossRef] [PubMed]

104. Klier, J.; Lindner, D.; Reese, S.; Mueller, R.S.; Gehlen, H. Comparison of four different allergy tests in equine asthma affected horses and allergen inhalation provocation test. *J. Equine Vet. Sci.* **2021**, *102*, 103433. [CrossRef] [PubMed]

105. White, S.J.; Moore-Colyer, M.; Marti, E.; Hannant, D.; Gerber, V.; Couetil, L.; Richard, E.A.; Alcocer, M. Antigen array for serological diagnosis and novel allergen identification in severe equine asthma. *Sci. Rep.* **2019**, *23*, 15170. [CrossRef]

106. Mutwiri, G.K.; Nichani, A.K.; Babiuk, S.; Babiuk, L.A. Strategies for enhancing the immunostimulatory effects of CpG oligodeoxynucleotides. *J. Control. Release* **2004**, *97*, 1–17. [CrossRef]

107. Rossi, H.; Virtala, A.-M.; Raekallio, M.; Rahkonen, E.; Rajamäki, M.M.; Mykkänen, A. Comparison of tracheal wash and bronchoalveolar lavage cytology in154 horses with and without respiratory signs in a referral hospital over 2009–2015. *Front. Vet. Sci.* **2018**, *5*, 61. [CrossRef]

108. Kündig, T.M.; Klimek, L.; Schendzielorz, P.; Renner, W.A.; Senti, G.; Bachmann, M.F. Is the Allergen Really Needed in Allergy Immunotherapy? *Curr. Treat. Options Allergy* **2015**, *2*, 72–82. [CrossRef]

109. Mahalingam-Dhingra, A.; Mazan, M.R.; Bedenice, D.; Ceresia, M.; Minuto, J.; Deveney, E.F. A CONSORT-guided, randomized, double-blind, controlled pilot clinical trial of inhaled lidocaine for the treatment of equine asthma. *Can. J. Vet. Res.* **2022**, *86*, 116–124.

110. Hartl, D.; Koller, B.; Mehlhorn, A.T.; Reinhardt, D.; Nicolai, T.; Schendel, D.J.; Griese, M.; Krauss-Etschmann, S. Quantitative and functional impairment of pulmonary CD4+ CD25hi regulatory T cells in pediatric asthma. *J. Allergy Clin. Immunol.* **2007**, *119*, 1258–1266. [CrossRef] [PubMed]

111. Stock, P.; Akbari, O.; Berry, G.; Freeman, G.; DeKruyff, R.H.; Umetsu, D.T. Induction of TH1-like regulatory cells that express Foxp3 and protect against airway hyperreactivity. *Nat. Immunol.* **2004**, *5*, 1149–1156. [CrossRef]

112. Kline, J.N.; Krieg, A.M. Oligodeoxynucleotides. In *New Drugs for Asthma, Allergy and COPD*; Hansel, T.T., Barnes, P.J., Eds.; Karger: Basel, Switzerland, 2001; Volume 31, pp. 229–232.

113. Von Garnier, C.; Nicod, L.P. Immunology taught by lung dendritic cells. *Swiss Med. Wkly.* **2009**, *139*, 186–192. [PubMed]

114. Liu, T.; Nerren, J.; Murrell, J.; Juillard, V.; Martens, R. CpG-induced stimulation of cytokine expression by peripheral blood mononuclear cells of foals and their dams. *J. Jevs.* **2008**, *28*, 419–426. [CrossRef]

115. Brzoska, M.; Langer, K.; Coester, C.; Loitsch, S.; Wagner, T.O.F.; Mallinckrodt, C. Incorporation of biodegradable nanoparticles into human airway epithelium cells—In vitro study of the suitability as a vehicle for drug or gene delivery in pulmonary diseases. *Biochem. Biophys. Res. Commun.* **2004**, *318*, 562–570. [CrossRef]

116. Montamat, G.; Leonard, C.; Poli, A.; Klimek, L.; Ollert, M. CpG Adjuvant in Allergen-Specific Immunotherapy: Finding the Sweet Spot for the Induction of Immune Tolerance. *Front. Immunol.* **2021**, *12*, 590054. [CrossRef] [PubMed]

Article

Upper and Lower Airways Evaluation and Its Relationship with Dynamic Upper Airway Obstruction in Racehorses

Chiara Maria Lo Feudo [1], Giovanni Stancari [2], Federica Collavo [3], Luca Stucchi [2], Bianca Conturba [2], Enrica Zucca [1] and Francesco Ferrucci [1,*]

1   Equine Sports Medicine Laboratory "Franco Tradati", Department of Veterinary Medicine and Animal Sciences, Università Degli Studi di Milano, 26900 Lodi, Italy; chiara.lofeudo@unimi.it (C.M.L.F.); enrica.zucca@unimi.it (E.Z.)
2   Veterinary Teaching Hospital, Department of Veterinary Medicine and Animal Sciences, Università Degli Studi di Milano, 26900 Lodi, Italy; giovanni.stancari@unimi.it (G.S.); luca.stucchi@unimi.it (L.S.); bianca.conturba@unimi.it (B.C.)
3   Freelance DVM, Rheinbacher Straße 24, 53919 Weilerswist, Germany; federicacollavo@gmail.com
*   Correspondence: francesco.ferrucci@unimi.it; Tel.: +39-02503-34146

Simple Summary: Dynamic upper airway obstructions (DUAO) are a common cause of poor performance in racehorses, including different forms which vary in severity. Previous studies reported contrasting results concerning the contribution of abnormal pharyngo-laryngeal appearance and airway inflammation to the pathogenesis of DUAO. The present study aimed to evaluate possible associations between the development of DUAO and resting airway endoscopic findings, epiglottis size, airways inflammation and exercise-induced pulmonary hemorrhage (EIPH). These relationships were statistically investigated retrospectively in 360 racehorses (Standardbreds and Thoroughbreds) with poor performance or abnormal respiratory noises. A flaccid appearance of the epiglottis was associated with the occurrence of the dorsal displacement of the soft palate, while no relationship was detected between DUAO and epiglottis length. Inflammation of the upper and lower airways was not related with the development of DUAO, nor were horses with DUAO more prone to experience EIPH. These results suggest that epiglottis may contribute to upper airway stability, while inflammation does not predispose horses to the onset of DUAO.

Abstract: Dynamic upper airway obstructions (DUAO) are common in racehorses, but their pathogenetic mechanisms have not been completely clarified yet. Multiple studies suggest that alterations of the pharyngo-laryngeal region visible at resting endoscopy may be predictive of the onset of DUAO, and the development of DUAO may be associated with pharyngeal lymphoid hyperplasia (PLH), lower airway inflammation (LAI) and exercise-induced pulmonary hemorrhage (EIPH). The present study aims to investigate the possible relationship between the findings of a complete resting evaluation of the upper and lower airways and DUAO. In this retrospective study, 360 racehorses (Standardbreds and Thoroughbreds) referred for poor performance or abnormal respiratory noises were enrolled and underwent a diagnostic protocol including resting and high-speed treadmill endoscopy, cytological examination of the bronchoalveolar lavage fluid and radiographic assessment of the epiglottis length. In this population, epiglottis flaccidity was associated with dorsal displacement of the soft palate, while no relationship was detected between DUAO and epiglottis length. No associations were detected between DUAO and PLH, LAI or EIPH. In conclusion, it is likely that epiglottis plays a role in upper airway stability, while airways inflammation does not seem to be involved in the pathogenesis of DUAO.

Keywords: DUAO; equine; horse; dynamic upper airway obstruction; upper respiratory tract; equine sports medicine; equine endoscopy; upper airway endoscopy

Citation: Lo Feudo, C.M.; Stancari, G.; Collavo, F.; Stucchi, L.; Conturba, B.; Zucca, E.; Ferrucci, F. Upper and Lower Airways Evaluation and Its Relationship with Dynamic Upper Airway Obstruction in Racehorses. *Animals* **2022**, *12*, 1563. https://doi.org/10.3390/ani12121563

Academic Editor: Chris W. Rogers

Received: 31 May 2022
Accepted: 16 June 2022
Published: 17 June 2022

Publisher's Note: MDPI stays neutral with regard to jurisdictional claims in published maps and institutional affiliations.

1. Introduction

Horses are obligatory nasal breathers and cannot avoid the high pressures occurring at the level of the nasopharynx during exercise by switching to oral breathing like other species. Moreover, the nasopharyngeal region is not supported by osseous or cartilaginous structures and relies only on muscle activity to maintain its stability [1,2]: therefore, during exercise, when airflow turbulence and negative pressures occur at the floor of the nasopharynx and within the larynx, these structures may collapse, and horses may develop different forms of dynamic upper airway obstruction (DUAO) [3]. As a consequence, respiratory function and gas exchanges at the alveolar-capillary level may be impaired, determining poor athletic performance, especially in racehorses working at supramaximal exercise [4–7].

Different types of DUAO have been described, including dorsal displacement of the soft palate (DDSP), medial deviation of aryepiglottic folds (MDAF), nasopharyngeal collapse (NPC), dynamic laryngeal collapse (DLC), epiglottis entrapment (EE) and epiglottis retroversion (ER) [8]. Although these conditions have been studied for a long time, their exact pathogenetic mechanisms are still not completely clarified and are likely to be multifactorial [8–10]. Most DUAOs are thought to result from neuromuscular dysfunction, fatigue, immaturity or conformational change of the structures controlling the patency of the nasopharyngeal region [8,9,11]. In particular, multiple researchers have hypothesized that epiglottic conformation may play an important role in the stability of upper airways [12]: in fact, the loss of epiglottis convexity has been associated with the development of DDSP [12,13] and MDAF [12,14] and is frequently reported as a flaccid appearance of the epiglottis. Abnormal epiglottis conformation was associated with lower earnings in young racehorses [15]: this result could be attributable to the development of DUAO in horses with a dysplastic epiglottis. Another study reported improved performance in horses after epiglottic augmentation, which may overcome deficits in neuromuscular function [16]. However, the findings of other studies do not support this hypothesis, as they failed to find an association between a flaccid appearance of epiglottis at rest and the detection of DDSP during exercise [17]. Moreover, some authors proposed that a short epiglottis, measured radiographically, may contribute to the pathogenesis of DDSP [18], but other studies reported normal epiglottis length in horses with DDSP [9,19]. Therefore, it is still debated whether a functional epiglottis with normal conformation is required to maintain soft palate stability. Other nasopharyngeal abnormalities, visible at resting endoscopy, have been proposed as predictors of DUAO, including episodes of DDSP, which can occur spontaneously or be induced by swallowing, and the presence of ulceration on the free margin of the soft palate [9,20–22]. Spontaneous DDSP during resting endoscopy has been reported as a highly specific but extremely insensitive test for DDSP during exercise [23]; however, the clinical significance of DDSP at rest remains uncertain, as it can also occur in horses with a normal function of the upper airway during exercise [17,22]. Palate ulceration may occur due to the friction between the ventral surface of the epiglottis and the free border of the soft palate [16]; however, in a study, only 16% of the horses with DDSP had palate ulceration [24], and vice versa—many horses with palate ulceration do not show DDSP during exercise [17]. It has been hypothesized that horses displacing quickly do not have enough abrasion between the epiglottis and palate to induce ulceration [16].

Different authors reported that inflammation of the nasopharyngeal region, visible at endoscopy as pharyngeal lymphoid hyperplasia (PLH), may impair the function of pharyngeal mechanoceptors [25] and contribute to the neuromuscular dysfunction and instability of the upper airway, predisposing horses to the onset of DUAOs, such as DDSP, NPC and ER [8,9,26–30]. Recently, the "one airway, one disease" concept, long known in human medicine and describing a relationship between the health of the upper and lower airways, has also been proposed and investigated in equine medicine [31]. Lower airway inflammation (LAI) has been associated with DUAO [9,28,30–33], probably because of the increased negative pressure driven by lower respiratory tract obstruction and increased respiratory impedance and work of breathing, which may impair upper airway patency and accelerate the onset of neuromuscular fatigue [30,31,33–35]. Another proposed theory is that DUAO may predispose horses to lower airway disease via unknown mechanisms [9,33]. Another lower airway disorder that has been associated with DUAO is exercise-induced pulmonary hemorrhage (EIPH) [9,32,36]. This may be due to the increase in the transmural pulmonary capillary pressure gradient resulting from upper airway obstruction and leading to the rupture of pulmonary capillaries and therefore EIPH [1,31,36,37]. In contrast, no association between DUAO and LAI inflammation or EIPH was detected in another study [38].

As the role of upper and lower airways resting evaluation has not been fully understood yet, the present study aims to investigate, in a wide population of racehorses, the clinical significance of upper airway anatomical and functional resting abnormalities and the possible contribution of upper and lower airway inflammation to the development of DUAO.

2. Materials and Methods

2.1. Horses

The clinical records of Standardbred and Thoroughbred racehorses referred to the Equine Sports Medicine Unit of the Veterinary Teaching Hospital of the University of Milan (Italy) between 2000 and 2021, with a history of poor performance or abnormal respiratory noises during exercise, were retrospectively reviewed. All horses ($n = 366$) were in full training upon admission. Each horse underwent a complete clinical examination, laboratory analyses and upper airway endoscopy at rest and during exercise on a high-speed treadmill. In 158 horses, the length of epiglottis was measured radiographically. In 339 horses, tracheobronchoscopy was performed 30 min after maximal exercise on a treadmill for EIPH evaluation. In 297 horses, lower airway endoscopy was performed at rest, at least 24 h after exercise, and bronchoalveolar lavage fluid (BALf) was collected in 265 horses and subjected to cytological examination.

2.2. Upper Airway Endoscopy at Rest

Following a complete clinical examination, resting upper airway endoscopy was performed in all of the horses ($n = 366$). With this aim, the horses were contained in a stock; to prevent interferences with upper airway function, no other physical or pharmacological restraint techniques were applied. A flexible videoendoscope (EC-530WL-P, Fujifilm, Tokyo, Japan) was passed through the left nasal passage, and the upper tract of the respiratory system was visualized. A 0–4 score was assigned to pharyngeal lymphoid hyperplasia (PLH) [39] observed at the level of the nasopharyngeal mucosa and/or dorsal pharyngeal recess. The larynx was visualized, and its function was assessed during spontaneous breathing and after the stimulation of laryngeal movements by inducing swallowing, performing nasal occlusion maneuvers and during the "slap test" (thoraco-laryngeal reflex). Epiglottis conformation was evaluated, and its alterations were recorded, including the loss of rigidity and convexity (epiglottis flaccidity) or the presence of an entrapping mucosal fold. Whenever dorsal displacement of the soft palate was observed spontaneously or after the induction of swallowing, this finding was recorded. The observation of an ulceration on the free margin of the soft palate was registered. Based on arytenoids abductive function, a 1–4 score was assigned to recurrent laryngeal neuropathy (RLN) [40]. Whenever an alteration suggestive of a previous surgical treatment of the upper airway was observed (i.e., laryngoplasty, ventriculectomy, cordectomy, staphylectomy, etc.), the horse was excluded from the study ($n = 6$).

2.3. High-Speed Treadmill Endoscopy

Before performing high-speed treadmill endoscopy (HSTE), the horses were conditioned to the high-speed treadmill (Sato I, Uppsala, Sweden) by at least two training sessions. All the horses ($n = 360$) were tacked with the same equipment used for racing and wore a heart rate (HR) meter (Polar, Equine Inzone FT1, Steinhausen, Switzerland) during exercise. To perform HSTE, the horses were first warmed up by a 4-min walk (1.5 m/s) and a 5-min trot (4.5 m/s for Thoroughbreds, 6 m/s for Standardbreds), with a 6° slope for Thoroughbreds and 3° slope for Standardbreds. After warm-up, the treadmill was temporarily stopped, and a videoendoscope (ETM PVG-325, Storz, Tuttlingen, Germany) was passed into the nasopharynx of the horse and held in position with Velcro® straps. Then, the treadmill was rapidly accelerated up to maximal speed (corresponding to HR $\geq$ 220 bpm) for a distance ranging from 1600 to 2100 m (based on the usual racing activity of each individual) or until the horse's fatigue [41]. The endoscopic images were visualized in real-time on a monitor and recorded on analogic (from 2000 to 2004) or digital supports (from 2005 to 2021) to allow for slow-motion analysis and storage. All the registered videos were later evaluated by the same operator, and the absence or presence of single or multiple DUAOs (DDSP, NPC, MDAF, EE, ER, DLC) was recorded. Based on the entity of airway obstruction, the DUAOs were classified as severe (DDSP, NPC, DLC and ER) or mild (MDAF and EE) [7].

2.4. Radiographic Measurement of Epiglottis Length

To measure the length of the epiglottis radiographically, a lateral view of the pharyngolaryngeal region was taken on standing horses (n = 158), with the head placed in a normal resting position. Radiodense markers of a known length were fixed on both sides of the neck and superimposed to the nasopharynx or the guttural pouches (Figure 1). The lengths of the markers were measured on the obtained radiographs and were used to determine the grade of radiographical magnification. Thyroepiglottic length was then measured on the radiograph, and a correction factor for the previously assessed magnification was applied to obtain the actual epiglottis length [18]. The reference values for epiglottis length were considered to be 8.76 $\pm$ 0.44 cm for Thoroughbreds [18,42] and 8.74 $\pm$ 0.38 cm for Standardbreds [20,42,43].

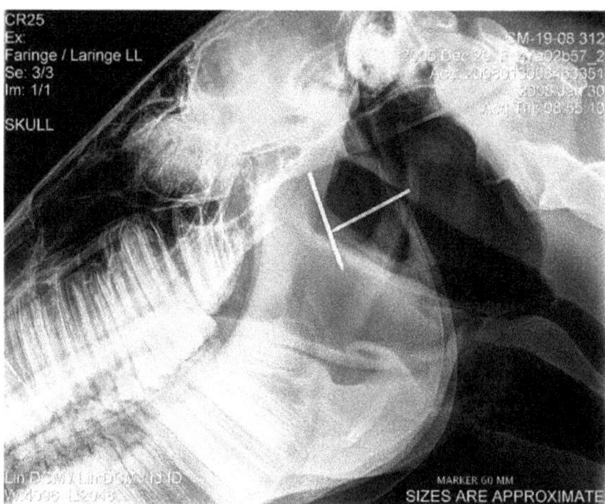

Figure 1. Example of how the radiographs were taken, with the markers of known length in place, to allow for epiglottis length measurements.

2.5. Post-Exercise Tracheobronchoscopy

Thirty minutes after the end of the HSTE, a tracheobronchoscopy was performed to verify whether exercise-induced pulmonary hemorrhage (EIPH) had occurred. The horses (n = 339) were contained in a stock and restrained with a twitch; endoscopy was performed as described above, and the lower tract of the respiratory system was examined. A 0–4 score was assigned to the presence of blood in the trachea and the mainstem bronchi [44].

2.6. Tracheobronchoscopy at Rest and BALf Collection

At least 24 h after HSTE, an endoscopy of the lower airways was performed as described above. With this aim, the horses (n = 297) were sedated with detomidine hydrochloride (0.01 mg/kg IV) and restrained with a twitch. A 0–5 score was assigned to tracheal mucus accumulation (TM) [45]. The BALf was collected in 265 horses as follows: 60 mL of a 0.5% lidocaine hydrochloride solution was sprayed at the level of the carena to inhibit a coughing reflex, and the endoscope was passed into the bronchial tree until it was wedged firmly within a segmental bronchus; here, a 300 mL sterile saline 0.9% was instilled, and the fluid immediately aspirated [46].

2.7. BALf Cytological Examination

The collected BALf was stored in sterile EDTA tubes and processed within 90 min. A few drops of pooled BALf were cytocentrifugated (Rotofix 32, Hettich Cyto System, Tuttlingen, Germany) at 500 rpm for 5 min. The slides were air dried, stained with May–Grünwald Giemsa and Perl's Prussian blue and observed under a light microscope at $400\times$ and $1000\times$ for 400-cell leukocyte differential count and the calculation of a simplified total hemosiderin score (THS) [47].

2.8. Statistical Analysis

The data were collected on an electronic sheet (Microsoft Excel, Redmond, WA, USA), analyzed with descriptive statistics and evaluated for normality by the Shapiro–Wilk test. Ages were compared between males and females and between Standardbreds and Thoroughbreds using the Mann–Whitney test. The associations between age and endoscopic scores (PLH, EIPH, TM), epiglottis length, BALf leukocyte populations and THS were evaluated by means of the Spearman correlation. The Mann–Whitney test was used to compare ages between horses with normal upper airways at rest and horses with any anatomical or functional alterations and between horses with and without DDSP at rest, epiglottis flaccidity and RLN. The same test was used to compare endoscopic scores, epiglottis lengths and BALf cytological findings between males and females and between Standardbreds and Thoroughbreds. The frequency of resting pharyngo-laryngeal alterations was compared between horses of different sexes and breeds using the Fisher's exact test. The horses were divided into four groups on the basis of the HSTE findings: no-DUAO, mild-DUAO, severe-DUAO or multiple-DUAOs. Ages was compared between the groups using the Kruskal–Wallis test and Dunn's multiple comparisons test; the distribution of sex and breed was compared between the groups by the Chi-square test. The Kruskal–Wallis test and the Dunn's multiple comparisons test were used to compare the endoscopic scores, epiglottis lengths, and BALf cytological results between the groups. The PLH score was also compared between the no-DUAO horses and the horses with dynamic DDSP, the horses with NPC and the horses with MDAF by the Mann–Whitney test; the same test was used to compare epiglottis lengths between the no-DUAO horses and the horses with dynamic DDSP. The frequency of pharyngo-laryngeal alterations at rest was compared between the groups by means of the Chi-square test and between the horses without DUAOs and the horses with dynamic DDSP, the horses with NPC and the horses with MDAF using the Fisher's exact test. The resting RLN grade was compared between the horses with and without DLC by means of the Mann–Whitney test. Moreover, the PLH score was compared between the horses with normal upper airway at rest and the horses with resting DDSP and between the horses with normal resting endoscopy and the horses with epiglottis flaccidity using the Mann–Whitney test. Finally, the frequency of DDSP at rest was compared between the horses with normal epiglottis and the horses with epiglottis flaccidity by means of the Fisher's exact test. The data are presented as the mean $\pm$ standard deviation (SD) if normally distributed and as the median and interquartile ranges (IQRs) if not normally distributed. The statistical significance was set at $p < 0.05$. The data were analyzed using a commercially available statistical software package (GraphPad Prism 9.3.1 for MacOS; GraphPad Software, San Diego, CA, USA).

3. Results

3.1. Horses

A total of 360 horses (312 Standardbreds and 48 Thoroughbreds) met the inclusion criteria. The population consisted of 233 males (203 stallions, 30 geldings) and 127 females, aged from 2 to 8 years (median 3 years, IQR 3–4 years). Among the Standardbred population, 200 horses were males (64.1%) and 112 were females (35.9%), while, among the Thoroughbred population, 33 subjects were males (68.75%) and 15 were females (31.25%). The median age was 3 years in both the Standardbred and Thoroughbred populations, with different IQRs (Standardbreds: IQR 3–4 years; Thoroughbreds: IQR 3–3 years). The horses were divided into the no-DUAO group (169 horses), mild-DUAO group (23 horses), severe-DUAO group (111 horses) and multiple-DUAOs group (57 horses).

3.2. Upper Airway Endoscopy at Rest

The frequency and grade distribution of PLH among the study population are shown in Table 1; the median PLH score was 2 (IQR 1–2). The normal appearance and function of the pharyngo-laryngeal region were observed in 160 horses (44.44%), while alterations were detected in 200 horses (55.56%); the distribution of upper airway alterations detected at resting endoscopy is displayed in Table 2.

Table 1. Frequency and distribution of pharyngeal lymphoid hyperplasia (PLH) grades among the study population.

PLH Grade	% of the Study Population
0	16.67%
1	17.22%
2	43.89%
3	20.83%
4	1.39%

Table 2. Distribution of upper airways endoscopic findings at rest.

Endoscopic Finding	% of the Study Population
Normal appearance and function	44.44%
Recurrent Laryngeal Neuropathy	30.27% 25.83% Grade II, 3.33% Grade III (0.28% III.a, 1.38% III.b, 1.67% III.c), 1.11% Grade IV
One or more episodes of dorsal displacement of the soft palate	27.22%
Epiglottis flaccidity	17.5%
Ulceration of the free margin of the soft palate	3.89%
Persistent epiglottis entrapment	1.11%

3.3. High-Speed Treadmill Endoscopy

During HSTE, no DUAO occurred in 169 horses (46.95%), one single DUAO was observed in 134 horses (37.22%) and multiple concomitant DUAOs were detected in 57 horses (15.83%). In order of frequency, the observed DUAOs included DDSP (107 horses, 29.72%), MDAF (63 horses, 17.5%), NPC (59 horses, 16.39%), DLC (19 horses, 5.28%), EE (7 horses, 1.94%) and ER (2 horses, 0.56%). The frequency of single or concomitant DUAOs is displayed in Table 3.

Table 3. Frequency of single or multiple dynamic upper airway obstructions (DUAOs) detected during high-speed treadmill endoscopy. DDSP = dorsal displacement of the soft palate; NPC = nasopharyngeal collapse; MDAF = medial deviation of aryepiglottic folds; DLC = dynamic laryngeal collapse; EE = epiglottis entrapment; ER = epiglottis retroversion.

DUAOs	% of the Study Population
DDSP	20.83%
NPC	7.22%
MDAF	6.11%
DLC	2.5%
EE	0.28%
ER	0.28%
MDAF + NPC	5.28%
MDAF + DDSP	2.22%
NPC + DDSP	1.94%
DDSP + EE	1.67%
MDAF + DLC	1.39%
DDSP + DLC	0.83%
MDAF + ER	0.28%
MDAF + NPC + DDSP	1.67%
MDAF + DDSP + DLC	0.28%
MDAF + NPC + DDSP + DLC	0.28%

3.4. Epiglottis Length

The epiglottis length was radiographically measured in 145 Standardbreds and 13 Thoroughbreds. The median epiglottis length in the investigated population was 8.2 cm (Standardbreds: IQR 7.88–8.7 cm; Thoroughbreds: IQR 7.5–8.8 cm). In the Standardbred population, the epiglottis length was within the reference ranges in 60 horses (41.67%), lower in 73 horses (50.69%) and higher in 11 horses (7.64%); in the Thoroughbreds population, the epiglottis length was within the normal limits in 2 horses (15.38%), lower in 8 horses (61.54%) and higher in 3 horses (23.08%).

3.5. Lower Airway Endoscopy and BALf Cytology

The tracheobronchoscopy performed 30 min after maximal exercise was negative for EIPH in 145 horses (42.77%) and positive in 194 horses (57.23%); the frequency and grade distributions of EIPH among the study population are shown in Table 4. The median EIPH grade was 1 (IQR 0–2). On the lower airway endoscopy performed at least after 24 h, tracheal mucus accumulation was observed in 296 horses (86.6%); the grade distribution is displayed in Table 5. The median TM score was 1 (IQR 1–2). The cytological examination of BALf showed a median of 46% (40–50.5%) macrophages, a mean of 35.72% ± 11.85% lymphocytes, a median of 10% (5–18%) neutrophils, a median of 1% (0–3%) eosinophils and a median of 4% (3–6%) mast cells. The median THS was 36 (12–66).

Table 4. Frequency and grade distribution of exercise-induced pulmonary hemorrhage (EIPH) among the study population.

EIPH Grade	Number of Horses	% of the Study Population
0	145	42.77%
1	75	22.12%
2	74	21.83%
3	34	10.03%
4	11	3.25%

3.6. Influence of Age, Breed and Sex

Among the study population, the Thoroughbreds were younger than the Standardbreds ($p = 0.0054$), and the females were younger than the males ($p < 0.0001$). Sex was equally distributed among the different breeds. Age was inversely correlated with PLH grade ($p < 0.0001$, r = −0.45) and TM score ($p = 0.0083$, r = −0.14) and was positively correlated with EIPH score ($p = 0.0053$, r = 0.15) and THS ($p = 0.0165$, r = 0.15); PLH score was significantly higher in females (median 2, IQR 2–3) than in males (median 2, IQR 1–2) ($p < 0.0001$). The horses showing pharyngo-laryngeal alterations at resting endoscopy were younger than the horses with normal appearance and function ($p = 0.0213$); moreover, pharyngo-laryngeal alterations were significantly more frequent in Thoroughbreds (83.33%) than in Standardbreds (51.28%) ($p < 0.0001$). In particular, DDSP at rest and grade > 1 RLN was more frequently observed in Thoroughbreds (DDSP 47.92%, RLN 63.83%) than in Standardbreds (DDSP 24.04%, RLN 25.32%) (DDSP $p = 0.0014$; RLN $p < 0.0001$). At BALf cytology, the neutrophils counts were higher in Thoroughbreds (median 17%, IQR 11–22.5%) than in Standardbreds (median 9%, IQR 4–17%) ($p = 0.0016$), while the lymphocyte counts were higher in Standardbreds (36.25% ± 11.78%) than in Thoroughbreds (30.92% ± 11.63%) ($p = 0.0293$). Among the different groups (no-DUAO, mild-DUAO, severe-DUAO and multiple-DUAOs), no differences were observed in age and sex distribution, while a significant difference in breed distribution was detected ($p = 0.0064$); in particular, in Thoroughbreds, DUAOs were significantly more frequent than they were in Standardbreds (Thoroughbreds 75%, Standardbreds 50.32%; $p = 0.001$).

Table 5. Frequency and grade distribution of tracheal mucus accumulation (TM) among the study population.

TM Grade	Number of Horses	% of the Study Population
0	45	13.2%
1	142	41.64%
2	117	34.31%
3	33	9.68%
4	4	1.17%

3.7. Upper Airway Endoscopy and Epiglottis Length vs. DUAOs

The PLH grade was significantly lower in the multiple-DUAOs group (median 1, IQR 0–2) compared to the no-DUAO group (median 2, IQR 1–3) ($p = 0.0066$) and the severe-DUAO group (median 2, IQR 1–2). No differences were observed in PLH between the horses with and without DDSP, MDAF and NPC. The epiglottis length did not differ between the groups, nor did it differ between the horses with and without DDSP. The frequency of abnormalities of pharyngo-laryngeal appearance and function detected at resting endoscopy differed between the groups ($p = 0.0003$) (Figure 2); in particular, alterations were more frequently observed in the severe-DUAO group (71.17%) compared to the no-DUAO group (44.97%) ($p < 0.0001$). While no differences were observed between the groups regarding the frequency of DDSP episodes observed during resting endoscopy, a trend was detected for different frequencies of epiglottis flaccidity among the different groups ($p = 0.0534$). In particular, a flaccid appearance of epiglottis was more frequently observed in the horses with DDSP diagnosed at HSTE (60.78%) compared to the horses without DDSP (39.22%) ($p = 0.0007$) (Figure 3). No differences were observed in the frequency of DDSP at rest between the horses with and without DDSP during HSTE, nor in the frequency of resting DDSP and epiglottis flaccidity between the horses with and without NPC and MDAF. The horses showing DDSP episodes at rest had a higher PLH grade (median 2, IQR 1–3) compared to the horses without resting DDSP (median 2, IQR 1–2) ($p = 0.0052$) and more frequently showed a flaccid appearance of the epiglottis (resting DDSP 31.63%, no resting DDSP 12.21%, $p < 0.0001$). Among the seven horses showing EE during exercise, three also presented persistent EE visible at resting endoscopy, two had a normal epiglottis and two had a flaccid appearance of the epiglottis; in four cases, an ulceration of the free margin of the soft palate was observed. Ulcerations of the margin of the soft palate were also observed in the four horses without DUAO, the two horses with MDAF + NPC, the two horses with DDSP and the two horses with NPC + DDSP. In the horses with DLC, the RLN grade was significantly higher (median 3, IQR 3–3) compared to the horses with normal dynamic arytenoids abduction (median 1, IQR 1–2) ($p < 0.0001$). In particular, all the horses with grade 1 (normal) RLN showed a normal laryngeal abduction during HSTE, and among the horses with grade 2 RLN, only two horses showed DLC (2.15%); in contrast, all the horses with grade 3 and grade 4 RLN showed DLC on HSTE.

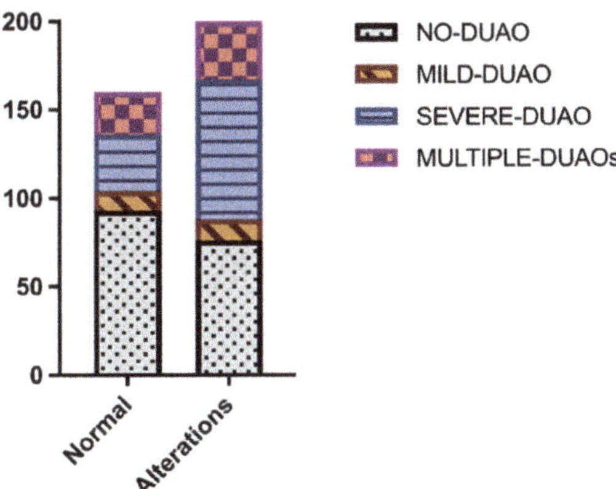

Figure 2. Stacked bars showing the frequency of alterations at upper airway resting endoscopy in the different groups (no-DUAO, mild-DUAO, severe-DUAO and multiple-DUAOs).

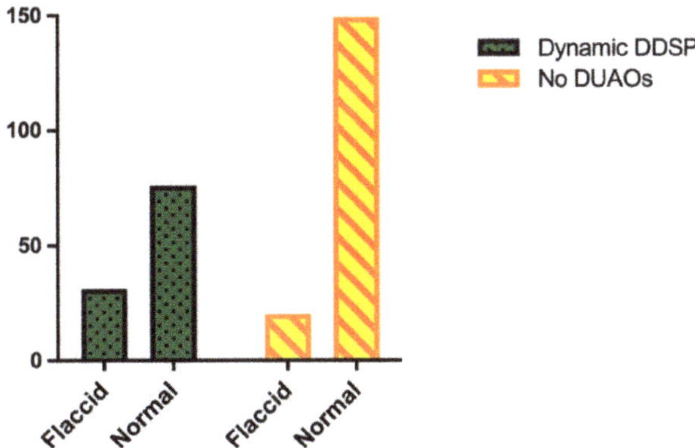

Figure 3. Bar graph showing the frequency of normal or flaccid appearance of epiglottis at upper airway resting endoscopy in the horses with no DUAOs and with DDSP diagnosed at HSTE.

3.8. Lower Airway Endoscopy and BALf Cytology vs. DUAOs

No differences between the groups were observed regarding the TM score, EIPH score, leukocyte differential count of BALf or THS.

4. Discussion

Although DUAOs are disorders that have long been known to cause poor performance in racehorses, their precise pathogenetic mechanisms have not been fully understood yet, and the possible role of structural or functional abnormalities of the nasopharynx has not been univocally clarified [8–11]. Moreover, some authors suggest that inflammation of the upper and lower respiratory tract may contribute to the development of DUAO, but contrasting results have been reported [26–33]. Therefore, the present study aimed to investigate the diagnostic role of upper and lower airways resting evaluation for a better understanding of DUAO development.

The value of this study relies on a very large and homogeneous population of racehorses, including a total of 360 patients, and on a highly standardized protocol for airways evaluation in both resting and dynamic conditions. Among the study population, 53.05% of the horses showed DUAO on HSTE; the prevalence of DUAO reported in the literature ranges from 22.6% to 80.1% [48,49], based on the kind of population (i.e., racehorses or other performance horses), inclusion criteria (randomized or horses with a history of poor performance and respiratory noises) and diagnostic technique (HSTE or overground endoscopy). Among reports including both Standardbreds and Thoroughbreds diagnosed by HSTE, DUAO prevalence varies from 22.6% [48] to 65% [22]. Interestingly, the highest frequencies have been reported by studies only including Thoroughbred racehorses, with percentages ranging from 72.6% to 80.1% [5,49–51]; similarly, in our study 75% of the Thoroughbred patients showed DUAO during HSTE, which was significantly more frequent than in Standardbreds (50.32%), suggesting a breed predisposition. However, in our population, most Standardbreds were referred for poor performance, while Thoroughbreds were more often referred for abnormal respiratory noises during exercise, with a consequent higher probability of being diagnosed with DUAO; this could have biased the results of our study. An apparent predisposition of Thoroughbreds to develop DUAO, and especially DDSP, in comparison with Standardbreds was also observed by previous researchers, but the reason for this breed discrepancy is unclear [11,52,53]. In our study, the Thoroughbreds more commonly showed upper respiratory tract alterations at rest; it must be noticed that the Thoroughbreds were younger than the Standardbreds, probably due to the shorter racing career, and that alterations at rest were more frequently observed in younger horses. Some authors suggest that young age may predispose horses to the development of DUAO [5,9] and that affected horses could spontaneously resolve this condition over time [10]; in fact, it has been hypothesized that increased palatal musculature in older horses may improve muscular activity and provide a better resistance against collapsing forces [5]. In contrast, other reports found no association between age and DUAO [11] or even a higher susceptibility in older horses [54].

In the present study, the most frequently observed forms of DUAO were DDSP (29.72%), MDAF (17.5%) and NPC (16.39%); likewise, most studies about racehorses report DDSP as the most common DUAO, with the prevalence ranging from 14% to 45.2% [5,14,22,48–51,55]. MDAF was the most common DUAO in a previous study [11], but its prevalence in the literature is highly variable, ranging from 4% to 40.4% [5,11,22,49–51]. Both in the present study and in the previous reports, MDAF is rarely diagnosed alone and is associated, in most cases, with other types of DUAO, especially NPC and DDSP; therefore, various authors have suggested that MDAF may either contribute or follow the development of associated DUAOs [5,8,11,12,38]. More generally, multiple concomitant DUAOs were detected relatively frequently in our population, with a prevalence of 15.83% of the total horses and of 29.84% of the horses with DUAO. Multiple DUAOs have also been commonly reported by previous studies, with a wide range of prevalence varying from 7% to 56.9% [4,5,11,15,22,49,50]. In the present report, NPC showed a prevalence of 16.39%, which was similar to those reported by some authors [4,14,50,51] but quite higher when compared to other studies [5,11,22,23,49]. The great variability of the prevalence of DUAOs, as reported by different studies, may be attributable to heterogeneous enrolled populations, diagnostic techniques and the lack of a consensus statement which could provide universal guidelines for the classification of different DUAOs [56]. The only upper respiratory tract for which a consensus statement exists is RLN detectable at rest [40]; in the present study, all of the horses with grade 1 RLN and 97.85% of the horses with grade 2 RLN showed a normal laryngeal function during exercise. In contrast, some authors have reported that low percentages (from 0.34% to 3.5%) of horses with grade 1 RLN and up to 11.9% of horses with grade 2 RLN may show DLC on HSTE [2,23,57]. Moreover, in our study, all horses with grade 3 and grade 4 RLN developed DLC on HSTE; conversely, previous studies reported a normal laryngeal function during exercise in 16% to 33% of the horses with grade 3 RLN [23,57], and it has been reported that 2% of horses with grade 4 RLN may still show a residual laryngeal activity during exercise without experiencing DLC [57]. Therefore, independently from the RLN grade at rest, HSTE should be always encouraged to evaluate dynamic laryngeal function.

It has been hypothesized that epiglottic hypoplasia may predispose horses to DUAOs, such as DDSP, MDAF and EE [12–15,18]; in the present study, epiglottis length was assessed radiographically, and alterations of its conformation visible at endoscopic examination were recorded. Interestingly, only 39.49% of our patients had a normal epiglottis length falling within the reference ranges [18,20,42,43]; a shorter epiglottis was observed in 51.59%, while the remaining 8.92% had a longer epiglottis. As the technique used for radiographic measurement was the same as that described by Linford (1983) [18], and our population of Standardbreds is by far the widest that underwent radiological assessment of epiglottis length, it is possible that the reference ranges may be reviewed, including more numerous populations of Standardbreds and Thoroughbreds. Unfortunately, only 158 horses in the present study underwent epiglottis length measurement, of which 145 were Standardbreds; the low number of Thoroughbreds subjected to this measurement represents a main limitation for the evaluation of the possible relationship between epiglottis length and the development of DUAOs. In our study epiglottis length did not influence the development of DUAO or the development of any resting alterations; this finding is in contrast with a previous one reporting a shorter epiglottis as a predisposing factor to DDSP [18]. However, similarly to our findings, other studies failed to detect any association between epiglottis length and DUAO [9,19]. Therefore, some authors suggested that the epiglottis may play a role in DUAO development—not for its length but for its conformation. In particular, the loss of rigidity and of the normal convex appearance has been associated with DDSP and MDAF [12,58]. Our findings partially confirm this hypothesis: among our population, the horses showing a flaccid appearance of epiglottis at endoscopic examination were more prone to develop DDSP both at rest and during exercise, while no relationship was observed with MDAF. As alterations at resting endoscopy were more frequent in the severe-DUAO group, we also evaluated whether DDSP detected at rest could be a good predictor of DDSP on HSTE; however, no association between DDSP at rest and dynamic DDSP was observed. Conversely, previous studies reported a low sensitivity but a high specificity of DDSP detected at rest [2,22,27]. In any case, upper respiratory tract endoscopy at rest is advised in cases of suspected DUAO, but dynamic endoscopy is confirmed to always be essential for a certain diagnosis.

It has been hypothesized that inflammation of the upper airways may predispose horses to DUAO [8,9,25–30]. In particular, PLH has been associated with DDSP [9,27,28], NPC [29,30] and ER [8]. Surprisingly, in our study, PLH was lower in horses belonging to the multiple-DUAOs group; a possible explanation for this may be that the horses in the multiple-DUAOs group were older than those in the other groups, even if a statistical significance was not detected. As the PLH grade was lower in older horses, a lower PLH grade in the multiple-DUAOs group may ensue, without implying any causative relation between a low PLH grade and the concomitant occurrence of multiple DUAOs. When individually evaluating different DUAOs, no association was detected between PLH and DDSP, MDAF or NPC. Unfortunately, in our study, it was not possible to individually evaluate the other types of DUAOs due to the low number of affected cases. Interestingly, the PLH grade was higher in horses experiencing DDSP at rest but not during exercise. It could be hypothesized that upper airway inflammation may have an influence on pharyngo-laryngeal movements at rest; however, during exercise, more forces come into play, and the contribution of pharyngeal inflammation may be too low to have an impact on pharyngeal stability.

Some authors have also reported an association between DUAO and LAI [9,28,30–33]; in particular, a relation between BALf neutrophilia and DDSP was reported in two studies from the same research group [9,28], while other studies detected an association between mild-moderate equine asthma and NPC [30,33]. However, multiple studies reported no association between upper and lower airways inflammation based on endoscopic and cytological findings [46,59–61], and other authors found no relationship between LAI and DUAO [38]. Similarly, in the present study, no differences between the groups were observed concerning tracheal mucus accumulation or the BALf cytological profile, suggesting that LAI is not associated with the onset of DUAO. Another lower airway disorder which has been associated with DUAO is EIPH [1,31,36,37]; however, analogously to a previous study [38], we found no relationship between DUAO and EIPH based on both post-exercise endoscopy and total hemosiderin score. Therefore, the findings of the present study do not support the theory of a possible contribution of upper and lower airways inflammation to the onset of DUAO, nor the cause–effect relationship between DUAO and EIPH; however, given the contrasting results reported by different studies, these hypotheses cannot be ruled out.

5. Conclusions

In conclusion, DUAOs were more commonly observed in Thoroughbred racehorses than in Standardbred racehorses, confirming a possible breed predisposition; conversely, age did not seem to influence the onset of DUAO on HSTE, as pharyngo-laryngeal alterations were detected more frequently in younger horses only at resting upper airway endoscopy. The detection, at resting endoscopy, of the abnormal appearance and function of the pharyngo-laryngeal region was associated with the development of DUAO: in particular, a flaccid appearance of the epiglottis, with a loss of convexity and rigidity, was associated with the occurrence of DDSP both at rest and on HSTE. In contrast, the epiglottis length measured radiographically was not associated with DUAO, suggesting that the epiglottis may be involved in maintaining the stability of the upper respiratory tract, not based on its dimensions but on its conformation. Finally, in the present study, no association was observed between DUAO and the inflammation of the upper and lower airways, nor between DUAO and EIPH.

Author Contributions: Conceptualization, C.M.L.F., G.S. and F.F.; methodology, C.M.L.F. and G.S.; formal analysis, C.M.L.F. and L.S.; investigation, C.M.L.F., G.S., F.C., L.S., B.C., E.Z. and F.F.; data curation, C.M.L.F., G.S. and F.C.; writing—original draft preparation, C.M.L.F., G.S., L.S. and F.F.; writing—review and editing, C.M.L.F., G.S., F.C., L.S., B.C., E.Z. and F.F.; visualization, C.M.L.F., G.S. and F.F.; supervision, F.F. All authors have read and agreed to the published version of the manuscript.

Funding: This research received no external funding.

Institutional Review Board Statement: Ethical review and approval were waived for this study, as it consists of a retrospective study and all the procedures were performed on clinical horses for diagnostic purposes and included informed owner consent for the use of clinical data. Moreover, all the procedures were performed according to relevant guidelines as a part of standard diagnostic protocols. The European directive no. 2010/63/UE about the protection of animals used for scientific purposes does not apply to retrospective studies about clinical patients that did not undergo any procedures for research aims, as long as informed consent for the use of clinical data was obtained from the animals' owners or holders.

Informed Consent Statement: Written informed consent was obtained from the owners—or authorized agents for the owners—of the animals included in the study for the use of data for research purposes.

Data Availability Statement: The data presented in this study are available on request from the corresponding author.

Acknowledgments: The authors wish to acknowledge the University of Milan for the support through the APC initiative.

Conflicts of Interest: The authors declare no conflict of interest.

References

1. Van Erck-Westergren, E.; Franklin, S.H.; Bayly, W.M. Respiratory diseases and their effects on respiratory function and exercise capacity. *Equine Vet. J.* **2013**, *45*, 376–387. [CrossRef]
2. Lane, J.G.; Bladon, B.; Little, D.R.M.; Naylor, J.R.J.; Franklin, S.H. Dynamic obstructions of the equine upper respiratory tract. Part 2: Comparison of endoscopic findings at rest and during high-speed treadmill exercise of 600 Thoroughbred racehorses. *Equine Vet. J.* **2006**, *38*, 401–408. [CrossRef] [PubMed]
3. Rakesh, V.; Ducharme, N.G.; Cheetham, J.; Datta, A.K.; Pease, A.P. Implications of different degrees of arytenoid cartilage abduction on equine upper airway characteristics. *Equine Vet. J.* **2018**, *40*, 629–635. [CrossRef] [PubMed]
4. Martin, B.B.; Reef, V.B.; Parente, E.J.; Sage, A.D. Causes of poor performance of horses during training, racing, or showing: 348 cases (1992–1996). *J. Am. Vet. Med. Assoc.* **2000**, *216*, 554–558. [CrossRef]
5. Lane, J.G.; Bladon, B.; Little, D.R.M.; Naylor, J.R.J.; Franklin, S.H. Dynamic obstructions of the equine upper respiratory tract. Part 1: Observations during high-speed treadmill endoscopy of 600 Thoroughbred racehorses. *Equine Vet. J.* **2006**, *38*, 393–399. [CrossRef] [PubMed]
6. Franklin, S.H.; Van Erck-Westergren, E.; Bayly, W.M. Respiratory responses to exercise in the horse. *Equine Vet. J.* **2012**, *44*, 726–732. [CrossRef]
7. Lo Feudo, C.M.; Stucchi, L.; Cavicchioli, P.; Stancari, G.; Conturba, B.; Zucca, E.; Ferrucci, F. Association between dynamic upper airway obstructions and fitness parameters in Standardbred racehorses during high-speed treadmill exercise. *J. Am. Vet. Med. Assoc.* **2022**. [CrossRef]
8. Franklin, S.H.; Allen, K.J. Assessment of dynamic upper respiratory tract function in the equine athlete. *Equine Vet. Educ.* **2017**, *29*, 92–103. [CrossRef]
9. Courouce-Malblanc, A.; Deniau, V.; Rossignol, F.; Corde, R.; Leleu, C.; Maillard, K.; Pitel, P.H.; Pronot, S.; Fortier, G. Physiological measurements and prevalence of lower airway diseases in Trotters with dorsal displacement of the soft palate. *Equine Vet. J.* **2010**, *42*, 246–255. [CrossRef]
10. Parente, E.J. Upper Airway Conditions Affecting the Equine Athlete. *Vet. Clin. N. Am. Equine Pract.* **2018**, *34*, 427–441. [CrossRef]
11. Tan, R.H.H.; Dowling, B.A.; Dart, A.J. High-speed treadmill videoendoscopic examination of the upper respiratory tract in the horse: The results of 291 clinical cases. *Vet. J.* **2005**, *170*, 243–248. [CrossRef] [PubMed]
12. Allen, K.; Franklin, S. Characteristics of palatal instability in Thoroughbred racehorses and their association with the development of dorsal displacement of the soft palate. *Equine Vet. J.* **2013**, *45*, 354–359. [CrossRef] [PubMed]
13. Haynes, P.F. Persistent dorsal displacement of the soft palate associated with epiglottic shortening in two horses. *J. Am. Vet. Med. Ass.* **1981**, *179*, 677–681.
14. Strand, E.; Skjerve, E. Complex dynamic upper airway collapse: Associations between abnormalities in 99 harness racehorses with one or more dynamic disorders. *Equine Vet. J.* **2012**, *44*, 524–528. [CrossRef]
15. Garrett, K.S.; Pierce, S.W.; Embertson, R.M.; Stromberg, A.J. Endoscopic evaluation of arytenoid function and epiglottic structure in Thoroughbred yearlings and association with racing performance at two to four years of age: 2954 cases (1998–2001). *J. Am. Vet. Med. Assoc.* **2010**, *236*, 669–673. [CrossRef]
16. Parente, E.J.; Martin, B.B.; Tulleners, E.P.; Ross, M.W. Dorsal Displacement of the Soft Palate in 92 Horses During High-Speed Treadmill Examination (1993–1998). *Vet. Surg.* **2002**, *31*, 507–512. [CrossRef]
17. Parente, E.J.; Martin, B.B. Correlation between standing endoscopic examinations and those made during high-speed exercise in horses—150 cases. *AAEP Proc.* **1995**, *41*, 170.
18. Linford, R.L.; O'Brien, T.R.; Wheat, J.D.; Meagher, D.M. Radiographic assessment of epiglottic length and pharyngeal and laryngeal diameters in the Thoroughbred. *Am. J. Vet. Res.* **1983**, *44*, 1660–1666.
19. Rehder, R.S.; Ducharme, N.G.; Hackett, R.P.; Nielan, G.J. Measurement of upper airway pressures in exercising horses with dorsal displacement of the soft palate. *Am. J. Vet. Res.* **1995**, *56*, 269–274.
20. Beard, W. Upper respiratory causes of exercise intolerance. *Vet. Clin. N. Am. Equine Pract.* **1996**, *12*, 435–455. [CrossRef]
21. Hobo, S.; Matsuda, Y.; Yoshida, K. Prevalence of Upper Respiratory Tract disorders detected with a flexible videoendoscope in Thoroughbred racehorses. *J. Vet. Med. Sci.* **1995**, *57*, 409–413. [CrossRef] [PubMed]
22. Kannegieter, N.J.; Dore, M.L. Endoscopy of the upper respiratory tract during treadmill exercise: A clinical study of 100 horses. *Aust. Vet. J.* **1995**, *72*, 101–107. [CrossRef] [PubMed]
23. Barakzai, S.Z.; Dixon, P.M. Correlation of resting and exercising endoscopic findings for horses with dynamic laryngeal collapse and palatal dysfunction. *Equine Vet. J.* **2011**, *43*, 18–23. [CrossRef]
24. Gille, D.; Lavoie, J.P. Review of seven cases of ulcers of the soft palate. *Equine Pract.* **1996**, *18*, 9–13.
25. Tessier, C. The equine nasopharynx in dynamic upper airway disorders: An update. *Pferdeheilkunde* **2006**, *22*, 565–569. [CrossRef]
26. Holcombe, S.J.; Derksen, F.J.; Stick, J.A.; Robinson, N.E. Effect of bilateral blockade of the pharyngeal branch of the vagus nerve on soft palate function in horses. *Am. J. Vet. Res.* **1998**, *59*, 504–508.
27. Tomaszewska, K.; Wierzbicka, M.; Masko, M.; Gajewski, Z. A resting and dynamic endoscopy as diagnostic tools in decreasing training capacity in horses—A review. *Pol. J. Nat. Sci.* **2019**, *34*, 595–605.
28. Courouce-Malblanc, A.; Pronost, S.; Fortier, G.; Corde, R.; Rossignol, F. Physiological measurements and upper and lower respiratory tract evaluation in French Standardbred Trotters during a standardised exercise test on the treadmill. *Equine Vet. J. Suppl.* **2002**, *34*, 402–407. [CrossRef]

29. Holcombe, S.J.; Derksen, F.J.; Stick, J.A.; Robinson, N.E. Pathophysiology of dorsal displacement of the soft palate in horses. *Equine Vet. J.* **1999**, *30*, 45–48. [CrossRef]

30. Van Erck, E. Dynamic respiratory videoendoscopy in ridden sport horses: Effect of head flexion, riding and airway inflammation in 129 cases. *Equine Vet. J.* **2011**, *43*, 18–24. [CrossRef]

31. Joo, K.; Povazsai, A.; Nyerges-Bohak, Z.; Szenci, O.; Kutasi, O. Asthmatic Disease as an Underlying Cause of Dorsal Displacement of the Soft Palate in Horses. *J. Equine Vet. Sci.* **2021**, *96*, 103308. [CrossRef] [PubMed]

32. Trope, G. Dynamic endoscopy of the equine upper airway—What is significant? *Vet. Rec.* **2013**, *172*, 499–500. [CrossRef] [PubMed]

33. Wysocka, B.; Klucinski, W. The occurrence of dynamic structural disorders in the pharynx and larynx, at rest and during exercise, in horses diagnosed with mild and moderate Equine Asthma (Inflammatory Airway Disease). *Pol. J. Vet. Sci.* **2018**, *21*, 203–211. [PubMed]

34. Couetil, L.L.; Rosenthal, F.S.; Denicola, D.B.; Chilcoat, C.D. Clinical signs, evaluation of bronchoalveolar lavage fluid, and assessment of pulmonary function in horses with inflammatory respiratory disease. *Am. J. Vet. Res.* **2001**, *62*, 538–546. [CrossRef]

35. Richard, E.A.; Fortier, G.D.; Denoix, J.M.; Art, T.; Lekeux, P.M.; Van Erck, E. Influence of subclinical inflammatory airway disease on equine respiratory function evaluated by impulse oscillometry. *Equine Vet. J.* **2009**, *41*, 384–389. [CrossRef]

36. Cook, W.R.; Williams, R.; Kirker-Head, C. Upper airway obstruction partial asphyxia as possible cause of exercise-induced pulmonary hemorrhage in the horse: An hypothesis. *Equine. Vet. J.* **1988**, *8*, 11–26. [CrossRef]

37. Ducharme, N.G.; Hackett, R.P.; Gleed, R.D.; Ainsworth, D.M.; Erb, H.N.; Mitchell, L.M.; Soderholm, L.V. Pulmonary capillary pressure in horses undergoing alteration of pleural pressure by imposition of various upper airway resistive loads. *Equine Vet. J.* **2010**, *31*, 27–33. [CrossRef]

38. Davidson, E.J.; Harris, M.; Martin, B.B.; Nolen-Watson, R.; Boston, R.C.; Reef, V. Exercising Blood Gas Analysis, Dynamic Upper Respiratory Tract Obstruction, and Postexercising Bronchoalveolar Lavage Cytology. A Comparative Study in Poor Performing Horses. *J. Equine Vet. Sci.* **2011**, *31*, 475–480. [CrossRef]

39. Holcombe, S.J.; Ducharme, N.G. Disorders of the Nasopharynx and Soft Palate. In *Equine Respiratory Medicine and Surgery*, 1st ed.; McGorum, B., Dixon, P., Robinson, E., Schumacher, J., Eds.; Saunders Elsevier: Philadelphia, PA, USA, 2007; pp. 437–457.

40. Robinson, N.E. Consensus statements on equine recurrent laryngeal neuropathy: Conclusions of the Havemeyer Workshop. September 2003, Stratford- upon-Avon. *Equine Vet. Educ.* **2004**, *16*, 333–336. [CrossRef]

41. Ferrucci, F.; Zucca, E.; Di Fabio, V.; Ferro, E. Treadmill Endoscopic Findings in 15 Racehorses Presented for Poor Performance. *Vet. Res. Commun.* **2003**, *27*, 395–397. [CrossRef]

42. Davenport-Goodall, C.L.; Parente, E.J. Disorders of the larynx. *Vet. Clin. N. Am. Equine Pract.* **2003**, *19*, 169–187. [CrossRef]

43. Butler, J.A.; Colles, C.M.; Dyson, S.J.; Kold, S.E.; Poulos, P.W. The Head. In *Clinical Radiology of the Horse*, 2nd ed.; Butler, J.A., Colles, C.M., Dyson, S.J., Kold, S.E., Poulos, P.W., Eds.; Blackwell Science Ltd.: Oxford, UK, 2000; pp. 327–402.

44. Hinchcliff, K.W.; Jackson, M.A.; Brown, J.A.; Dredge, A.F.; O'Callaghan, P.A.; McCaffrey, J.P.; Morley, P.S.; Slocombe, R.E.; Clarke, A.F. Tracheobronchoscopic assessment of exercise-induced pulmonary hemorrhage in horses. *Am. J. Vet. Res.* **2005**, *66*, 596–598. [CrossRef] [PubMed]

45. Gerber, V.; Straub, R.; Marti, E.; Hauptman, J.; Herholz, C.; King, M.; Imhof, A.; Tahon, L.; Robinson, N.E. Endoscopic scoring of mucus quantity and quality: Observer and horse variance and relationship to inflammation, mucus viscoelasticity and volume. *Equine Vet. J.* **2004**, *36*, 576–582. [CrossRef] [PubMed]

46. Lo Feudo, C.M.; Stucchi, L.; Alberti, E.; Stancari, G.; Conturba, B.; Zucca, E.; Ferrucci, F. The Role of Thoracic Ultrasonography and Airway Endoscopy in the Diagnosis of Equine Asthma and Exercise-Induced Pulmonary Hemorrhage. *Vet. Sci.* **2021**, *8*, 276. [CrossRef] [PubMed]

47. Lo Feudo, C.M.; Stucchi, L.; Stancari, G.; Alberti, E.; Conturba, B.; Zucca, E.; Ferrucci, F. Associations between Exercise-Induced Pulmonary Hemorrhage (EIPH) and Fitness Parameters Measured by Incremental Treadmill Test in Standardbred Racehorses. *Animals* **2022**, *12*, 449. [CrossRef] [PubMed]

48. Morris, E.A.; Seeherman, H.J. Clinical evaluation of poor performance in the racehorse: The results of 275 evaluations. *Equine Vet. J.* **1991**, *23*, 169–174. [CrossRef]

49. Davison, J.A.; Lumsden, J.M.; Boston, R.C.; Ahern, B.J. Overground endoscopy in 311 Thoroughbred racehorses: Findings and correlation to resting laryngeal function. *Aust. Vet. J.* **2017**, *95*, 338–342. [CrossRef]

50. Mirazo, J.E.; Page, P.; Rubio-Martinez, L.; Marais, H.J.; Lyle, C. Dynamic upper respiratory abnormalities in Thoroughbred racehorses in South Africa. *J. S. Afr. Vet. Assoc.* **2014**, *85*, 1140. [CrossRef]

51. Oellers, S.; Barton, A.K.; Ohnesorge, B. Frequencies of dynamic obstructions of the upper respiratory tract in 135 Thoroughbreds during on-board exercise endoscopy. *Pferdeheilkunde* **2018**, *3*, 223–231. [CrossRef]

52. Durando, M.M.; Martin, B.B.; Hammer, E.J.; Langsam, S.P.; Birks, E.K. Dynamic upper airway changes and arterial blood gas parameters during treadmill exercise. *Equine Vet. J. Suppl.* **2002**, *34*, 408–412. [CrossRef]

53. Durando, M.M.; Martin, B.B.; Davidson, E.J.; Birks, E.K. Correlations between exercising arterial blood gas values, tracheal wash findings and upper respiratory tract abnormalities in horses presented for poor performance. *Equine Vet. J. Suppl.* **2006**, *36*, 523–528. [CrossRef] [PubMed]

54. Ordidge, R.M. The treatment of dorsal displacement of the soft palate by thermal cautery: A review of 252 cases. In Proceedings of the 2001 World Equine Airways Symposium, Edingurgh, UK, 19–23 July 2001.

55. Priest, D.T.; Cheetham, J.; Regner, A.L.; Mithcell, L.; Soderholm, L.V.; Tamzali, Y.; Ducharme, N.G. Dynamic respiratory endoscopy of Standardbred racehorses during qualifying races. *Equine Vet. J.* **2012**, *44*, 529–534. [CrossRef] [PubMed]
56. Barnett, T.P.; Smith, L.C.R.; Cheetham, J.; Barakzai, S.Z.; Southwood, L.; Marr, C.M. A call for consensus on upper airway terminology. *Equine Vet. J.* **2015**, *47*, 505–507. [CrossRef]
57. Elliot, S.; Cheetham, J. Meta-analysis evaluating resting laryngeal endoscopy as a diagnostic tool for recurrent laryngeal neuropathy in the equine athlete. *Equine Vet. J.* **2019**, *51*, 167–172. [CrossRef] [PubMed]
58. Kelly, P.G.; Reardon, R.J.; Johnston, M.S.; Pollock, P.J. Comparison of dynamic and resting endoscopy of the upper portion of the respiratory tract in 57 Thoroughbred yearlings. *Equine Vet. J.* **2013**, *45*, 700–704. [CrossRef] [PubMed]
59. Koblinger, K.; McDonald, N.K.; Wasko, A.; Logie, N.; Weiss, M.; Leguillette, R. Endoscopic Assessment of Airway Inflammation in Horses. *J. Vet. Intern. Med.* **2011**, *25*, 1118–1126. [CrossRef] [PubMed]
60. Holcombe, S.J.; Robinson, N.E.; Derksen, F.J.; Bertold, B.; Genovese, R.; Miller, R.; De Feiter Rupp, H.; Carr, E.A.; Eberhardt, S.W.; Boruta, D.; et al. Effect of tracheal mucus and tracheal cytology on racing performance in Thoroughbred racehorses. *Equine Vet. J.* **2006**, *38*, 300–304. [CrossRef]
61. Holcombe, S.J.; Jackson, C.; Gerber, V.; Jefcoat, A.; Berney, C.; Eberhardt, S.; Robinson, N.E. Stabling is associated with airway inflammation in young Arabian horses. *Equine Vet. J.* **2001**, *33*, 244–249. [CrossRef]

Article

Giant Multinucleated Cells Are Associated with Mastocytic Inflammatory Signature Equine Asthma

Ilaria Basano [1,†], Alessandra Romolo [1,†], Giulia Iamone [1], Giulia Memoli [1], Barbara Riccio [1], Jean-Pierre Lavoie [2], Barbara Miniscalco [1] and Michela Bullone [1,*]

[1] Department of Veterinary Sciences, University of Turin, 10095 Grugliasco, Italy; ilaria.basano@unito.it (I.B.); alessandra.romolo@edu.unito.it (A.R.); giulia.iamone@unito.it (G.I.); giulia.memoli@unito.it (G.M.); barbara.riccio@unito.it (B.R.); barbara.miniscalco@unito.it (B.M.)
[2] Department of Clinical Sciences, Faculty of Veterinary Medicine, Université de Montréal, Saint-Hyacinthe, QC J2S 2M2, Canada; jean-pierre.lavoie@umontreal.ca
* Correspondence: michela.bullone@unito.it
† These authors contributed equally to this work.

Simple Summary: Lung inflammation is commonly assessed in asthmatic horses by bronchoalveolar lavage fluid (BALF) cytology. Among the cell types commonly found in equine BALF samples, macrophages are the most abundant. The clinical significance of the abnormal cytological appearance of macrophages in samples from diseased horses is largely disregarded. The present work focuses on cytological alterations observed in macrophages during a chronic inflammatory disease such as equine asthma. Our data, although limited in number, support macrophage fusion (resulting in multinucleated giant cells) as a mechanism of interest in the classification of equine asthma phenotypes, as it was significantly associated with the inflammatory signature and chronicity of the disease.

Abstract: Equine asthma is currently diagnosed by the presence of increased neutrophil (>5%), mast cell (>2%), and/or eosinophil (>1%) differential cell count. Macrophages are normal resident cells within the alveoli. Their presence in BALF is considered normal, but the clinical implication of the presence of activated or fused macrophages (giant multinucleated cells, GMC) is currently overlooked. We aimed to assess the prevalence, cytological determinants, and clinical significance of increased GMC counts in BALF of 34 asthmatic horses compared to 10 controls. Counts were performed on 15 randomly selected high magnification fields per cytospin slide (40×), and expressed as GMC:single macrophage (GMC:M) ratio. Regression models were used for statistical analysis. GMC was frequently observed in both asthmatic and control horses, with an increased prevalence of equine asthma ($p = 0.01$). GMC:M ratio was significantly higher in severe vs. mild to moderate equine asthmatic and control horses. In asthmatic horses, an increased GMC:M ratio was significantly associated with BALF mastocytosis ($p = 0.01$), once adjusting for age and the presence and severity of clinical signs of the horses. Tachypnea was the only clinical sign that tended to be positively associated with GMC:M ratio after adjustment ($p = 0.08$). In conclusion, our data suggest that a relationship might exist between molecular mechanisms regulating GMC formation and mast cell recruitment in the equine lung. The same mechanisms could lead to tachypnea even in the absence of respiratory effort at rest. We suggest including GMC count in the basic cytological assessment of BALF samples to gain more insights into their role in equine asthma.

Keywords: equine asthma; macrophage fusion; giant multinucleated cells; mast cells; bronchoalveolar lavage; cytology

Citation: Basano, I.; Romolo, A.; Iamone, G.; Memoli, G.; Riccio, B.; Lavoie, J.-P.; Miniscalco, B.; Bullone, M. Giant Multinucleated Cells Are Associated with Mastocytic Inflammatory Signature Equine Asthma. *Animals* 2022, 12, 1070. https://doi.org/10.3390/ani12091070

Academic Editor: Chris W. Rogers

Received: 10 March 2022
Accepted: 18 April 2022
Published: 20 April 2022

Publisher's Note: MDPI stays neutral with regard to jurisdictional claims in published maps and institutional affiliations.

1. Introduction

Equine asthma is a highly prevalent disease in horses of all ages, characterized by chronic lower airway non-septic inflammation [1]. Equine asthma can manifest as different disease entities or phenotypes, based on severity and recurrence of clinical signs. The severe asthma phenotype (severe equine asthma, SEA) describes adult horses presenting with recurrent and reversible bouts of bronchospasm and respiratory effort at rest, often associated with cough, mucus nasal discharge, tachypnoea, and exercise intolerance. Severe asthma is typically associated with bronchoalveolar lavage fluid (BALF) neutrophilia >25% [2]. Sporadically, concomitant mastocytosis is also reported [3]. Mild to moderate equine asthma (MEA) describes a much broader clinical entity with no age predilection and clinical signs lasting for more than 3 weeks. Clinical signs span from poor performance in racehorses to more obvious cough, nasal discharge, or dyspnoea during or after exercise in show horses. The MEA phenotype is currently classified based on the inflammatory signature observed in BALF as neutrophilic (if neutrophils are >5% at differential cell counts), mastocytic (>2%), eosinophilic ($\geq$1%), or mixed [1]. There is substantial agreement on the fact that both genetics and environmental factors play a role in equine asthma pathogenesis and pathophysiology [4–6]. A large amount of evidence supports the primary role of straw and hay dust (and/or pollen) exposure as triggers for SEA exacerbations [7–14]. On the other hand, information on the etiology of MEA cases is scarce. Whether an initial trigger is needed for disease development is unknown in both conditions.

Cytological assessment of BALF is considered the gold standard technique for the diagnosis of equine asthma, together with relevant and compatible clinical signs [1]. The presence of qualitative and/or quantitative alterations of cytological entities other than neutrophils, mastocytes, and eosinophils is often disregarded in equine cytopathology. Nevertheless, it deserves attention as it might have clinical implications. Macrophages are normal resident cells within the lung, particularly within the alveoli. Their presence in equine BALF is physiological and their quantity has been shown to remain unchanged or decrease with aging [15,16]. It must be noticed that available data on age-associated changes in equine BALF differential cell counts are limited, and often restricted to secondary outcomes in relevant studies. Macrophages can exhibit various phenotypical features of activation, phagocytosis, and fusion, whose significance is currently overlooked. Monocyte/macrophage fusion results in giant multinucleated cells (GMC) and is considered a marker of chronic inflammation. In tissues, they are assumed to represent a specialization for improved phagocytosis (i.e., during a foreign body response) [17]. Molecular pathways leading to GMC formation are many and represent an emerging research field [17]. It is overall recognized that interleukin (IL)-4, a Th-2 cytokine whose role has been highlighted also in SEA [6,18,19], is likely to play a major role in this process [20]. Of note, IL-4 is implicated in mast cell recruitment and activation as well [21].

With the current work, we aimed at investigating the prevalence of GMCs in the BALF of asthmatic horses, as well as their association with relevant clinical or cytological features of equine asthma. Our hypothesis was that GMCs are increased in horses with equine asthma vs. controls, and their number is associated with disease chronicity (namely, with a diagnosis of SEA) and BALF mastocytosis.

2. Materials and Methods

2.1. Study Design

Cases referred to the Veterinary Teaching Hospital of the University of Turin with a diagnosis of equine asthma as reported in the electronic medical record file were searched retrospectively on April 2021. Additionally, newly referred cases diagnosed with equine asthma were prospectively included starting from April 2021. Cases were included in the study if the information on history, first examination, bronchoalveolar lavage procedure, and results were reported, and if BAL cytology slides were still available for re-assessment. Records were reviewed and cases with incomplete information or with uncertain diagnoses were excluded. Additionally, horses with concomitant positive bacterial culture from transtracheal wash samples and horses that had corticosteroids or antimicrobial treatments in the 2-week period preceding BALF sampling were excluded. Asthma cases (SEA and MEA) were defined following the current revised ACVIM Consensus Statement [1]. Briefly, SEA cases were defined based on relevant history and clinical signs; mild to moderate equine asthma cases were defined as those with relevant clinical signs and BALF neutrophil >5%, eosinophils >1%, and/or mast cells >2%. Mastocytic mild to moderate asthma (M-MEA) was defined based on the finding of >2% mast cells in BALF, in the absence of any other cytological abnormality [1], in horses with relevant respiratory signs. The following data were obtained from medical records concerning the horses studied: age, sex, relevant clinical signs (in particular: cough, nasal discharge, dyspnea, and poor performance), and time of the year when the exam was performed (month). Cytospins of all cases were reviewed for a blind reassessment of differential cell counts by the same operator and GMC quantification. The study was approved by the Local Animal Ethical Committee of the Department of Veterinary Medicine (Prot. N. 711, 17 March 2021), University of Turin. Written informed consent was obtained from the owners to use clinical data for research purposes.

Further asthmatic and control horses were retrospectively enrolled among equine patients referred to the Veterinary Teaching Hospital of the University of Montreal, Canada. Data and BAL cytospins slides of this second population of horses were retrieved by the Equine Respiratory Tissue Bank (https://btre.ca/, accessed on 1 November 2021). Horses were randomly selected among those available based on their diagnosis (5 SEA, 5 MEA, 10 controls defined as healthy or with musculoskeletal disease).

2.2. Bronchoalveolar Lavage Procedure and Processing

The BAL procedure was performed as described by Hoffman [22] on standing sedated horses restrained in a stock. Briefly, two boluses of 250 mL warm sterile saline solution were instilled in the distal airways and lung parenchyma by means of a Bivona catheter and gently withdrawn. A 5 mL-aliquot obtained after gentle resuspension of the entire solution withdrawn was immediately submitted for cytology in an EDTA tube. Cytospin slides (n.2) were prepared using 200 µL of unfiltered fresh samples centrifugated at 500 RPM for 5 min at room temperature. Slides were air-dried and stained with May-Grunwald-Giemsa using an automated slide stainer (Mirastainer^{TM} II System, EMD Chemicals Inc., Darmstadt, Germany).

The methods used in Turin and Montreal [2] for BALF fluid withdrawal and analysis differed only for the use of the Bivona catheter (3 m in length, 11 mm outer diameter) or Olympus SIF-Q140 videoendoscope (working length 250 cm, 10.5 mm outer diameter) to reach the sampling site.

2.3. Cytology

Cytospins slides were available for all horses studied. Slides were assessed by the same operator (AR), blinded to the animal ID, who performed: (1) differential cell counts for neutrophils, lymphocytes, macrophages, mast cells, and eosinophils; and (2) GMC counts. We defined GMC as cells with 2 or more nuclei and with evidence of unique cytoplasm.

(1) Differential cell counts were performed at 40× magnification on a minimum of 500 cells and 5 high power fields (HPF) [23]. Epithelial cells were excluded, when present and GMC were counted as macrophages (i.e., a GMC with two nuclei accounted for two macrophages).

(2) Counts of GMC were performed at 40× magnification over 15 HPF and expressed as GMC:single macrophages (GMC:M) ratio. They were obtained from:

$$GMC : M_{horse} = \frac{\sum_{i=1}^{n=15}(GMC)_i}{\sum_{i=1}^{n=15}(M)_i}$$

where i represents HPF and n is the minimum number of HPF per slide that must be assessed for a reliable estimate. To determine n, the absolute number of GMC and single macrophages were initially counted in all non-empty HPF of a subset of 10 samples (from 5 MEA and 5 SEA horses).

2.4. Statistical Analysis

Data were analyzed with GraphPad Prism version 8.0.0 for Windows (GraphPad Software, San Diego, CA, USA) and with STATA 15 software (StataCorp 2017, Stata Statistical Software: Release 15, College Station, TX, USA; StataCorp LLC.). Data distribution was assessed with the Saphiro Wilk normality test before and after the log transformation of the data. As log transformation did not yield significant advantages in data distribution, raw data were used for analysis. Groups' mean values were compared using the ANOVA test. Association between GMC:M ratio and other clinical or cytological/inflammatory parameters were assessed with linear or logistic regression models, depending on the nature of the independent variable (continuous or binomial, respectively). Data were adjusted for age (years) and for the presence and severity of clinical signs (for each clinical sign observed among cough, nasal discharge, tachypnea, increased respiratory effort at rest, this variable was increased by 1 point; the final variable could range from 0 to 4). Normalized cumulative average charts were used to determine the minimum number of HPF needed to accurately estimate the GMC:M ratio and its difference among the groups studied, under the assumption that an average 15% error is considered normal in manual cell counting. Cumulative average charts were normalized with respect to the grand average of the sample (equal to 1).

3. Results

3.1. Horses

A total of 42 horses were included in the study. Their clinical details are reported in Table 1.

Table 1. Details of the horses studied.

	Turin Cohort (N = 22)	Montreal Cohort (N = 20)
Controls		
N	0	10
Sex (M:F)	-	5:5
Age (years)	-	10.4 ± 3.6
Mild to moderate equine asthma (MEA)		
N	14	5
Sex (M:F)	8:6	3:2
Age (years)	7.4 ± 5.7	8.2 ± 1.6
Severe equine asthma (SEA)		
N	8	5
Sex (M:F)	6:2	3:2
Age (years)	15.6 ± 5.4	16.0 ± 4.5
Mastocytic asthma (M-MEA)		
N	7	1
Sex (M:F)	6:1	M
Age (years)	6.5 ± 7.5	8

MEA: mild to moderate equine asthma; M-MEA mastocytic mild to moderate equine asthma; SEA: severe equine asthma. M: male; F: female.

Briefly, there were 32 asthmatic and 10 control horses. There were 18 Warmblood horses, 7 Standardbred horses (none of them actively racing, 6 used for leisure activities, and 1 stallion used only for breeding), 1 draft horse, 5 ponies, 2 Friesian horses, and 9 Quarter Horse or related breeds. The Turin cohort was composed mainly of horses with MEA (n = 14). Horses with MEA with no previous treatments presented mostly with mastocytic inflammation at BALF (n = 7), with a small percentage having pure neutrophilic (n = 2) or mixed neutrophilic and mastocytic (n = 5) inflammation. The Montreal cohort was composed of 10 control horses (of which, 5 were healthy and 5 with concomitant orthopedic problems), 5 MEA (3 neutrophilic, 1 mastocytic, and 1 paucigranulocytic), and 5 severe asthmatic horses (Table 2). Five out of ten control horses presented mild increases in inflammatory cell counts at BALF (Table 2). The paucigranulocytic IAD case was a 10-year-old female Warmblood used for dressage, diagnosed with MEA in Montreal. Medical records reported differential cell counts as: 5% neutrophils, 1% mast cells, 0% eosinophils, 41% macrophages and 54% lymphocytes. This mare was referred for chronic cough at the beginning of exercise and at rest in box, and prolonged recovery after exercise. The owner reported no exercise intolerance, nor was noticed the presence of respiratory effort at rest during a clinical examination. The horse had been showing these signs for 2–3 years before referral, and the owner reports clinical improvements while feeding wet hay and following bronchodilator and anti-inflammatory therapies in the past. Horses with MEA were presented for: cough (8/19), increased respiratory effort during or after exercise (8/19), nasal discharge (5/19), tachypnea (5/19), and poor performance (3/19). Two horses diagnosed with MEA from the Turin cohort (2 M-MEA and 1 mixed) and 1 horse from the Montreal cohort did not present any relevant clinical signs at the time of examination. They were however included due to the relevant history and clinical complaint.

Table 2. Inflammatory signature of the horses studied.

	Paucigranulocytic	Neutrophilic [1]	Mastocytic [2]	Mixed
Controls	5	3	2	0
MEA	1	5	8	5
SEA	1	8	1	3

[1] Cutoff for neutrophilia in controls and MEA = 5%, in SEA = 25%, respectively [1,2]. [2] Cutoff mastocytosis in all groups studied = 2%.

3.2. Giant Multinucleated Cell Counts

3.2.1. Counting Method

The normalized cumulative average chart of GMC:M values shows that starting from 13 HPF, the average error falls below 15%, suggesting that the assessment of 15 HPF per sample provides precise estimates (Figure 1A). Of note, initial variability was higher for SEA vs. MEA samples, possibly because GMC are unevenly distributed across the microscope slide or because of a subjective bias towards reading first the fields with the highest number of GMC. Further tests specifically assessing cell distribution on the slide and comparing subjective vs. computer-based randomized reads on the same slides might ascertain these points.

The counting method employed allowed the detection of significant differences between SEA and MEA horses using a subset of randomly selected 10 horses (5 per group). Data suggest that a reliable estimate of the difference between these two groups can be achieved by counting 5 HPF (Figure 1B). For the aim of the present study, all GMC:M counts were performed on a total of 15 HPF per slide, to increase the accuracy of the measurement.

3.2.2. Disease Effect and External Validity of the Study

Giant multinucleated cells were observed in 7/10 (70%) control horses, 17/19 (89%) mild to moderate, and 13/13 (100%) severe asthmatic horses. Overall, a progressive increase in GMC:M ratio was noticed in MEA and SEA horses compared to controls. Considering both cohorts in a two-way ANOVA model, a significant effect of disease ($p < 0.0001$) and of the cohort ($p = 0.02$) were observed on GMC:M ratio (Figure 2), as well as a significant interaction among these covariates ($p = 0.004$). Post-tests revealed significantly higher values of GMC:M ratio in SEA and MEA compared to control horses ($p < 0.0001$ for both). When the two cohorts were assessed separately, GMC:M ratio was higher in SEA vs. MEA only in the Montreal cohort, due to the presence of an outlier in the Turin cohort. This outlier was a 9-year-old female horse with mixed BALF inflammation (6% neutrophil, 3% mast cell count in BALF) referred for chronic mucous nasal discharge and tachypnea at rest. Severe equine asthma in this horse was ruled out based on the absence of response to bronchodilators and no response to a 10-day course of oral dexamethasone treatment (in the absence, however, of environmental improvements). The mare had normal lung radiography and ultrasound, tested negative for EHV5 on both trans-tracheal wash (TTW) and BALF samples, and TTW yielded no bacterial growth. Removal of this outlier from the analysis revealed a significantly higher GMC:M ratio in SEA vs. MEA also in the Turin cohort ($p < 0.01$). Based on the available data, we cannot exclude that the mare had SEA, but the lack of response to corticosteroid does not support this hypothesis. It is possible that a longer treatment time could have shown an effect. A higher GMC:M ratio was also noticed in SEA horses from Montreal vs. SEA horses from Turin, and this was not associated with a different clinical presentation, based on the information available in medical records. No difference was observed in the GMC:M counts between healthy controls and controls with orthopedic problems (data not shown).

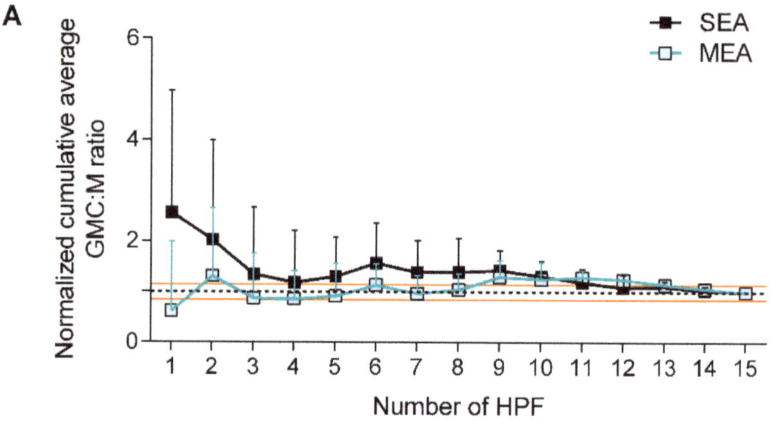

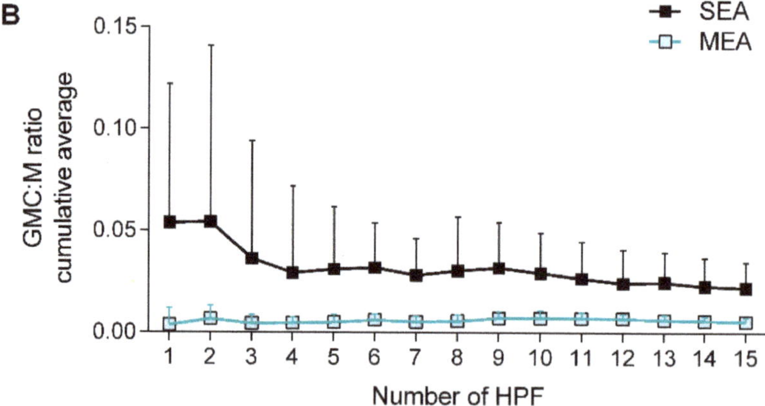

Figure 1. Validation of the methodology proposed for GMC counting in BALF cytological slides. (**A**) Normalized cumulative average curve shows the percentage variation of each cumulative measure against the total average (constrained to 1 and identified with the dotted line). Red lines identify the limits of the acceptable 15% error above or below the mean. (**B**) Cumulative average of GMC:M ratio over 15 measurements performed on 15 high power fields (HPF). SEA: severe equine asthma. MEA: mild to moderate equine asthma.

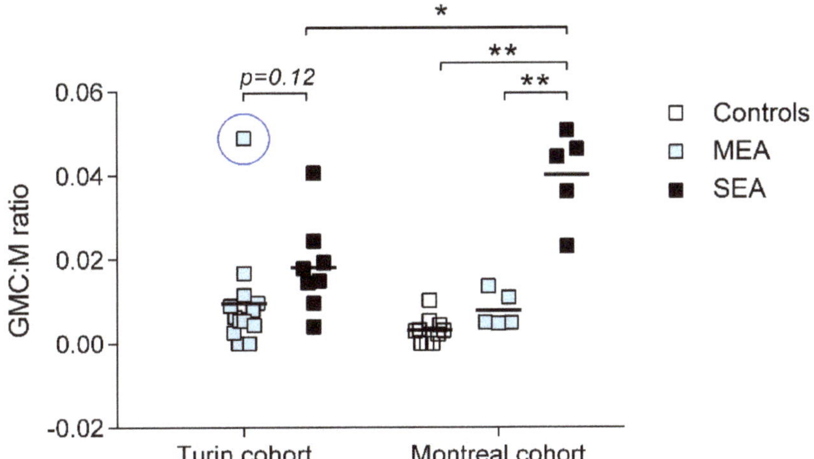

Figure 2. Effect of disease and cohort on GMC:M ratio (N = 42). *: significantly different ($p < 0.001$); **: significantly different ($p < 0.0001$). SEA: severe equine asthma. MEA: mild to moderate equine asthma. The blue circle identifies the outlier, whose clinical description is provided within the text.

Based on these results, horses from both cohorts were included in subsequent analysis to increase the power of the study.

3.3. Giant Multinucleated Cell Counts Relationship with Clinical Parameters and Inflammatory Signatures

GMC:M ratio and the percentage of macrophages in BALF were both significantly associated with the age of the horses. Specifically, GMC:M ratio increased while the percentage of macrophages decreased with aging ($p = 0.01$ and $p = 0.001$, respectively). The association between GMC:M and macrophage percentage in BAL also showed a significant trend towards an inverse association ($p = 0.05$). Once adjusted for clinical signs, however, the association of GMC:M with both age and macrophages in BALF became not significant ($p = 0.85$ and $p = 0.34$, respectively). Contrarily, the adjusted approach revealed a significant decrease in the percentage of alveolar macrophages with aging (β coefficient = -1.06; 95% CI -2.00, -0.11; $p = 0.03$).

GMC:M ratio was not associated with any cytological variable in an unadjusted univariate analysis restricted to asthmatic subjects. Adjusting for the presence and severity of clinical signs and for age revealed a significant association between GMC:M ratio and the percentage of mast cells in BALF (Table 3). Given the presence of an outlier in the Turin cohort with a very high GMC:M ratio (0.0489) and 3% mast cells in BALF, analyses were repeated in the absence of this animal. Exclusion of the outlier confirmed the observed results concerning the association between GMC:M ratio and BALF mastocytosis (β coefficient = -0.0034826; 95% CI -0.0053072, -0.0016579; $p = 0.001$), and revealed a previously undetected significant association between GMC:M ratio and BALF neutrophilia (β coefficient = 0.0004044; 95% CI 0.0001487, 0.0006601; $p = 0.004$).

A graphical representation of the relationship between GMC:M and BALF mastocytosis is provided in Figure 3. This graph suggests the existence of two distinct cytological clusters that follow the same pattern (inverse relationship), possibly related to disease chronicity. Concerning the clinical signs presented by the horses studied, those significantly associated with the presence and quantity of GMC (expressed as GMC:M ratio) were found to be the presence of respiratory effort at rest ($p = 0.01$) and tachypnea ($p = 0.007$). Adjusting for the presence and severity of clinical signs completely abolished the significant association with an increased respiratory effort at rest, while a tendency to significance was still appreciable for tachypnea ($p = 0.08$).

Table 3. Results of linear regression model used to investigate the relationship between GMC:M ratio and BALF cytological parameters, adjusting for age and for the presence and severity of clinical signs. Analyses were run on all asthmatic subjects (N = 32).

	β Coefficient	95% CI		*p* Value
BALF Neutrophil %	0.0002534	−0.0001471	0.0006538	0.20
BALF Mast cell %	−0.0037398	−0.0065036	−0.0009760	0.01
BALF Macrophage %	−0.0000191	−0.0004785	0.0004402	0.93
BALF Lymphocyte %	−0.0001442	−0.0005176	0.0002291	0.43
BALF Eosinophil %	−0.0003903	−0.0072857	0.0065051	0.91

BALF: bronchoalveolar lavage fluid. CI: confidence interval.

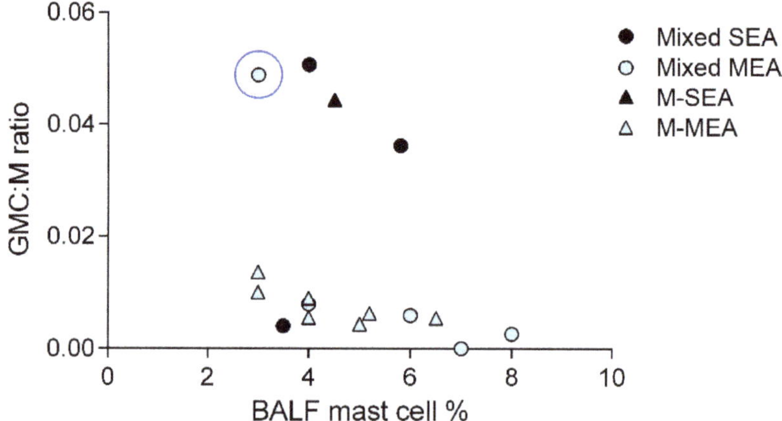

Figure 3. Relationship between GMC:M ratio and BALF mastocytosis in equine asthma (N = 17). Two clusters can be distinguished based on GMC:M ratio. In both, the relationship between GMC:M ratio and BALF mastocytosis is inverse. The blue circle identifies the outlier already identified in Figure 2, with a 1-year-history of tachypnea and mucous nasal discharge. Circles represent horses with mixed neutrophilic-mastocytic inflammation. Triangles represent horses with pure mastocytic inflammation. Black identifies SEA. Light blue identifies horses with MEA. Horses with no clinical signs at examination (n = 3) are not depicted to better represent the statistical model employed.

4. Discussion

Increased attention has been paid to the role of alveolar macrophages in equine asthma in recent years, as they appear to behave as crucial regulators of both acute inflammatory response and its subsequent resolution [24]. Alveolar macrophages are normal inhabitants of the lower airways, and their fusion produces GMC [17]. Giant multinucleated cells are commonly observed in BALF samples from asthmatic horses, thus often disregarded. Like macrophages, different phenotypes of GMC are recognized in health and disease, some of which can be distinguished based on morphological features while others do not [17,25,26]. A deeper understanding of the role and activity of GMC in equine respiratory health and diseases might provide additional information in clinical settings. It might also help in the comprehension of physiopathological mechanisms driving disease development and maintenance, with potential implications for therapeutic advances. Our work provides the first insight into this complex panorama, suggesting that the presence of GMC in increased quantities is a feature of equine asthma, especially in its severe form. Moreover, the significant association with mastocytes suggests the interleukin (IL)-4 signaling pathway is involved, in accordance with previous studies identifying a central role of this cytokine in equine asthma [18,19,27].

The prevalence of horses with GMC in their BALF was high in our study, reaching 70% in control horses and almost 100% in asthmatic horses. This was unexpected but easily explained by the environment horses are exposed to and to the criteria we used to define GMC. Horses are naturally exposed to dusts of biological, vegetal, or mineral origin. It is reasonable to hypothesize that a dust-rich environment stimulates GMC formation in the lungs even in the absence of pulmonary diseases. Our definition of GMC as cells with two or more nuclei of monocyte origin was also loose. Data in human beings highlights that GMC with a low number of nuclei (>three) are highly prevalent in respiratory samples from healthy people and patients with chronic lung diseases. Contrarily, GMC with >10 nuclei were observed in only 10% of human pathological BALF samples, most often associated with granulomatous diseases (sarcoidosis, pneumoconiosis, interstitial lung diseases) or chronic non-interstitial lung diseases [25]. Equine asthma is not a granulomatous disease. However, cases characterized by miliary granulomatous lesions of unknown origin have been reported [28], which would be consistent with the finding of GMC in BALF. Idiopathic asthmatic granulomatosis has been described also in human beings [29,30]. Further work on alveolar GMC in healthy and asthmatic horses might be of value for establishing, respectively, normal values and expected ranges in this common disease. This approach could facilitate the identification of less common granulomatous or GMC-associated pulmonary conditions in horses.

Control horses included in our study did present cytological abnormalities. This is in line with previous observations in a population of actively performing show horses [16], and could be linked with the stabling environment [7,31]. We acknowledge that control animals were available from only one site in our study, and this is a limitation of our current approach. However, data from MEA and SEA horses from both sites are concordant, and there is no reason to expect that the lack of control horses from one site would have introduced a bias in our data. A further aspect that might have helped in understanding the causes of GMC formation in our population is the availability of environmental particle loads by air sampling. This is a piece of information that can be obtained quite easily, although it remains unspecific. A thorough characterization of the exogenous particles within alveolar macrophages and GMC would provide relevant information concerning the origin of GMC formation in our horses.

Macrophage fusion and GMC formation require cell priming and expression of adhesion molecules called fusogens [17]. Depending on the stimulus and mechanisms underlying macrophage fusion, different GMC phenotypes are currently recognized. Broadly, GMC are classified as classically activated and alternatively activated. Classically activated GMC are those typical of acute infective processes (typically mycobacterial ones), its development is driven by interferon (IFN)-γ, lipopolysaccharide (LPS), and tumor necrosis factor (TNF)-α. The major fusogen of this phenotype is the dendrocyte-expressed seven-transmembrane protein (DC-STAMP). Alternatively activated GMC are a wider family typical of sarcoidosis, foreign body, parasite reactions, and chronic tuberculosis. Alternatively activated GMC are mainly linked with a Th-2 type immune response (IL-4, IL-13 rich environment), but sub-phenotypes responsive to IL-10 or IL-6 are also recognized. They typically express fusogens as E-caderhin or a mannitol receptor (CD206) [17]. Recent work has shown that macrophage polarization occurs in equine alveolar macrophages following the same patterns described in human beings, with CD206 expression being significantly enhanced by IL-4 stimulation [27]. The same work, however, failed to show a significant increase in CD206 expression in asthmatic horses, either in remission or exacerbation of the disease, compared to control horses. Another study reports that CD206 expression assessed by flow cytometry was increased in alveolar macrophages of asthmatic horses during disease exacerbation vs. remission of the disease, although values of asthmatic horses were not different from controls [32]. Whether GMC formation was assessed and accounted for, was not specified in both papers. The results observed are in line with previous studies showing activation of different immune (T cell) responses in equine asthma cases [33], and highlight the need to find clinically applicable markers of such equine asthma endotypes that can be used together with BALF cell counts.

Quantification and interpretation of BALF cell counts, per se, remains challenging. BALF cell counts are commonly expressed in percentage, as the absolute number is dependent on many uncontrollable factors. This is especially relevant in equine medicine, in which the BAL technique is only partly standardized and data exists in support of significant effects on BALF cell counts induced by the volume of fluid instilled or by the counting method used [22,23,32,34]. The same is true for GMC quantification, for which only relative measures are available or commonly used [35]. In this perspective, an increase in GMC:M ratio as the one we observed in SEA could be driven by an increased fusion of macrophages or by a decreased number of alveolar macrophages. In turn, decreased macrophage count might have been linked with increases in other cell types, given their expression in relative units. The measuring unit we used did not permit the assessment of a parallel increase in both singular alveolar macrophages and GMC. GMC:M ratio was not associated with BALF macrophage percent count in our population, while significant associations were observed with neutrophils and mastocytes. Among several cytokines involved in neutrophil or mast cell recruitment, IL-4 is of particular interest as it has been shown to play a role in SEA-associated neutrophilia [19], it induces mast cell proliferation and survival through several pathways [36], and it is synthesized by mast cells [21]. The negative relationship observed between mast cell counts and GMC:M ratio should be interpreted cautiously, as a decrease in relative count might not reflect an absolute decrease, but only a milder decrease compared to other cell populations. It is likely that a larger number of macrophages, and/or neutrophils are recruited compared to mast cells following an acute inflammatory event, even in the presence of a positive trend for all cell types. Absolute counts of BALF inflammatory cells as well as the knowledge of the volumes of saline-infused and recovered during the BAL procedure might help in the interpretation of relative counts and should always be considered in the future. Lastly, we acknowledge that most mastocytic MEA cases were from Turin. It is possible that the environment near Turin is characterized by pollutants or natural antigens driving BALF mastocytosis and GMC formation in the absence of disease in horses. This could have introduced a bias in our results. However, data from Montreal and Turin were coherent concerning the mast cell and GMC relationship observed. The only mastocytic MEA case in Montreal presented with increased GMC compared to neutrophilic MEA cases from the same site.

Alveolar macrophage counts (expressed in percentage) appeared to be negatively associated with age in our population, after appropriate adjustments, while no age-related effect was observed on GMC. Alveolar macrophage counts reduction with aging is coherent with immunosenescence, an age-associated deterioration in immunity characterized by a diminished reaction toward host pathogens [37]. It is described also in human beings [38]. Previous studies failed to detect any age-related decrease in alveolar macrophages in horses [37]. Further large-scale work is warranted to clarify this point. To the authors' knowledge, whether GMC prevalence, phenotype or function is altered by age or by the aging environment is an unexplored research field. Age is a recognized risk factor for some GMC-mediated human diseases [39]. Age is also a well-recognized risk factor for SEA [40]. In addition, whether macrophage recruitment and its fusion (GMC formation) into the alveoli are driven by the same molecular pathways is largely undefined. Our data suggest that, in equine asthma, the two processes are regulated differently, and the assessment of these two cytological entities could provide complementary information from a clinical perspective.

5. Conclusions

In conclusion, our data support a possible role of GMC in equine asthma that deserves further attention. Their increased prevalence in SEA and their significant association with mast cells and, to a lesser extent, neutrophil relative cell counts in BALF, suggest they play a regulatory role in disease. This is in line with recent evidence showing alveolar macrophages, the precursors of GMC, as important orchestrators in equine asthma physiopathology. It must be remembered, however, that our results were obtained using relative ratios and this requires caution in data interpretation. Relative counts can indeed cause independently under or over-sized effects.

Alveolar GMC might be increased in equine asthma in association with the increased burden of inhaled dust, due to altered mucociliary clearance, decreased respiratory flow, air trapping, or also other causes. A better knowledge of alveolar GMC prevalence, morphology, etiology, and function in asthma might disclose pathological mechanisms still undefined and deserves attention. Due to the high number of variables and biases that should be considered for this aim, large-scale work is warranted to gain meaningful insights and applicable results in the future.

Author Contributions: Conceptualization, M.B. and B.M.; investigation: I.B., G.I., G.M., B.R.; data curation, A.R., G.I., J.-P.L.; methodology: A.R., M.B.; project administration: M.B.; resources: J.-P.L., B.M., M.B.; formal analysis, M.B., A.R.; writing—original draft preparation, M.B., A.R.; writing—review and editing, I.B., G.I., G.M., B.R., B.M.; supervision, M.B., B.M., J.-P.L. All authors have read and agreed to the published version of the manuscript.

Funding: This research received no external funding.

Institutional Review Board Statement: The animal study protocol was approved by Ethics Committee of the Department of Veterinary Sciences, University of Turin (protocol code n. 711, date of approval 17 March 2021). Montréal data were obtained from the Equine Respiratory Tissue Bank.

Informed Consent Statement: Written informed consent was obtained from the owners of the horses included in the study.

Data Availability Statement: Not applicable. All data produced in the study are presented in this article.

Acknowledgments: The authors would like to thank Laura Starvaggi Cucuzza and Sophie Manguy-Seers for their precious help with laboratory procedures and Montréal case selection, respectively. We would also like to thank the Quebec Respiratory Health Network (RHN) for providing financial support to the Equine Respiratory Tissue Bank.

Conflicts of Interest: The authors declare no conflict of interest.

References

1. Couetil, L.L.; Cardwell, J.M.; Gerber, V.; Lavoie, J.P.; Leguillette, R.; Richard, E.A. Inflammatory airway disease of horses-revised consensus statement. *J. Vet. Intern. Med./Am. Coll. Vet. Intern. Med.* **2016**, *30*, 503–515. [CrossRef]
2. Jean, D.; Vrins, A.; Beauchamp, G.; Lavoie, J.P. Evaluation of variations in bronchoalveolar lavage fluid in horses with recurrent airway obstruction. *Am. J. Vet. Res.* **2011**, *72*, 838–842. [CrossRef]
3. Lavoie, J.P.; Maghni, K.; Desnoyers, M.; Taha, R.; Martin, J.G.; Hamid, Q.A. Neutrophilic airway inflammation in horses with heaves is characterized by a th2-type cytokine profile. *Am. J. Respir. Crit. Care Med.* **2001**, *164*, 1410–1413. [CrossRef]
4. Klukowska-Rotzler, J.; Swinburne, J.E.; Drogemuller, C.; Dolf, G.; Janda, J.; Leeb, T.; Gerber, V. The interleukin 4 receptor gene and its role in recurrent airway obstruction in swiss warmblood horses. *Anim. Genet.* **2012**, *43*, 450–453. [CrossRef]
5. Leclere, M.; Lavoie-Lamoureux, A.; Lavoie, J.P. Heaves, an asthma-like disease of horses. *Respirology* **2011**, *16*, 1027–1046. [CrossRef]
6. Jost, U.; Klukowska-Rotzler, J.; Dolf, G.; Swinburne, J.E.; Ramseyer, A.; Bugno, M.; Burger, D.; Blott, S.; Gerber, V. A region on equine chromosome 13 is linked to recurrent airway obstruction in horses. *Equine Vet. J.* **2007**, *39*, 236–241. [CrossRef]
7. Pirie, R.S.; Collie, D.D.; Dixon, P.M.; McGorum, B.C. Evaluation of nebulised hay dust suspensions (hds) for the diagnosis and investigation of heaves. 2: Effects of inhaled hds on control and heaves horses. *Equine Vet. J.* **2002**, *34*, 337–342. [CrossRef]
8. Pirie, R.S.; Collie, D.D.; Dixon, P.M.; McGorum, B.C. Inhaled endotoxin and organic dust particulates have synergistic proinflammatory effects in equine heaves (organic dust-induced asthma). *Clin. Exp. Allergy J. Br. Soc. Allergy Clin. Immunol.* **2003**, *33*, 676–683. [CrossRef]
9. Pirie, R.S.; Dixon, P.M.; Collie, D.D.; McGorum, B.C. Pulmonary and systemic effects of inhaled endotoxin in control and heaves horses. *Equine Vet. J.* **2001**, *33*, 311–318. [CrossRef]
10. Pirie, R.S.; Dixon, P.M.; McGorum, B.C. Endotoxin contamination contributes to the pulmonary inflammatory and functional response to aspergillus fumigatus extract inhalation in heaves horses. *Clin. Exp. Allergy J. Br. Soc. Allergy Clin. Immunol.* **2003**, *33*, 1289–1296. [CrossRef]

11. Costa, L.R.; Johnson, J.R.; Baur, M.E.; Beadle, R.E. Temporal clinical exacerbation of summer pasture-associated recurrent airway obstruction and relationship with climate and aeroallergens in horses. *Am. J. Vet. Res.* **2006**, *67*, 1635–1642. [CrossRef]

12. Bullone, M.; Murcia, R.Y.; Lavoie, J.P. Environmental heat and airborne pollen concentration are associated with increased asthma severity in horses. *Equine Vet. J.* **2016**, *48*, 479–484. [CrossRef]

13. McGorum, B.C.; Dixon, P.M.; Halliwell, R.E. Phenotypic analysis of peripheral blood and bronchoalveolar lavage fluid lymphocytes in control and chronic obstructive pulmonary disease affected horses, before and after 'natural (hay and straw) challenges'. *Vet. Immunol. Immunopathol.* **1993**, *36*, 207–222. [CrossRef]

14. Schmallenbach, K.H.; Rahman, I.; Sasse, H.H.; Dixon, P.M.; Halliwell, R.E.; McGorum, B.C.; Crameri, R.; Miller, H.R. Studies on pulmonary and systemic aspergillus fumigatus-specific ige and igg antibodies in horses affected with chronic obstructive pulmonary disease (copd). *Vet. Immunol. Immunopathol.* **1998**, *66*, 245–256. [CrossRef]

15. Christmann, U.; Buechner-Maxwell, V.A.; Witonsky, S.G.; Hite, R.D. Role of lung surfactant in respiratory disease: Current knowledge in large animal medicine. *J. Vet. Intern. Med./Am. Coll. Vet. Intern. Med.* **2009**, *23*, 227–242. [CrossRef]

16. Gerber, V.; Robinson, N.E.; Luethi, S.; Marti, E.; Wampfler, B.; Straub, R. Airway inflammation and mucus in two age groups of asymptomatic well-performing sport horses. *Equine Vet. J.* **2003**, *35*, 491–495. [CrossRef]

17. Brooks, P.J.; Glogauer, M.; McCulloch, C.A. An overview of the derivation and function of multinucleated giant cells and their role in pathologic processes. *Am. J. Pathol.* **2019**, *189*, 1145–1158. [CrossRef]

18. Cordeau, M.E.; Joubert, P.; Dewachi, O.; Hamid, Q.; Lavoie, J.P. Il-4, il-5 and ifn-gamma mrna expression in pulmonary lymphocytes in equine heaves. *Vet. Immunol. Immunopathol.* **2004**, *97*, 87–96. [CrossRef]

19. Lavoie-Lamoureux, A.; Moran, K.; Beauchamp, G.; Mauel, S.; Steinbach, F.; Lefebvre-Lavoie, J.; Martin, J.G.; Lavoie, J.P. Il-4 activates equine neutrophils and induces a mixed inflammatory cytokine expression profile with enhanced neutrophil chemotactic mediator release ex vivo. *Am. J. Physiology. Lung Cell. Mol. Physiol.* **2010**, *299*, L472–L482. [CrossRef]

20. Moreno, J.L.; Mikhailenko, I.; Tondravi, M.M.; Keegan, A.D. Il-4 promotes the formation of multinucleated giant cells from macrophage precursors by a stat6-dependent, homotypic mechanism: Contribution of e-cadherin. *J. Leukoc. Biol.* **2007**, *82*, 1542–1553. [CrossRef]

21. Caslin, H.L.; Kiwanuka, K.N.; Haque, T.T.; Taruselli, M.T.; MacKnight, H.P.; Paranjape, A.; Ryan, J.J. Controlling mast cell activation and homeostasis: Work influenced by bill paul that continues today. *Front. Immunol.* **2018**, *9*, 868. [CrossRef] [PubMed]

22. Hoffman, A.M. Bronchoalveolar lavage: Sampling technique and guidelines for cytologic preparation and interpretation. *Vet. Clin. N. Am. Equine Pract.* **2008**, *24*, 423–435. [CrossRef] [PubMed]

23. Fernandez, N.J.; Hecker, K.G.; Gilroy, C.V.; Warren, A.L.; Leguillette, R. Reliability of 400-cell and 5-field leukocyte differential counts for equine bronchoalveolar lavage fluid. *Vet. Clin. Pathol./Am. Soc. Vet. Clin. Pathol.* **2013**, *42*, 92–98. [CrossRef] [PubMed]

24. Karagianni, A.E.; Lisowski, Z.M.; Hume, D.A.; Scott Pirie, R. The equine mononuclear phagocyte system: The relevance of the horse as a model for understanding human innate immunity. *Equine Vet. J.* **2021**, *53*, 231–249. [CrossRef]

25. Kern, I.; Kecelj, P.; Kosnik, M.; Mermolja, M. Multinucleated giant cells in bronchoalveolar lavage. *Acta Cytol.* **2003**, *47*, 426–430. [CrossRef]

26. Lemaire, I.; Yang, H.; Lauzon, W.; Gendron, N. M-csf and gm-csf promote alveolar macrophage differentiation into multinucleated giant cells with distinct phenotypes. *J. Leukoc. Biol.* **1996**, *60*, 509–518. [CrossRef]

27. Wilson, M.E.; McCandless, E.E.; Olszewski, M.A.; Robinson, N.E. Alveolar macrophage phenotypes in severe equine asthma. *Vet. J.* **2020**, *256*, 105436. [CrossRef]

28. Bullone, M.; Joubert, P.; Gagne, A.; Lavoie, J.P.; Helie, P. Bronchoalveolar lavage fluid neutrophilia is associated with the severity of pulmonary lesions during equine asthma exacerbations. *Equine Vet. J.* **2018**, *50*, 609–615. [CrossRef]

29. Wenzel, S.E.; Vitari, C.A.; Shende, M.; Strollo, D.C.; Larkin, A.; Yousem, S.A. Asthmatic granulomatosis: A novel disease with asthmatic and granulomatous features. *Am. J. Respir. Crit. Care Med.* **2012**, *186*, 501–507. [CrossRef]

30. Umezawa, H.; Naito, Y.; Ogasawara, T.; Takeuchi, T.; Kasamatsu, N.; Hashizume, I. Idiopathic bronchocentric granulomatosis in an asthmatic adolescent. *Respir. Med. Case Rep.* **2015**, *16*, 134–136. [CrossRef]

31. Leclere, M.; Lavoie-Lamoureux, A.; Gelinas-Lymburner, E.; David, F.; Martin, J.G.; Lavoie, J.P. Effect of antigenic exposure on airway smooth muscle remodeling in an equine model of chronic asthma. *Am. J. Respir. Cell Mol. Biol.* **2011**, *45*, 181–187. [CrossRef] [PubMed]

32. Kang, H.; Bienzle, D.; Lee, G.K.C.; Piche, E.; Viel, L.; Odemuyiwa, S.O.; Beeler-Marfisi, J. Flow cytometric analysis of equine bronchoalveolar lavage fluid cells in horses with and without severe equine asthma. *Vet. Pathol.* **2022**, *59*, 91–99. [CrossRef] [PubMed]

33. Bullone, M.; Lavoie, J.P. Asthma "of horses and men"-how can equine heaves help us better understand human asthma immunopathology and its functional consequences? *Mol. Immunol.* **2015**, *66*, 97–105. [CrossRef] [PubMed]

34. Leclere, M.; Desnoyers, M.; Beauchamp, G.; Lavoie, J.P. Comparison of four staining methods for detection of mast cells in equine bronchoalveolar lavage fluid. *J. Vet. Intern. Med./Am. Coll. Vet. Intern. Med.* **2006**, *20*, 377–381. [CrossRef]

35. Trout, K.L.; Holian, A. Macrophage fusion caused by particle instillation. *Curr. Res. Toxicol.* **2020**, *1*, 42–47. [CrossRef] [PubMed]

36. McLeod, J.J.; Baker, B.; Ryan, J.J. Mast cell production and response to il-4 and il-13. *Cytokine* **2015**, *75*, 57–61. [CrossRef]

37. Hansen, S.; Baptiste, K.E.; Fjeldborg, J.; Horohov, D.W. A review of the equine age-related changes in the immune system: Comparisons between human and equine aging, with focus on lung-specific immune-aging. *Ageing Res. Rev.* **2015**, *20*, 11–23. [CrossRef]

38. Zissel, G.; Schlaak, M.; Muller-Quernheim, J. Age-related decrease in accessory cell function of human alveolar macrophages. *J. Investig. Med. Off. Publ. Am. Fed. Clin. Res.* **1999**, *47*, 51–56.

39. Mohan, S.V.; Liao, Y.J.; Kim, J.W.; Goronzy, J.J.; Weyand, C.M. Giant cell arteritis: Immune and vascular aging as disease risk factors. *Arthritis Res. Ther.* **2011**, *13*, 231. [CrossRef]

40. Couetil, L.L.; Ward, M.P. Analysis of risk factors for recurrent airway obstruction in north american horses: 1,444 cases (1990–1999). *J. Am. Vet. Med. Assoc.* **2003**, *223*, 1645–1650. [CrossRef]

Review

The Immune Mechanisms of Severe Equine Asthma—Current Understanding and What Is Missing

Joana Simões [1,2,3,*], Mariana Batista [1,2,4] and Paula Tilley [1,2]

[1] CIISA-Centre for Interdisciplinary Research in Animal Health, Faculty of Veterinary Medicine, University of Lisbon, 1300-477 Lisbon, Portugal; marianabatista@fmv.ulisboa.pt (M.B.); paulatilley@fmv.ulisboa.pt (P.T.)

[2] Associate Laboratory for Animal and Veterinary Sciences (AL4Animals), Faculty of Veterinary Medicine, University of Lisbon, 1300-477 Lisbon, Portugal

[3] Equine Health and Welfare Academic Division, Faculty of Veterinary Medicine, Lusófona University, Campo Grande 376, 1749-024 Lisbon, Portugal

[4] Basic Sciences Academic Division, Faculty of Veterinary Medicine, Lusófona University, Campo Grande 376, 1749-024 Lisbon, Portugal

* Correspondence: joana.simoes@ulusofona.pt

Simple Summary: Severe equine asthma (sEA) is a highly prevalent respiratory disease affecting adult horses. Affected horses present with cough, nasal discharge and increased respiratory effort at rest. Although a complex diversity of genetic and immunological pathways contribute to the disease, these remain to be fully understood. Several studies have reported the role of inflammatory mediators and of some cells found in sEA airway inflammation. However, the reported results revealed some inconsistencies between studies. A better understanding of sEA's genetics and detailed immunology is fundamental in order to characterize the underlying mechanisms involved in the disease's occurrence and to establish an adequate therapy and a precise prognosis. This review examines some literature findings on the genetic and immunology of sEA and discusses further research areas.

Citation: Simões, J.; Batista, M.; Tilley, P. The Immune Mechanisms of Severe Equine Asthma—Current Understanding and What Is Missing. *Animals* **2022**, *12*, 744. https://doi.org/10.3390/ani12060744

Academic Editors: Francesco Ferrucci, Chiara Maria Lo Feudo and Luca Stucchi

Received: 7 January 2022
Accepted: 14 March 2022
Published: 16 March 2022

Publisher's Note: MDPI stays neutral with regard to jurisdictional claims in published maps and institutional affiliations.

Abstract: Severe equine asthma is a chronic respiratory disease of adult horses, occurring when genetically susceptible individuals are exposed to environmental aeroallergens. This results in airway inflammation, mucus accumulation and bronchial constriction. Although several studies aimed at evaluating the genetic and immune pathways associated with the disease, the results reported are inconsistent. Furthermore, the complexity and heterogeneity of this disease bears great similarity to what is described for human asthma. Currently available studies identified two chromosome regions (ECA13 and ECA15) and several genes associated with the disease. The inflammatory response appears to be mediated by T helper cells (Th1, Th2, Th17) and neutrophilic inflammation significantly contributes to the persistence of airway inflammatory status. This review evaluates the reported findings pertaining to the genetical and immunological background of severe equine asthma and reflects on their implications in the pathophysiology of the disease whilst discussing further areas of research interest aiming at advancing treatment and prognosis of affected individuals.

Keywords: severe equine asthma; immunology; genetic; neutrophils; horse

1. Introduction

Severe equine asthma (sEA) is a naturally occurring chronic respiratory disease [1], affecting up to 20% of adult horses in the Northern hemisphere [2]. Disease develops upon exposure of genetically susceptible individuals to environments with high concentrations of airborne respirable particles, capable of inducing airway inflammation [3]. A vast array of antigens have been implicated in the etiology of sEA and it is thought that airway inflammation results from the synergistic effect of multiple allergens [4], to which individuals are susceptible in unique ways. This disease has been mostly associated with hay feeding and stabling, being termed as stable-associated sEA, but summer pasture-associated sEA also occurs [1,5,6]. Fungal spores, bacterial endotoxins, forage and storage mites, microbial toxins, peptidoglycans, proteases, pollen and plant debris, as well as inorganic particles trigger clinical signs of disease [5–10]. Several fungi (>50 species), especially *Aspergillus fumigatus*, have been widely recognised as significant risk factors for sEA [11–13]. Recent research by White and colleagues has uncovered the potential role of novel allergens including new species of fungi, mites, pollen, and arthropods, but also that of latex proteins [5], which hadn't yet been clearly associated with the disease.

During disease exacerbation affected horses develop cough, nasal discharge and increased respiratory effort at rest [1,14,15], due to neutrophil recruitment, mucus plugging, bronchospasm and airway remodeling [16,17].

Severely asthmatic horses are usually managed through antigen avoidance and the use of corticosteroids and bronchodilators to reduce airway inflammation, bronchoconstriction and improve lung function [18]. However, some horses are unresponsive to corticosteroid treatment posing a challenge to clinicians. Thus, the identification of causal antigens and the development of antigen screening tests is fundamental and will enable a personalized treatment approach using specific immunotherapy [5,19].

Disease diagnosis is mostly based on history, clinical signs, and bronchoalveolar lavage fluid (BALF) differential cytology. Although lung function testing can accurately detect sEA, such equipment is unavailable to most field practitioners [1,14,20]. The genetic and immunological mechanisms associated with this disease are complex and heterogenous, implicating the activation of different inflammatory pathways [21–23]. Currently there is a need to better characterize the immune events leading to the occurrence and persistence of airway inflammation as this will help clinicians in determining the best treatment approach and in providing an accurate prognosis. Moreover, the development of novel ancillary diagnostic tests and therapeutic targets are required for early diagnosis of sEA and total resolution of airway inflammation in refractory cases.

Because sEA shares many similarities to its human counterpart, the horse is considered a good model for the study of the non-allergic and late on-set asthma phenotypes, since disease occurs naturally and sample collection can be easily performed [24]. Thus, further contributions to the disease's characterization will benefit both horses and humans alike.

In the present systematic review, the intention was to combine the data published over the last twenty years on the immune mechanisms which have been identified, described, and associated with sEA, in order to help researchers and clinicians to better understand this highly prevalent respiratory disease. We must recognize the limitations of this systematic review, as it does not reanalyze available data as a metanalysis would. Nevertheless, we believe that, by assembling the existing information, we can contribute to the identification of knowledge gaps to address in future scientific discussions and research projects, hopefully leading to further enlightening of the immune mechanisms of sEA.

2. Genetic Background

Although sEA's heritability has been shown in several horse breeds, and a familial aggregation has long been ascertain, external factors, such as environment, increase the likelihood of expressing the disease [3,21,22].

The chromosome region ECA13 has been associated with sEA in one family of Swiss Warmbloods, while region ECA15 has been implicated in a different family of the same breed. The inheritance mode differed between both families, being autosomal recessive in the first family and autosomal dominant in the second [25]. Additionally, in the first family of horses the Interleukin 4 receptor (IL-4R) gene and its neighboring regions in ECA13 appeared to contribute to disease in some individuals [26–28]. In humans, polymorphic differences in the Interleukin 4 receptor α chain (IL4Rα) gene play an important role in the development of asthma, since they induce the isotopic switch to immunoglobulin E (IgE) and the differentiation of T-helper type 2 (Th2) lymphocytes [29,30].

Racine and colleagues described an interaction between IL-4R and products of the *SOCS5* gene, which may influence the molecular cascades involving nuclear factor (NF)-κB [31]. The gene coding for SOCS5 is located in the ECA15 chromosome region, and it is predominantly expressed by Th1 cells while further inhibiting Th2 differentiation. The inhibitory effect of SOCS5 on IL-4 signaling contributes to the non-Th2 cytokine profile observed in human non-allergic asthma [32], and may explain further similarities between both species.

In a genome wide association study (GWAS), the gene responsible for the TXNDC11 protein, also located in the ECA13 region, has been linked to sEA [33]. In humans, TXNDC11 controls the production of hydrogen peroxide in the respiratory epithelium [34], as well as the expression of MUC5AC mucin, which has been shown to play a significant role on airway hyperreactivity in mice [35]. In sEA-affected horses MUC5AC is upregulated, thus contributing to the mucus plugging observed in the disease [36].

The analysis of genomic copy number variants did not reveal any relevant variant regions which could be associated with the sEA, although a copy number loss was reported on chromosome 5 involving the gene NME7 [37]. The expression of this gene is necessary for ciliary function in the lungs and may be involved in sEA, since in knockout mice it induces primary ciliary dyskinesia [38]. Also, using RNA sequencing technique, a single point substitution was detected in the PACRG and RTTN genes in asthmatic horses, predictively altering their proteins, which are related to ciliary function [39].

In a gene set enrichment analysis of the bronchial epithelium after hay dust exposure, asthmatic horses presented upregulated genes of the E2F transcription factor family, which contribute to cell cycle regulation. Thus, asthmatic horses may suffer from impaired bronchial epithelial regeneration associated to subepithelial remodeling [40].

These recent studies have shown that the respiratory epithelium contributes to the immunological response observed in severely asthmatic horses.

Furthermore, an analysis of expression quantitative trait loci (eQTLs) allied with GWAS did not find a significant association between observed genetic variants and sEA, except for a disease-genetic variant in CLEC16A gene, which regulates gene expression of dexamethasone-induced protein (DEXI) [41]. This is of special importance in comparative pathology, as DEXI has also been reported in human asthma [42], although in horses it appears to not be a reliable indicator of sEA [41].

The identification and differential expression analysis of microRNAs (miRNAs) present in the serum of sEA-affected horses, showed a downregulation of miR-128 and miR-744. These findings suggest that a Th2/Th17 immunological response may characterize sEA [10,43].

Additionally, a recent work on Polish Konik horses aimed to detect the effects of inbreeding on sEA, however no effects were observed at the individual level [44].

Although most of these findings relate to certain families of Swiss Warmblood horses, they illustrate the complex genetic heterogeneity of sEA, which most likely results from the interaction of different genes. However, the use of such specific horse families and the likelihood of high variety of genetic background mechanisms contributing to the disease limits the application of these findings to the general equine population.

3. Immunological Phenotypes and Endotypes

Phenotype is the term used to describe the observable clinical characteristic of a disease, whereas endotype, a subclass of phenotype, refers to its molecular and genetic mechanism or treatment response [45].

As stated in the 2016 consensus, equine asthma (EA) is currently defined by two major phenotypes, which differ according to disease onset, clinical presentation and its severity—mild/moderate EA (mEA) and the already described sEA, which is the focus of this review [1]. However, phenotypes are insufficient when deciding upon the appropriate therapeutic management or determining the prognosis, which mainly depend on the immunological mechanisms of the disease.

Human asthma is usually considered to be a type 1 hypersensitivity, due to increased levels of IgE associated with a Th2 response, resulting in the recruitment of eosinophils into the airways [46]. However, an endotype which does not appear to be associated to a Th2 response has also been identified. As such, human asthma is divided into two major endotypes according to cytokine profile: Th2 and non-Th2 type asthma [47]. The Th2 type asthma is considered an allergic phenotype with the aforementioned eosinophil involvement and because its cytokine profile has been thoroughly described, several biomarkers are available for characterizing the disease and will be addressed bellow.

However, sEA is typically characterized by a neutrophilic response [1,48], and appears to not have the typical presentation of a type 1 hypersensitivity [49]. Although a Th2 cytokine profile has been described in sEA-affected horses, these animals do not display an early phase response [50,51]. However, a late phase response leading to neutrophilic bronchiolitis, associated with an increase in CD4+ T cells in the bronchoalveolar lavage fluid (BALF), has been described [50,52,53]. These features have led to the hypothesis that a type 3 hypersensitivity response, resulting in antibody-antigen complexes and activation of complement cascade, were involved in the disease's immunology [54]. However, because sEA does not possess most of the features described in type 3 hypersensitivities, it is unlikely that this type of response accounts for the main immunological features of the disease [55].

Still the precise cytokine profile of sEA remains unclear, with a multitude of reports pointing to either a Th1, a Th2, a Th17 or a mixed mediated response. Table 1 illustrates the cytokines reported in sEA-affected horses.

Table 1. Cytokines reported in sEA-affected horses according to T helper subtype [10,43,49,52,56–62].

Th2	Th17	Th1/Th2	Th1/Th17	Th2/Th17	Undefined
↑ IL-4 [1(r)]	↑ CXCL13 [2(r)]	↑ IL-4 [1(r)]	↑ IL-1β [1(r)]	↓ miR-197 [2(r)]	↓ IFN-γ [1(r)]
↑ IL-5 [1(r)]		↑ IFN-γ [1(r)]	↑ IL-8 [1(r); 3(r); 3(p)]	↑ miR-744 [2(r)]	↓ IL-4 [1(r)]
↓ IFN-γ [1(r)]			↑ IFN-γ [1(r)]	↓ miR-26a [4(r)]	↓ IL-5 [1(r)]
			↑ TNF-α [1(r)]	↑ miR-31 [4(r)]	↓ IL-13 [1(r)]
			↑ IL-17 [1(r)]	↓ TNF-α [4(r)]	
				↑ IL-4R [4(r)]	

[1]—BALF recovered cells; [2]—Peripheral blood; [3]—bronchial epithelial biopsy; [4]—Lung tissue (post-mortem). r—RNA detected; p—protein detected. ↑—increased expression; ↓—decreased expression. IL—interleukin; IL-4R—interleukin 4 receptor; IFN-γ—gamma-interferon; TNF-α—tumor necrosis factor-α; miR—microRNA.

Using immunohistochemistry and in situ hybridization, the expression of IL-4 and IL-5 was observed in the BALF lymphocytes of sEA-affected horses [49,53]. However, the Th2 cytokine profile of these animals was accompanied by airway neutrophilia, but not eosinophilia.

Other authors have reported an increased expression of IL-1β, IL-8, gamma-interferon (IFN-γ), tumor necrosis factor (TNF)-α, and IL-17, mainly suggesting a Th1 and/or Th17 mixed mediated response [59–61].

Gene expression analysis of BALF cells and bronchial epithelium of severely asthmatic horses, using reverse transcription polymerase chain reaction (RT-PCR), revealed that the expression of IL-1β, IL-8, NF-κB and toll-like receptor (TLR)4 was upregulated in these animals. Furthermore, authors reported that these findings correlated with the neutrophil percentage detected in the BALF [59].

Ainsworth and colleagues reported that during remission severely asthmatic horses exhibited an increased expression of IL-13 and despite BALF neutrophilia no differences in cytokine expression were observed 24 h after environmental challenge. However, after 5 weeks of chronic exposure to aeroallergens asthmatic horses presented increased IFN-γ and IL-8 gene expression [60].

After antigen challenge, the BALF cells of sEA-affected horses showed elevated gene expression of IL-17, IL-8 and TLR4. Gene expression of IL-8 was also increased in the bronchial epithelium, and using immunohistochemistry was tracked to the ciliated epithelium of affected horses. Additionally, stimulated peripheral blood neutrophils of asthmatic horses incubated with lipopolysaccharide (LPS) and formyl-methionyl-leucine phenylalanine (fMLP), two potent pro-inflammatory agents associated with sEA, revealed upregulated gene expression of IL-17 and TLR4 [61].

The presence of a mixed Th1/Th2 cytokine profile, involving mediators such as IL-4 and IFN-γ, has also been reported [57,58]. Disease exacerbation, post-antigen challenge, was also accompanied by elevated expression of IL-1β, TNF-α, IL-8 and IFN-γ, and treatment with fluticasone decreased mRNA expression of TNF-α [57]. Similarly, horses diagnosed with summer pasture-associated sEA developed disease exacerbation during the summer months, with increased expression of IL-13 and IFN-γ by BALF lymphocytes and CD4+ lymphocytes from peripheral blood. Furthermore, during disease remission, in the winter, these animals exhibited increased IL-4 mRNA expression [58].

The possibility of a mixed Th2/Th17 response has also been postulated [43], associated with a dysregulated Th17 cell differentiation pathway [62]. Eleven differentially expressed miRNAs (DEmiRs) were reported in the serum of asthmatic horses, compared to healthy individuals. Also, a shift towards the maturation of Th2 cells was proposed, supported by decreased levels of miR-128, which in association with decreased miR-197 and increased levels of miR-744 affects the maturation of Th cells towards a Th17 profile [43].

The analysis of the miRNAs and mRNA found in the lung tissue of sEA-affected horses supports the hypothesis of a Th17 mediated response, but also of a Th2 immune response [62]. Additionally, the upregulated miRNAs miR-142-3p and miR-223 found in asthmatic horses are also associated with severe neutrophilic asthma in humans, and with increased expression of IL-1β, IL-6 and IL-8 [63] cytokines, some of which have been associated with sEA [57,59].

Contrarily, Kleiber and colleagues reported neither a specific Th1 nor a Th2 specific response, but a downregulation of expressed cytokines (IL-4, IL-5, IL-13 and IFN-γ) in the CD4 and CD8 populations of the peripheral blood and BALF of sEA-affected horses [56], which could implicate the involvement of other pathways in the disease.

Thus, the reported results may reflect the heterogeneity of the cytokine profile involved in sEA and may imply the existence of different disease endotypes. However the interpretation of these results must necessarily take into consideration the described methodologies of the above-mentioned studies. For example, cytokine expression was investigated using distinct samples, namely BALF, bronchial and lung tissue, as well as peripheral blood. As such, results may not only reflect the inflammatory response of the examined cells, but also differences between local and systemic inflammatory responses.

With the development of transcriptomics, novel techniques for assessing the existence and relative prevalence of several RNA species have been introduced to the scientific community. This is portrayed in the reported methodologies of the aforementioned studies, where recent experiments sequence mRNA and miRNA, contrasting with the less comprehensive/detailed methods, such as traditional targeted immunohistochemistry, in situ hybridization and RT-PCR.

Additionally, the experimental design of most studies involved the exacerbation of the disease by exposing the asthmatic horse to an intense pro-inflammatory environment, using hay dust and/or by stabling the affected horses. It cannot be excluded that the experimental induction of airway inflammation may interfere to some extent with the expressed cytokine profile, especially considering individual susceptibilities to specific allergens. Therefore, this factor also needs to be taken into account when interpreting reported results.

As in human asthma, it is highly likely that sEA possesses multiple endotypes, and considering the neutrophil recruitment observed in affected horses, a Th17 mediated response is probably part of the inflammatory pathways involved in this disease. Nevertheless, more encompassing studies involving genomics, transcriptomics and proteomics are required to better define the cytokine profile of sEA and to determine therapeutic targets in affected horses, and although further confirmation is required, the reported DEmiRs may constitute novel therapeutic targets for sEA.

Severely asthmatic horses may also present an altered response to allergens, since ex vivo and in vivo hay dust stimulation revealed upregulation of several genes participating in the inflammation [10,64]. Pacholewska and colleagues reported an increased expression of CXCL13 chemokine [10] which may indicate a Th17 mediated response [65], but no evidence of a Th1 nor a Th2 response was found. Additionally, in a murine model of allergic airway inflammation increased expression of CXCL13 has been reported and its neutralization reduced allergic inflammation by decreasing lymphocytes, eosinophils, as well as the recruitment of CXCR5-bearing cells [66]. Accordingly, in humans, IL-17 expression has been associated with severe neutrophilic asthma and in horses this cytokine is responsible for the activation and persistence of neutrophils in the airways. Also, IL-17 was shown to be associated with reduced response to corticosteroids, with post-treatment persistence of IL-8 [67,68].

The described heterogeneity also occurs in human asthma, where one could consider the existence of three distinct phenotypes: allergic asthma, non-allergic asthma and late-onset asthma [22]. These phenotypes may also be applicable when describing sEA. In this sense the allergic asthma phenotype would be characterized by a Th2 mediated response and usually associated with other allergic diseases. In general, horses affected with sEA may also suffer from other allergic diseases such as insect bite hypersensitivity or atopy [69,70]. Interestingly, Lo Feudo and colleagues have reported the presence of a type 1 hypersensitivity in sEA-affected horses in response to intradermal allergen test, which may further support the hypothesis of an allergic phenotype [71]. Additionally, the use of skin prick tests has previously been used to identify allergic sensitization in severely asthmatic horses [19]. Similarly, evidence of a type 1 hypersensitivity to different allergens has also been described in severely asthmatic horses using allergen inhalation [72].

The non-allergic phenotype in humans is usually associated with the presence of neutrophils in the airways, a hallmark of sEA. This phenotype also reflects the involvement of a Th1 and of a Th17 response [62], which has also been described as contributing to the immune response in asthmatic horses.

Finally, the late-onset asthma is age-associated and also occurs in sEA where affected individuals are mature adults [2,73]. This age association is thought to be the consequence of immunosenescence and inflammaging, which describe the immune and inflammatory modifications observed in geriatric patients [74–76], a subject extensively revised by Bullone et al. [75]. Immunosenescence is essentially a disfunction of the immune system. In horses it is usually characterized by a dysregulation of adaptative immunity associated with a lower proliferative response of T lymphocytes [77], and a decrease in mean percentage of regulatory T cells [78].

On the other hand, the term inflammaging defines the chronic inflammatory state observed in older individuals accompanied by an increased expression of inflammatory cytokines [74]. A correlation between age and IL-6 has also been described in healthy geriatric horses [79]. Also, compared to young adults, geriatric horses with colitis had higher levels of IL-6 and TNF-α [80]. It has also been described that older horses exhibit increased expression of IL-1β, IL-15, IL-18, IFN-γ and TNF-α mRNA [77,81,82]. These studies confirm that inflammaging and immunosenescence occur in geriatric horses both systemically and locally [74], and are most likely involved in the immunology of sEA, although further research is needed to clarify the age-associated changes and how they affect airway inflammatory response.

The reported differences in the immunological pathways contributing to sEA illustrate the complexity of this disease and suggest the existence of several endotypes, which converge into the same clinical phenotype. One must also consider that the methodological differences of the above mentioned studies, such as time of sample collection, natural vs. stimulated inflammatory response and duration of the disease, may have contributed to the reported variations. It is therefore fundamental that holistic studies, encompassing more exhaustive and complementary approaches, and preferably large multi-center studies can be performed to unravel sEA's different immunological responses.

4. The Epithelium

The bronchial epithelium is a complex tissue composed of a single layer of ciliated columnar or cuboidal cells that intercalate with secretory cells, namely club and goblet cells [83]. These epithelial cells act as a protective barrier against foreign particles and microbes, while the mucus secreted by both epithelial secretory cells and submucosal glands, comprising a mixture of ions, proteins, lipids and large amounts of mucin glycoproteins, namely MUC5AC [36], actively contribute to this protection [84]. The relative amount of the cells that constitute this tissue is also dynamic. One example of this plasticity is the fact that the relative number of secretory cells is known to be increased in asthmatic horses in exacerbation phases when compared to controls [85]. Also, the composition of the produced mucus is disrupted in inflammatory conditions, such as sEA, and it is known that asthmatic horses, whether or not in exacerbation, have significantly decreased Salivary Scavenger and Agglutinin (SALSA) production, thus impairing innate antibacterial abilities [86].

Also present are a number of immunologically active cells, such as neutrophils, macrophages and lymphocytes. The relative number of these immune cells also varies according to inflammation status and it is known that mononuclear leukocytes [87], and mast cells [88] are increased in asthmatic patients. Airway epithelial cells are supported by a loose connective tissue [83], which is also involved in immune response and in reactive airway remodeling. In fact, recent studies have determined that the interactions between fibroblasts and the epithelium can influence airway remodeling [89].

A wide array of studies have found that the protective role of the respiratory epithelium is not exclusively mechanic, as these cells are capable of responding to offensive stimuli by secreting immunomodulatory molecules, such as chemokines, cytokines, and host defense molecules, including acute phase proteins and complement proteins [90], that regulate respiratory innate immune response. As such, studies have found that healthy bronchiolar epithelium transcribes genes of all *TLR* [90], and that *TLR3* is not only the most transcribed *TLR* in equine bronchial epithelial cells [91], but is also particularly active in response to stimulation [90,91].

Unsurprisingly, several studies have found that the epithelium of asthmatic horses responds differently to offensive stimuli than that of healthy horses. In a transcriptome analysis of endoscopically obtained biopsies, Tessier et al. found many differentially expressed genes, among which were the neutrophil chemotaxis (GO:0030593) gene set which was overrepresented in asthmatic derived samples and can explain the observed marked airway neutrophilia [64]. This study also identified the increase in *MMP-1, MMP-9, TLR4* and *IL-8* transcription as a central player in the inflammation process observed in asthmatic horses. Another relevant epithelial produced chemokine, likely involved in asthma pathophysiology and uncovered by in vitro hay dust exposure assays, is *CXCL2* [92]. CXCL2 is related to IL-8 and shares its ability to potentiate early airway neutrophilia.

Interestingly, the differences found between the airway epithelium of asthmatic and healthy horses can account for more than the pathophysiological development of the condition, impacting also its treatment. RNA-seq analysis found a decreased expression of the rhythmic process (GO:0048511) gene set in asthmatic horses, namely *CIART*, which could disrupt the natural glucocorticoid response and promote treatment resistance [64].

As the first barrier against foreign respirable particles and microbes, the airway epithelium plays an essential role in the defense of the lung and is an integral part of the regulatory mechanisms that drive lung inflammatory processes. The continuation of the study of this tissue's immunomodulatory properties and ways in which they can be impaired in sEA will undoubtably contribute to advance our knowledge of this pathology and find better and more efficient treatment approaches.

5. Alveolar Macrophages

Alveolar macrophages (AMs) are the most common immune cells found in the lungs of healthy horses. By releasing cytokines and chemokines, such as IL-8, CXCL2 (also known as macrophage inflammatory protein-2), and TNF-α, AMs act as first respondents in the host's defense [57,93–96]. Therefore, it is likely that these cells contribute to the pathomechanisms of sEA.

Macrophages present different characteristics depending on the tissue where they are located [97]. Compared to peritoneal macrophages (PMs), AMs appear to possess increased responsiveness and phagocytic capacity [98]. When exposed to LPS, an important antigen implicated in sEA, AMs presented upregulation of the MyD88 and TRIF pathways [98], further highlighting the importance of these cells in the innate immune response. In comparison, only the MyD88 pathway was upregulated in PMs [98], further illustrating the differentiated role of AMs.

Depending on local microenvironment, macrophages can modify their phenotype [99,100]. The pro-inflammatory phenotype (M1) is induced by pathogen-associated molecular patterns (PAMPS) and IFN-γ [100], while the anti-inflammatory (M2) phenotype is related to wound healing and tissue repair, thus playing an important role in controlling neutrophilic inflammation through efferocytosis [101,102]. The cytokines IL-4 and IL-13 have been reported as modulators of the latter phenotype [103,104].

However, sEA-affected horses may lack this dynamic modulation. A recent study found that asthmatic horses at pasture had an increased expression of IL-10 in comparison to healthy controls. The authors also reported a simultaneous increased expression of IL-10 (M1 phenotype) and CD206 (M2 phenotype), suggesting the presence of a non-canonical phenotype. Furthermore, AMs maintained their responsiveness to LPS and expressed an increase in IL-8, even in the presence of IL-10, known as an inhibitor of LPS response. As such, an impaired response to IL-10 may contribute to sEA's pathogenesis [105]. Additionally, the AMs of sEA-affected horses exposed to moldy hay presented not only increased expression of CD206 markers, but also of CD163 markers [106], further reenforcing the anti-inflammatory profile of these cells.

AMs may also contribute to the dysregulation of apoptosis described in sEA [105,107]. Apoptosis is a physiological mechanism for programed cell death and thus fundamental for inflammatory control. In asthmatic horses a delay in the apoptosis of BALF neutrophils has been reported [107–109]. Furthermore, Niedzwiedz and colleagues also described an increase in the early apoptotic rate of the BALF AMs [109].

Current knowledge on the role of macrophages is quite limited, and advances in genomic and transcriptomic analysis may help to enlighten how these cells contribute to the disease. In particular we need to understand if these cells are in fact the main agent responsible for the neutrophil recruitment and understand how they influence the inflammatory pathways associated with equine asthma. One must also consider that the M1/M2 phenotype nomenclature of macrophages is a rather simplistic concept, since it categorizes the activity of these cells into two extreme opposites—pro and anti-inflammatory. This concept is mostly based on in vitro studies (i.e., stimulation with LPS) and fails to take into consideration the local microenvironment found in vivo. Furthermore, studies have shown that macrophages can simultaneously exhibit characteristics of both phenotypes [97,110], urging the creation of a new nomenclature based on their ontogeny which more clearly encompasses the recent findings on this subject.

AMs derive from embryonic precursors and blood monocytes, which migrate to the lung and differentiate into AMs. A recent work by Evren and colleagues describes the pathways involved in this differentiation and further defines populations of AMs based on surface cell proteins, using a humanized mouse model [111]. However, compared to humans and mice, knowledge about the precise origin of equine AMs is still vague and several conclusions are extrapolated from in vitro and in vivo studies conducted in other species.

By expressing a non-canonical phenotype associated with an altered apoptotic rate, AMs may contribute to the persistence of airway inflammation. Nonetheless, a better characterization of the equine lung macrophage population is fundamental to understand how these cells influence the inflammatory response in EA.

6. The Role of Neutrophils

Neutrophil recruitment, and consequent airway infiltration, is a hallmark of sEA [48,112] and has been extensively discussed in the literature [113,114]. Neutrophils contribute to the innate immune response through the phagocytosis of pathogens and the production of cytokines, chemokines and proteases, as well as reactive oxygen species (ROS) and neutrophil extracellular traps (NETs) [115,116]. Their apoptosis serves as a mechanism of controlling the action of these cells and limiting secondary tissue damage [117].

The airway neutrophilia observed in sEA and in human severe neutrophilic asthma is typically not associated with a septic inflammation. The exposure to aeroallergens and irritants results in the activation of pattern recognition receptors, namely TLR2, TLR4 and NOD2 [118,119], which interact with adaptor protein MyD88 and result in a consequent increase of cytokines and chemokines, such as IL-17, IL-8, CXCL2 and CXCL10, promoting the migration of neutrophils into the airways [120–123]. Although several studies have reported an increased expression of IL-8 in asthmatic horses during disease exacerbation [57,60,124,125], the IL-17 cytokine, whose role is further upstream in the cytokine cascade, appears to play a more significant role in neutrophil recruitment contributing to the chronicity of sEA [61,67,125,126]. IL-17 stimulates the production of CXCL1, CXCL2 [127–129], IL-8 and granulocyte macrophage-colony stimulating factor (GM-CSF) [127,130] and decreases neutrophil apoptosis [68].

Additionally, the increased expression of *TLR4* mRNA in asthmatic horses after antigen challenge [61] correlates with *IL-8* mRNA expression [131] and may further contribute to neutrophil inflammation.

NETosis is one of the mechanisms employed by neutrophils to impair infectious agents [132]. NETs are composed by nuclear DNA associated with nuclear and granule proteins and enzymes [133–135]. However, they are also cytotoxic and can themselves promote lung injury [135–137]. The IL-8 chemokine, which has been shown to be upregulated in sEA-affected horses, induces NETosis in severe human asthma [138]. Furthermore, NETs were increased in the BALF of asthmatic horses during exacerbation [139,140] and low density neutrophils (LDNs), a subpopulation of neutrophils with a greater capacity for producing NETs, have been found to be increased in the peripheral blood of humans and horses with severe asthma [141,142].

ROS are formed by inflammatory cells, such as neutrophils, involving NADPH oxidase [143] and contribute to cell injury and airway remodeling [144–146]. They are also responsible for the activation of transcription factors [147,148] and the expression of inflammatory cytokines [149]. In order to prevent oxidative injury, cells produce antioxidants. However, horses with sEA show signs of oxidant/antioxidant imbalance [150,151], including a reduction in ascorbic acid and an increase in elastase concentrations in the BALF [152,153]. Oxidative stress may also contribute to corticosteroid insensitivity in asthmatic horses. These animals maintain neutrophilic inflammation even after treatment with corticosteroids, which may be caused by the expression of the chemoattractant IL-8. In vitro it was demonstrated that oxidative stress increases the mRNA expression of *IL-8* and *IL-1β* by peripheral blood neutrophils of both healthy and asthmatic horses and that in spite treatment with dexamethasone, the upregulation of *IL-8* persisted, whilst *IL-1β* became downregulated [154]. In vivo research about the precise role of IL-8, and IL-17 will help determine if the IL-8 pathway is a suitable target for immunotherapy in asthmatic horses and humans with corticosteroid insensitivity. Additionally, sEA-affected horses may benefit from the correction of oxidative stress, although research using a more encompassing model, illustrating the microenvironment of the lungs in vivo, should be considered to evaluate the impact of oxidative stress in the inflammatory pathway.

Moreover, the bronchial epithelium is susceptible to the cytotoxic effects of neutrophil byproducts, such as ROS, exosomes and proteases [155,156], and in humans NETs are also able to induce the expression of pro-inflammatory cytokines by the epithelium [157]. In asthmatic horses the production of secretoglobulin 1A1 (SCGB1A1), a protein produced by club cells with anti-inflammatory functions, is compromised. This could be due to the decrease in the number of club cells or to a depletion of SCGB1A1 in response to chronic inflammation [139,158].

As previously mentioned, dysregulation of neutrophil apoptosis may also contribute to sEA [107,109]. This can occur through several mechanisms, such as (1) the expression of a non-canonical phenotype by AMs, which may compromise their response to efferocytes [105,159], and (2) the presence of IL-17 which increases neutrophil viability [68], thus perpetuating neutrophilic inflammation.

Neutrophils play a significant role in sEA and thus limiting their activation and increasing their clearance can improve disease resolution and limit potential complications associated with tissue injury.

7. Inflammatory Biomarkers

Several biological molecules have been implicated in the inflammatory response of sEA and their identification and the knowledge of their interactions could ultimately contribute to a personalized diagnosis and disease monitoring. However, in order for a biomarker to have clinical applicability it must meet several requirements, which are summed up by the "SAVED" model. "SAVED" stands for "Superior", "Actionable", "Valuable", "Economical", and "clinically Deployable", indicating that the new biological marker must improve current practice and patient management, as well as patient outcome, but also be cost-effective while using technology available in clinical laboratories [160]. This criteria is also being used in the development of novel biomarkers for human asthma [161].

Research on the biomarkers for equine asthma focuses mostly on two major sampling methods–BALF and peripheral blood. BALF has the advantage of better portraying the degree of airway inflammation, thus it is considered to be more representative of the disease [162]. However, BALF collection is an invasive procedure unfit to be used routinely to obtain repeated measurements or in horses with severe respiratory distress. Serum biomarkers require sampling of peripheral blood which is a far less invasive process and is usually well tolerated by horses. Unlike BALF markers, peripheral blood markers indicate systemic inflammation, rendering their application less disease-specific and making the interpretation of the obtained results more challenging [163]. Comparatively human medicine has other non-invasive alternatives, such as sputum induction [164], which is an unfeasible option in horses, and exhaled breath condensate (EBC). The applicability of exhaled breath condensate (EBC) is currently being investigated in equine asthma, although it requires specific equipment [165].

As previously mentioned, Th2 type human asthma has been thoroughly described and several biomarkers are available, thus better aiding the definition of a suitable therapeutic approach [161,166]. As such, serum IgE, fractional exhaled nitric oxide (FeNO) and blood eosinophilia are used in a clinical context to characterize disease and predict response to corticosteroids [166]. Current research is focusing on novel biomarkers which have shown promise for clinical application and require minimally invasive procedures [161], such as sputum mRNA analysis [167], serum periostin [168], exhaled breath volatile organic compounds [169], dipeptidyl peptidase-4 [170] and urinary leukotriene E4 [171]. Contrastingly, considerably less biomarkers are described for the non-Th2 type human asthma, where cytokine profile studies based on genetic, transcriptomic and proteomic analysis are considered fundamental [166,167]. Ultimately, these studies will allow optimization of personalized therapeutic targets and ensure a good clinical outcome.

Currently, only one biomarker is widely used for the diagnosis of sEA, requiring the sampling of BALF (Table 2). Since BALF neutrophilia is a hallmark of severely asthmatic horses, cutoff values have been established and are commonly used in everyday practice [1,14,172]. Unlike human Th2 type asthma, where eosinophilia occurs both in the BALF and in the peripheral blood [173], in sEA blood neutrophilia isn't observed [1] and therefore cannot be used as a diagnostic tool. As such, current research focuses on alternative biomarkers which can substitute BALF collection, as well as consensual cutoff values, which would prove useful in identifying severely asthmatic horses in remission, monitor treatment response and contribute to precision medicine.

Acute phase proteins (APP), such as haptoglobin or serum amyloid A (SAA), are being investigated as potential indicators of disease [174–176]. The expression of haptoglobin was decreased in horses with summer pasture-associated sEA, compared to healthy controls [175]. This protein has been associated with airway remodeling [177], and in asthmatic children serum haptoglobin is reported to have decreased immediately after antigen challenge, although its levels did increase 24 h after exposure [178]. A different study reported that mean serum haptoglobin values were increased in severely asthmatic horses. However, authors reported that after antigen exposure an increase was also observed in the control group, suggesting that airway inflammation is reflected systemically [176].

SAA has been associated with neutrophil recruitment [179] and is increased in the serum and sputum of asthmatic people [180]. Similarly, seven days after antigen challenge sEA-affected horses presented a higher concentration of SAA [176].

Table 2. Biomarkers described for sEA.

Sampling Method	Biomarker	Reported Results	References
BALF			
	Neutrophils (>25% as cutoff for sEA)	Marked neutrophilia	[1]
	Haptoglobin	Decreased	[175]
	IFN-γ	Increased	[172]
	MMP-8	Increased	[181]
	MMP-9	Increased	[181,182]
Peripheral blood			
	Serum amyloid A	Increased	[176]
	Haptoglobin	Increased	[176]
	Circulating immune complexes	Conflicting results	[183,184]
Exhaled breath condensate			
	Methanol	Increased	[185]
	Ethanol	Increased	[185]

BALF—Bronchoalveolar lavage fluid; IFN-γ—gamma-interferon; MMP—matrix metalloproteinase.

Although APP can help in identifying local and systemic inflammation associated with sEA, they are not disease specific, and while SAA could potentially be associated with neutrophil recruitment, the role of these proteins in sEA needs to be better investigated.

The applicability of circulating immune complexes (CIC), which are formed by the union of an antigen and an antibody, has also been studied in sEA. These complex molecules were found to be increased in this disease [183,184] and, although conflicting reports question their diagnostic power, they may contribute to the monitoring of treatment response [184].

Similarly, matrix metalloproteinases (MMPs), tissue inhibitors of metalloproteinases (TIMPs) and MMPs/TIMPs ratio can be used to monitor disease severity and response to corticosteroids [181,182,186–188]. In general, MMPs are responsible for tissue destruction through collagen degradation [189], whilst TIMPs lead to the formation of fibrosis [190], and as such are thought to contribute to airway remodeling and fibrosis in chronic inflammation. The concentration of MMP-2, MMP-9, TIMP-1 and TIMP-2 decrease in response to treatment with corticosteroids [186], and with cytosine-phosphate-guanosine-oligodeoxynucleotides (CpG-ODN), an immunostimulatory drug [187]. However, since these biomarkers require invasive sampling, as they are measured in BALF, they are not suitable for evaluating treatment response.

IFN-γ has also been proposed as a biomarker of sEA, since it is increased in the BALF of these animals and is capable of distinguishing severely asthmatic horses from healthy individuals [172]. However, similarly to what is seen with MMP, it requires BALF sampling, rendering it unsuitable for repeated measurements.

In mEA-affected horses with airway neutrophilia (presenting more than 15% neutrophils in BALF), serum concentration of surfactant protein D (SPD) was reported to be significantly increased [191]. SPD could be a relevant biomarker for the diagnosis of EA, but additional research is required in severely asthmatic horses.

EBC is a non-invasive method which allows sampling of airway material and access to information about the metabolic status of the patient, even during disease exacerbation [165,185,192]. Preez et al. found that horses with lower airway inflammation have a higher pH and an increased hydrogen peroxide (H_2O_2) concentration in the air they exhale [192]. However, a different study reported no variations in the pH nor in the concentration of H_2O_2 between healthy horses and those with lower airway inflammation [165]. Both these studies did not focus exclusively on severely asthmatic horses and so it remains unclear whether these parameters could be of use as biomarkers of the disease. An additional study of metabolites in the EBC revealed that sEA-affected horses had increased concentrations of methanol and ethanol, compared to healthy controls [185]. The study of metabolomics in EBC has the potential to be a non-invasive approach to sEA diagnosis. Nonetheless, further research is necessary to better understand if this method has limitations, particularly if it can adequately distinguish between sEA-affected horses during remission and healthy individuals.

Current knowledge on sEA biomarkers is still limited and their use in a clinical context requires further research, since a noticeable benefit must be associated with these molecules in order for them to be included in the clinical guidelines for disease diagnosis and monitoring.

8. Microbiome

The term microbiome refers to the community of microbes, such as bacteria, fungi, virus and archaea of a particular biological location [193,194]. These microorganisms interact functionally and metabolically, playing an important role in modulating the host's innate immune response [195] and contributing to the inhibition of potential pathogens [196–198].

For many years it was mistakenly thought that the lung environment was sterile. However, the identification of microbiota in the lower respiratory tract of healthy humans has since discredited this belief [199–202].

Similarly, studies on equine microbiome have revealed that the same organizational taxonomic units (OTUs) can be found in the upper and lower airways of healthy individuals, although the latter anatomic region possessed an inferior biomass with decreased richness and diversity [203,204].

The phyla Proteobacteria, Firmicutes, Bacteroidetes, and Actinobacteria were found to be highly represented in healthy horses [203]. These phyla are also dominant in the lungs of healthy humans along with two other—Fusobacteria and Acidobacteria [199,200,205].

The microbiome is highly dependent on the host's interaction with its environment. Thus, geographical location, housing conditions, diet, interactions with other individuals, and also treatment with antimicrobials and corticosteroids make each individual's microbiome unique [204,206–212].

In humans, dysbiosis during infancy is considered a significant risk-factor for respiratory diseases such as asthma [213–215]. However, this relation has yet to be described in the equine population.

The lung microbiome of adult asthmatic people has been demonstrated to differ in both number and composition from that of healthy individuals, and these differences have been associated with airway hyperresponsiveness and obstruction [216,217]. Thus, the study of the lower airway microbiome in equine asthma stems from the hypothesis that these microbes play an important role in modulating the innate immune response of the host and may, therefore, contribute to the immunology of sEA [195].

The comparison of lung, nasal, and oral microbiomes of healthy and asthmatic horses showed that significant differences between groups could only be found in the lung microbiome at the taxonomic family level, with an overrepresentation of the Pasteurellacea family, but not at the phylum or OTU level, leading to the hypothesis that these differences were not inherent, but rather a consequence of inflammation [204].

Additionally, the bacteria *Nicoletella semolina*, a Pasteurellacea, has been detected in the upper and lower airways of both healthy and severely asthmatic horses [218–220]. Although its prevalence was increased in asthmatic horses, as detected by quantitative polymerase chain reaction (qPCR), no specific functional association between this bacteria and sEA was found [220], suggesting that it may be an opportunistic agent perpetuating airway inflammation in asthmatic animals.

Inversely, *Corynebacterium* spp. was commonly found in the trachea of a group of healthy horses, but its presence was decreased in the evaluated asthmatic group. As such, this microorganism could be a part of the normal microbiota of healthy horses, and might be one of the populations affected by the inflammatory changes in the airways [221].

Recent studies further point to the occurrence of dysbiosis in the lower airway microbiome of mEA-affected horses, although it is still unclear whether this results from persistent inflammation or chronic treatment with corticosteroids [203,221]. Furthermore, the relative abundance of *Streptococcus* was increased in mildly asthmatic horses, suggesting that the presence of this genus might be a risk-factor for mEA [203,222]. Similarly, infections with *Streptococcus pneumoniae*, and other opportunistic agents such as *Haemophilus influenzae* and *Moraxella catarrhalis*, are associated with acute asthma exacerbations in humans [223,224].

Fungi are often implicated in sEA exacerbation and are also considered as risk factors in human asthma [5,6]. Despite their relevance in disease pathophysiology, research focusing on the lung mycobiota of healthy and asthmatic horses is very limited. In healthy humans agents such as *Davidiellaceae* and *Cladosporium* [225], as well as *Eremothecium*, *Systenostrem*, and *Malassezia* [202], are the main contributors to the lung mycobiome. On the other hand, Charlson et al. have found that *Malassezia pachydermatis* is exclusively found in the BALF of asthmatic patients, which also showed significantly increased populations of *Termitomyces clypeatus* and *Psathyrella candollean* [225]. As for equine species, Bond and colleagues reported that the mycobiota of a group of mildly asthmatic horses comprised mainly of two phyla - Ascomycota and Basidiomycota [226]. Although a better characterization of healthy and asthmatic equine lung mycobiome is necessary, the reported results differ from those described in healthy humans [202,225]. Nonetheless, significant differences between the two species are to be expected, since stabling and hay feeding promote an environment rich in fungi [6].

To mitigate signs of airway inflammation and improve lung function, asthmatic patients are usually prescribed long-term treatment with corticosteroids. Because of their immunomodulatory effect, the use of corticosteroids can promote microbiome dysbiosis. In mEA-affected horses, the use of systemic dexamethasone affected the microbiota of the lower respiratory tract of healthy and asthmatic horses, increasing the relative abundance of 9 OTUs, including the abundance of *Streptococcus* spp. in the asthmatic group [203]. Similarly, the nebulization of dexamethasone resulted in an increase in the genera *Alysiella* and *Bordetella* in the lower respiratory tract, however this treatment had no effect on the population of *Streptococcus* found in the airways [226].

The relation between microbiome and corticosteroid is not unidirectional, since response to treatment is also influenced by lung microbiome. The microbiome of asthmatic humans diagnosed with corticosteroid resistance showed differences at the genus level compared to that of responsive patients. Furthermore, BALF AMs from asthmatic patients stimulated with *Haemophilus parainfluenzae*, a potential pathogen found in asthmatics with corticosteroid resistance, resulted in inhibition of response to corticosteroids, along with increased p38 mitogen-activated kinase phosphatase (MAPK) activation and increased *IL-8* and *mitogen-activated kinase phosphatase 1 (MKP1)* mRNA expression. On the other hand, exposure to commensal *Prevotella melaninogenica* did not have a similar effect [227]. Reports also indicate that *H. parainfluenzae* can convert a Th2-type allergic asthma sensitive to corticosteroid treatment to a Th1 neutrophilic profile, with Il-17 expression [228]. Interestingly, Goleva and colleagues have also reported that inhibition of the transforming growth factor-β-associated kinase-1 (TAK1) in monocytes collected from the peripheral blood of asthmatic patients restored cellular sensitivity to corticosteroids [227], which could represent a novel therapeutic approach in patients which are refractory to these drugs. As such, the study of the microbiome can be useful in determining the response to therapy with corticosteroids.

Several studies have described how the gut microbiome of people with respiratory disease differs from that of healthy individuals [229–231], highlighting an immunological relationship between the lungs and gut. This interaction is termed the gut-lung axis, illustrating how these two distant anatomical sites appear to communicate. Not only does gut microbiota have a local immunological effect by interacting with the mucosal immune system [232], but it also produces pro and anti-inflammatory metabolites, such as biogenic amines (i.e., histamine), oxilipins, and short-chain fatty acids (SCFAs), which modulate the inflammatory response both locally and in the lung [233–236]. SCFAs reduce allergic response and airway inflammation in both humans and mice [237–241]. Trompette and colleagues reported that in an ovalbumin (OVA)-model, where mice were challenged with ovalbumin protein to induce an allergic inflammation in the lung, treatment with SCFAs increased the presence of dendritic cells with high phagocytic ability but with impaired capacity of activating Th2 effector cells in the lungs [240].

Allergic diseases are also associated with a lower fecal microbial diversity in humans, and, although these findings have mostly been reported in infants, they have also been described in adults [242].

Differences in the fecal microbiome of asthmatic (mEA) and healthy horses appeared to occur mostly during disease exacerbation [207]. The reported dysbiosis was observed using an OTU analysis approach and was found to be accompanied by an increased representation of the Firmicutes phylum, namely Clostridia class [207]. These authors hypothesized that lung inflammation and compromised oxygenation would induce changes in the gut microenvironment, since fewer differences were observed when both groups of horses were at pasture, a less pro-inflammatory environment. However, no causative relation was established and disease remission could also be secondary to the changes induced in the gut microbiome by leaving the horses out to pasture. Dysbiosis of bacteria belonging to the phylum Firmicutes has also been documented in several studies on gut and respiratory microbiome of asthmatic humans [243,244] and further research could prove to be of interest in equine asthma.

Conversely, Kaiser-Thom and colleagues compared the fecal microbiota of horses diagnosed with either sEA, culicoides hypersensitivity or both to that of healthy individuals using a Divisive Amplicon Denoising Algorithm (DADA2) approach to analyze microbial taxonomy, and found no significant differences between the microbe populations [206].

Research is currently focused on alternative therapies which could revert the dysbiosis observed in asthmatic individuals and thus impacting immune responses. Since treatment with antibiotics is not a viable option, supplementation with probiotics and soluble fiber are currently under investigation [238,245]. In an OVA-induced mouse model of allergic airway inflammation, the use of probiotics induced regulatory T cells differentiation and suppressed Th2 allergic response [245]. Further studies are necessary to understand if horse gut microbiome could also benefit from such treatments and whether they would in fact result in the modulation of the inflammatory response associated with sEA.

The microbiome of asthmatic humans has been associated with specific disease endotypes. For example, increased representation of Proteobacteria is found in severe asthma with neutrophilic exacerbation [246] and Th17-related gene expression [247]. As such, study of the microbiome can further enable the practice of precision medicine, and increase the likelihood of a good prognosis in asthmatic patients since it will allow a personalized therapeutic approach. However, these interactions have yet to be described in equine medicine.

Studies on the microbiome of asthmatic horses are mostly descriptive, portraying the microbial populations found in the respiratory and intestinal tract. Although some studies do characterize the degree of airway inflammation of these horses by using BALF cytology and lung function evaluation, cross-referenced data with existing cytokine profile is, nonetheless, currently missing. Whether equine pulmonary microbiome will differ according to asthma endotype and inflammatory cell population remains to be ascertained.

Another limitation is the small number of animals enrolled in each study which limits the statistical significance of the results, influences the conclusions and may further impair the establishment of causality. Additionally, in most studies disease exacerbation was achieved by altering the horses' environment and diet which inevitably influences the microbiome of the studied horses and works as a confounding factor in the interpretation of results. Furthermore, it is still not clear to which extent the altered microbiome causes, perpetuates or is secondary to airway inflammation in equine and human asthmatic patients.

Further research on the microbiome of equine asthmatics will undoubtably contribute to the elucidation of the current loopholes in this subject.

9. Conclusions and Future Directions

Further characterization of the disease's genetic background is fundamental to improve current knowledge of the pathways involved in the heritability and expression of sEA. The reported genetic research focuses mostly on a well characterized subpopulation of Swiss Warmblood horses which limits the applicability of these findings to the general horse population. Nonetheless, the reported genetic heterogeneity and complexity observed in the above mentioned families likely occurs in other individuals, although, potentially, other genes and pathways may be involved. Thus, research on a large multi-center population of client owned sEA-affected horses with a detailed genetic background is needed to further contribute to the description of the genetic events that take place in these animals.

Although the clinical phenotype of sEA has been thoroughly described [1], the immunological mechanisms which lead to inflammation and structural changes in the airways (mucus accumulation, bronchial constriction and bronchial wall thickening) still lack clarification. Several cytokines, chemokines and inflammatory cells participate in the pathogenesis of asthma; however their precise characterization is still unclear. The inconsistencies found between reported studies may arise from differences in their experimental design and methodologies. Furthermore, the limited number of animals included in these works may hinder the attainment of significant results.

Thus, cooperation between research groups and research based on large multi-center populations of client owned sEA-affected horses could potentially solve some of these limitations. Also, uniformization of methodologies and protocols will enable the comparison of reported results, allowing a better definition of the genetic and immune mechanisms associated with sEA.

Additionally, more encompassing studies using genomic, transcriptomic, proteomic and metabolomic analysis will undoubtably enhance the scientific knowledge of the disease. This will enable an understanding of how genetics can determine cytokine and chemokine expression and how these proteins and metabolites influence disease expression. Furthermore, the impact of lung and gut microbiome also needs to be assessed, since these microbes regulate the immune innate response from an early age and can also promote airway hyperreactivity and inflammation in asthmatic individuals.

Although sEA shares many similarities with its human counterpart, an understanding of this disease based on the extrapolation of reported data for other species is unfeasible. Thus understanding the origin of equine AMs and the characterization of this population is necessary to recognize how these cells influence the inflammatory pathways of asthmatic horses.

Immunological and genetic characterization will likely assist in the identification of disease endotypes and more importantly contribute to the development of novel therapeutic targets. For example, anti-interleukin targeted therapies, using monoclonal antibodies, could help manage the disease, especially in horses with resistance to corticosteroids. This is currently being researched in human asthma, where identification of disease endotypes associated with specific cytokine profiles has led to the development of monoclonal antibodies. One such example is Tralokinumab, a human anti-IL-13 monoclonal antibody for uncontrolled asthma [248].

Since the precise cytokine profiles of equine asthma are not fully understood, current research is focusing on identifying the causal allergens which trigger airway inflammation [5], and how immunotherapy can help modulate the inflammatory response with promising results [9,187]. Also, determining specific allergen susceptibility can contribute to the development of specific immunotherapy and, in theory, help devise environmental management protocols for affected horses.

Novel diagnostic tools based on genetics or disease biomarkers would prove of significant value to equine medical practitioners, especially if they are able to positively identify a severely asthmatic horse during remission. Current diagnosis relies mostly on invasive methods which are not suitable for evaluating treatment response, since it will require repeated measurements. Thus, systemic blood biomarkers and exhaled breath condensate are attractive alternatives to BALF sampling. Research should focus on defining cutoff levels and constructing a panel of biomarkers which could substitute BALF cytology when monitoring treatment response.

In conclusion, current research shows that the genetic and inflammatory pathways involved in sEA are complex and variations are to be expected between subsets of individuals. A deeper knowledge of the disease's immunological pathways will allow the definition of endotypes, the detection of inflammatory biomarkers of diagnostic value, and a personalized therapeutic approach targeting the inflammatory pathways involved in the disease.

Author Contributions: All authors have contributed equally to this work. All authors have read and agreed to the published version of the manuscript.

Funding: The research developed by the authors is supported by CIISA and FCT–Fundação para a Ciência e Tecnologia through project UIDB/00276/2020.

Institutional Review Board Statement: Not applicable.

Informed Consent Statement: Not applicable.

Data Availability Statement: Not applicable.

Conflicts of Interest: The authors declare no conflict of interest.

References

1. Couëtil, L.L.; Cardwell, J.M.; Gerber, V.; Lavoie, J.P.; Léguillette, R.; Richard, E.A. Inflammatory Airway Disease of Horses-Revised Consensus Statement. *J. Vet. Intern. Med.* **2016**, *30*, 503–515. [CrossRef] [PubMed]
2. Hotchkiss, J.W.; Reid, S.W.J.; Christley, R.M. A survey of horse owners in Great Britain regarding horses in their care. Part 1: Horse demographic characteristics and management. *Equine Vet. J.* **2007**, *39*, 294–300. [CrossRef]
3. Ramseyer, A.; Gaillard, C.; Burger, D.; Straub, R.; Jost, U.; Boog, C.; Marti, E.; Gerber, V. Effects of Genetic and Environmental Factors on Chronic Lower Airway Disease in Horses. *J. Vet. Intern. Med.* **2007**, *21*, 149–156. [CrossRef] [PubMed]
4. Pirie, R.S.; Collie, D.D.S.; Dixon, P.M.; McGorum, B.C. Inhaled endotoxin and organic dust particulates have synergistic proinflammatory effects in equine heaves (organic dust-induced asthma). *Clin. Exp. Allergy* **2003**, *33*, 676–683. [CrossRef] [PubMed]
5. White, S.J.; Moore-Colyer, M.; Marti, E.; Hannant, D.; Gerber, V.; Coüetil, L.; Richard, E.A.; Alcocer, M. Antigen array for serological diagnosis and novel allergen identification in severe equine asthma. *Sci. Rep.* **2019**, *9*, 15171. [CrossRef] [PubMed]
6. Moore-Colyer, M.J.S.; Taylor, J.L.E.; James, R. The Effect of Steaming and Soaking on the Respirable Particle, Bacteria, Mould, and Nutrient Content in Hay for Horses. *J. Equine Vet. Sci.* **2016**, *39*, 62–68. [CrossRef]

7. Niedzwiedz, A.; Jaworski, Z.; Kubiak, K. Serum concentrations of allergen-specific IgE in horses with equine recurrent airway obstruction and healthy controls assessed by ELISA. *Vet. Clin. Pathol.* **2015**, *44*, 391–396. [CrossRef] [PubMed]

8. McGorum, B.C.; Ellison, J.; Cullen, R.T. Total and respirable airborne dust endotoxin concentrations in three equine management systems. *Equine Vet. J.* **1998**, *30*, 430–434. [CrossRef]

9. Klier, J.; Geis, S.; Steuer, J.; Geh, K.; Reese, S.; Fuchs, S.; Mueller, R.S.; Winter, G.; Gehlen, H. A comparison of nanoparticullate CpG immunotherapy with and without allergens in spontaneously equine asthma-affected horses, an animal model. *Immun. Inflamm. Dis.* **2018**, *6*, 81–96. [CrossRef]

10. Pacholewska, A.; Jagannathan, V.; Drögemüller, M.; Klukowska-Rötzler, J.; Lanz, S.; Hamza, E.; Dermitzakis, E.T.; Marti, E.; Leeb, T.; Gerber, V. Impaired cell cycle regulation in a natural equine model of asthma. *PLoS ONE* **2015**, *10*, e0136103. [CrossRef]

11. Pirie, R.S.; Dixon, P.M.; McGorum, B.C. Endotoxin contamination contributes to the pulmonary inflammatory and functional response to Aspergillus fumigatus extract inhalation in heaves horses. *Clin. Exp. Allergy* **2003**, *33*, 1289–1296. [CrossRef] [PubMed]

12. Schmallenbach, K.H.; Rahman, I.; Sasse, H.H.L.; Dixon, P.M.; Halliwell, R.E.W.; McGorum, B.C.; Crameri, R.; Miller, H.R.P. Studies on pulmonary and systemic Aspergillus fumigatus-specific IgE and IgG antibodies in horses affected with chronic obstructive pulmonary disease (COPD). *Vet. Immunol. Immunopathol.* **1998**, *66*, 245–256. [CrossRef]

13. Morán, G.; Folch, H.; Araya, O.; Burgos, R.; Barria, M. Detection of reaginic antibodies against Faenia rectivirgula from the serum of horses affected with Recurrent Airway Obstruction by an in vitro bioassay. *Vet. Res. Commun.* **2010**, *34*, 719–726. [CrossRef] [PubMed]

14. Tilley, P.; Sales Luis, J.P.; Branco Ferreira, M. Correlation and discriminant analysis between clinical, endoscopic, thoracic X-ray and bronchoalveolar lavage fluid cytology scores, for staging horses with recurrent airway obstruction (RAO). *Res. Vet. Sci.* **2012**, *93*, 1006–1014. [CrossRef]

15. Simões, J.; Sales Luís, J.; Tilley, P. Contribution of lung function tests to the staging of severe equine asthma syndrome in the field. *Res. Vet. Sci.* **2019**, *123*, 112–117. [CrossRef]

16. Pirie, R.S.; Collie, D.D.S.; Dixon, P.M.; McGorum, B.C. Evaluation of nebulised hay dust suspensions (HDS) for the diagnosis and investigation of heaves. 2: Effects of inhaled HDS on control and heaves horses. *Equine Vet. J.* **2002**, *34*, 337–342. [CrossRef]

17. Leclere, M.; Lavoie-Lamoureux, A.; Gélinas-Lymburner, É.; David, F.; Martin, J.G.; Lavoie, J.P. Effect of antigenic exposure on airway smooth muscle remodeling in an equine model of chronic asthma. *Am. J. Respir. Cell Mol. Biol.* **2011**, *45*, 181–187. [CrossRef]

18. Simões, J.; Sales Luís, J.P.; Tilley, P. Owner Compliance to an Environmental Management Protocol for Severe Equine Asthma Syndrome. *J. Equine Vet. Sci.* **2020**, *87*, 102937. [CrossRef]

19. Tilley, P.; Sales Luis, J.P.; Branco Ferreira, M. Comparison of Skin Prick Tests with In Vitro Allergy Tests in the Characterization of Horses with Recurrent Airway Obstruction. *J. Equine Vet. Sci.* **2012**, *32*, 719–727. [CrossRef]

20. Stucchi, L.; Ferrucci, F.; Bullone, M.; Dellacà, R.L.; Lavoie, J.P. Within-breath oscillatory mechanics in horses affected by severe equine asthma in exacerbation and in remission of the disease. *Animals* **2022**, *12*, 4. [CrossRef]

21. Gerber, V.; Baleri, D.; Klukowska-Rötzler, J.; Swinburne, J.E.; Dolf, G. Mixed inheritance of equine recurrent airway obstruction. *J. Vet. Intern. Med.* **2009**, *23*, 626–630. [CrossRef] [PubMed]

22. Marti, E.; Gerber, H.; Essich, G.; Oulehla, J.; Lazary, S. The genetic basis of equine allergic diseases 1. Chronic hypersensitivity bronchitis. *Equine Vet. J.* **1991**, *23*, 457–460. [CrossRef] [PubMed]

23. Couetil, L.; Cardwell, J.M.; Leguillette, R.; Mazan, M.; Richard, E.; Bienzle, D.; Bullone, M.; Gerber, V.; Ivester, K.; Lavoie, J.P.; et al. Equine Asthma: Current Understanding and Future Directions. *Front. Vet. Sci.* **2020**, *7*, 450. [CrossRef] [PubMed]

24. Bullone, M.; Lavoie, J.P. Asthma "of horses and men" - How can equine heaves help us better understand human asthma immunopathology and its functional consequences? *Mol. Immunol.* **2015**, *66*, 97–105. [CrossRef]

25. Swinburne, J.E.; Bogle, H.; Klukowska-Rötzler, J.; Drögemüller, M.; Leeb, T.; Temperton, E.; Dolf, G.; Gerber, V. A whole-genome scan for recurrent airway obstruction in Warmblood sport horses indicates two positional candidate regions. *Mamm. Genome* **2009**, *20*, 504–515. [CrossRef]

26. Klukowska-Rötzler, J.; Swinburne, J.E.; Drögemüller, C.; Dolf, G.; Janda, J.; Leeb, T.; Gerber, V. The interleukin 4 receptor gene and its role in recurrent airway obstruction in Swiss Warmblood horses. *Anim. Genet.* **2012**, *43*, 450–453. [CrossRef]

27. Shakhsi-Niaei, M.; Klukowska-Rötzler, J.; Drögemüller, C.; Swinburne, J.; Ehrmann, C.; Saftic, D.; Ramseyer, A.; Gerber, V.; Dolf, G.; Leeb, T. Replication and fine-mapping of a QTL for recurrent airway obstruction in European Warmblood horses. *Anim. Genet.* **2012**, *43*, 627–631. [CrossRef]

28. Jost, U.; Klukowska-Rötzler, J.; Dolf, G.; Swinburne, J.E.; Ramseyer, A.; Bugno, M.; Burger, D.; Blott, S.; Gerber, V. A region on equine chromosome 13 is linked to recurrent airway obstruction in horses. *Equine Vet. J.* **2007**, *39*, 236–241. [CrossRef]

29. Ober, C.; Leavitt, S.A.; Tsalenko, A.; Howard, T.D.; Hoki, D.M.; Daniel, R.; Newman, D.L.; Wu, X.; Parry, R.; Lester, L.A.; et al. Variation in the interleukin 4-receptor α gene confers susceptibility to asthma and atopy in ethnically diverse populations. *Am. J. Hum. Genet.* **2000**, *66*, 517–526. [CrossRef]

30. Youn, J.; Hwang, S.H.; Cho, C.S.; Min, J.K.; Kim, W.U.; Park, S.H.; Kim, H.Y. Association of the interleukin-4 receptor α variant Q576R with Th1/Th2 imbalance in connective tissue disease. *Immunogenetics* **2000**, *51*, 743–746. [CrossRef]

31. Racine, J.; Gerber, V.; Miskovic Feutz, M.; Riley, C.P.; Adamec, J.; Swinburne, J.E.; Couetil, L.L. Comparison of genomic and proteomic data in recurrent airway obstruction affected horses using ingenuity pathway analysis®. *BMC Vet. Res.* **2011**, *7*, 48. [CrossRef] [PubMed]

32. Seki, Y.I.; Hayashi, K.; Matsumoto, A.; Seki, N.; Tsukada, J.; Ransom, J.; Naka, T.; Kishimoto, T.; Yoshimura, A.; Kubo, M. Expression of the suppressor of cytokine signaling-5 (SOCS5) negatively regulates IL-4-dependent STAT6 activation and Th2 differentiation. *Proc. Natl. Acad. Sci. USA* **2002**, *99*, 13003–13008. [CrossRef] [PubMed]

33. Schnider, D.; Rieder, S.; Leeb, T.; Gerber, V.; Neuditschko, M. A genome-wide association study for equine recurrent airway obstruction in European Warmblood horses reveals a suggestive new quantitative trait locus on chromosome 13. *Anim. Genet.* **2017**, *48*, 691–693. [CrossRef] [PubMed]

34. Pacquelet, S.; Lehmann, M.; Luxen, S.; Regazzoni, K.; Frausto, M.; Noack, D.; Knaus, U.G. Inhibitory action of NoxA1 on dual oxidase activity in airway cells. *J. Biol. Chem.* **2008**, *283*, 24649–24658. [CrossRef]

35. Evans, C.M.; Raclawska, D.S.; Ttofali, F.; Liptzin, D.R.; Fletcher, A.A.; Harper, D.N.; McGing, M.A.; McElwee, M.M.; Williams, O.W.; Sanchez, E.; et al. The polymeric mucin Muc5ac is required for allergic airway hyperreactivity. *Nat. Commun.* **2015**, *6*, 6281. [CrossRef]

36. Gerber, V.; Robinson, N.E.; Venta, P.J.; Rawson, J.; Jefcoat, A.M.; Hotchkiss, J.A. Mucin genes in horse airways: MUC5AC, but not MUC2, may play a role in recurrent airway obstruction. *Equine Vet. J.* **2003**, *35*, 252–257. [CrossRef]

37. Ghosh, S.; Das, P.J.; McQueen, C.M.; Gerber, V.; Swiderski, C.E.; Lavoie, J.P.; Chowdhary, B.P.; Raudsepp, T. Analysis of genomic copy number variation in equine recurrent airway obstruction (heaves). *Anim. Genet.* **2016**, *47*, 334–344. [CrossRef]

38. Šedová, L.; Buková, I.; Bažantová, P.; Petrezsélyová, S.; Prochazka, J.; Školníková, E.; Zudová, D.; Včelák, J.; Makovický, P.; Bendlová, B.; et al. Semi-lethal primary ciliary dyskinesia in rats lacking the nme7 gene. *Int. J. Mol. Sci.* **2021**, *22*, 3810. [CrossRef]

39. Tessier, L.; Côté, O.; Bienzle, D. Sequence variant analysis of RNA sequences in severe equine asthma. *PeerJ* **2018**, *2018*, e5759. [CrossRef]

40. Tessier, L.; Côté, O.; Clark, M.E.; Viel, L.; Diaz-Méndez, A.; Anders, S.; Bienzle, D. Gene set enrichment analysis of the bronchial epithelium implicates contribution of cell cycle and tissue repair processes in equine asthma. *Sci. Rep.* **2018**, *8*, 16408. [CrossRef]

41. Mason, V.C.; Schaefer, R.J.; McCue, M.E.; Leeb, T.; Gerber, V. eQTL discovery and their association with severe equine asthma in European Warmblood horses. *BMC Genom.* **2018**, *19*, 581. [CrossRef] [PubMed]

42. Ferreira, M.A.R.; Matheson, M.C.; Tang, C.S.; Granell, R.; Ang, W.; Hui, J.; Kiefer, A.K.; Duffy, D.L.; Baltic, S.; Danoy, P.; et al. Genome-wide association analysis identifies 11 risk variants associated with the asthma with hay fever phenotype. *J. Allergy Clin. Immunol.* **2014**, *133*, 1564–1571. [CrossRef] [PubMed]

43. Pacholewska, A.; Kraft, M.F.; Gerber, V.; Jagannathan, V. Differential expression of serum MicroRNAs supports CD4+ t cell differentiation into Th2/Th17 cells in severe equine asthma. *Genes* **2017**, *8*, 383. [CrossRef] [PubMed]

44. Borowska, A.; Wolska, D.; Niedzwiedz, A.; Borowicz, H.; Jaworski, Z.; Siemieniuch, M.; Szwaczkowski, T. Some genetic and environmental effects on equine asthma in polish konik horses. *Animals* **2021**, *11*, 2285. [CrossRef]

45. Kuruvilla, M.E.; Lee, F.E.H.; Lee, G.B. Understanding Asthma Phenotypes, Endotypes, and Mechanisms of Disease. *Clin. Rev. Allergy Immunol.* **2019**, *56*, 219–233. [CrossRef]

46. Holgate, S.T. Innate and adaptive immune responses in asthma. *Nat. Med.* **2012**, *18*, 673–683. [CrossRef]

47. Kuo, C.H.S.; Pavlidis, S.; Loza, M.; Baribaud, F.; Rowe, A.; Pandis, I.; Sousa, A.; Corfield, J.; Djukanovic, R.; Lutter, R.; et al. T-helper cell type 2 (Th2) and non-Th2 molecular phenotypes of asthma using sputum transcriptomics in U-BIOPRED. *Eur. Respir. J.* **2017**, *49*, 1602135. [CrossRef]

48. Rossi, H.; Virtala, A.M.; Raekallio, M.; Rahkonen, E.; Rajamäki, M.M.; Mykkänen, A. Comparison of tracheal wash and bronchoalveolar lavage cytology in 154 horses with and without respiratory signs in a referral hospital over 2009–2015. *Front. Vet. Sci.* **2018**, *5*, 61. [CrossRef]

49. Cordeau, M.E.; Joubert, P.; Dewachi, O.; Hamid, Q.; Lavoie, J.P. IL-4, IL-5 and IFN-γ mRNA expression in pulmonary lymphocytes in equine heaves. *Vet. Immunol. Immunopathol.* **2004**, *97*, 87–96. [CrossRef]

50. McGorum, B.C.; Dixon, P.M.; Halliwell, R.E.W. Phenotypic analysis of peripheral blood and bronchoalveolar lavage fluid lymphocytes in control and chronic obstructive pulmonary disease affected horses, before and after "natural (hay and straw) challenges. *Vet. Immunol. Immunopathol.* **1993**, *36*, 207–222. [CrossRef]

51. Deaton, C.M.; Deaton, L.; Jose-Cunilleras, E.; Vincent, T.L.; Baird, A.W.; Dacre, K.; Marlin, D.J. Early onset airway obstruction in response to organic dust in the horse. *J. Appl. Physiol.* **2007**, *102*, 1071–1077. [CrossRef]

52. Kleiber, C.; Grünig, G.; Jungi, T.; Schmucker, N.; Gerber, H.; Davis, W.C.; Straub, R. Phenotypic Analysis of Bronchoalveolar Lavage Fluid Lymphocytes in Horses with Chronic Pulmonary Disease. *J. Vet. Med. Ser. A Physiol. Pathol. Clin. Med.* **1999**, *46*, 177–184. [CrossRef]

53. Lavoie, J.P.; Maghni, K.; Desnoyers, M.; Taha, R.; Martin, J.G.; Hamid, Q.A. Neutrophilic airway inflammation in horses with heaves is characterized by a Th2-type cytokine profile. *Am. J. Respir. Crit. Care Med.* **2001**, *164*, 1410–1413. [CrossRef]

54. Moran, G.; Folch, H.; Henriquez, C.; Ortloff, A.; Barria, M. Reaginic antibodies from horses with Recurrent Airway Obstruction produce mast cell stimulation. *Vet. Res. Commun.* **2012**, *36*, 251–258. [CrossRef] [PubMed]

55. Felippe, M.J.B. *Equine Clinical Immunology*, 1st ed.; John Wiley & Sons, Ltd.: Hoboken, NJ, USA, 2016; ISBN 9781119086512.

56. Kleiber, C.; McGorum, B.C.; Horohov, D.W.; Pirie, R.S.; Zurbriggen, A.; Straub, R. Cytokine profiles of peripheral blood and airway CD4 and CD8 T lymphocytes in horses with recurrent airway obstruction. *Vet. Immunol. Immunopathol.* **2005**, *104*, 91–97. [CrossRef] [PubMed]

57. Giguère, S.; Viel, L.; Lee, E.; MacKay, R.J.; Hernandez, J.; Franchini, M. Cytokine induction in pulmonary airways of horses with heaves and effect of therapy with inhaled fluticasone propionate. *Vet. Immunol. Immunopathol.* **2002**, *85*, 147–158. [CrossRef]

58. Horohov, D.W.; Beadle, R.E.; Mouch, S.; Pourciau, S.S. Temporal regulation of cytokine mRNA expression in equine recurrent airway obstruction. *Vet. Immunol. Immunopathol.* **2005**, *108*, 237–245. [CrossRef] [PubMed]

59. Padoan, E.; Ferraresso, S.; Pegolo, S.; Castagnaro, M.; Barnini, C.; Bargelloni, L. Real time RT-PCR analysis of inflammatory mediator expression in recurrent airway obstruction-affected horses. *Vet. Immunol. Immunopathol.* **2013**, *156*, 190–199. [CrossRef]

60. Ainsworth, D.M.; Grünig, G.; Matychak, M.B.; Young, J.; Wagner, B.; Erb, H.N.; Antczak, D.F. Recurrent airway obstruction (RAO) in horses is characterized by IFN-γ and IL-8 production in bronchoalveolar lavage cells. *Vet. Immunol. Immunopathol.* **2003**, *96*, 83–91. [CrossRef]

61. Ainsworth, D.M.; Wagner, B.; Franchini, M.; Grünig, G.; Erb, H.N.; Tan, J.Y. Time-dependent alterations in gene expression of interleukin-8 in the bronchial epithelium of horses with recurrent airway obstruction. *Am. J. Vet. Res.* **2006**, *67*, 669–677. [CrossRef] [PubMed]

62. Hulliger, M.F.; Pacholewska, A.; Vargas, A.; Lavoie, J.P.; Leeb, T.; Gerber, V.; Jagannathan, V. An integrative mirna-mrna expression analysis reveals striking transcriptomic similarities between severe equine asthma and specific asthma endotypes in humans. *Genes* **2020**, *11*, 1143. [CrossRef] [PubMed]

63. Maes, T.; Cobos, F.A.; Schleich, F.; Sorbello, V.; Henket, M.; De Preter, K.; Bracke, K.R.; Conickx, G.; Mesnil, C.; Vandesompele, J.; et al. Asthma inflammatory phenotypes show differential microRNA expression in sputum. *J. Allergy Clin. Immunol.* **2016**, *137*, 1433–1446. [CrossRef]

64. Tessier, L.; Côté, O.; Clark, M.E.; Viel, L.; Diaz-Méndez, A.; Anders, S.; Bienzle, D. Impaired response of the bronchial epithelium to inflammation characterizes severe equine asthma. *BMC Genom.* **2017**, *18*, 708. [CrossRef]

65. Takagi, R.; Higashi, T.; Hashimoto, K.; Nakano, K.; Mizuno, Y.; Okazaki, Y.; Matsushita, S. B Cell Chemoattractant CXCL13 Is Preferentially Expressed by Human Th17 Cell Clones. *J. Immunol.* **2008**, *181*, 186–189. [CrossRef] [PubMed]

66. Baay-Guzman, G.J.; Huerta-Yepez, S.; Vega, M.I.; Aguilar-Leon, D.; Campillos, M.; Blake, J.; Benes, V.; Hernandez-Pando, R.; Teran, L.M. Role of CXCL13 in asthma: Novel therapeutic target. *Chest* **2012**, *141*, 886–894. [CrossRef] [PubMed]

67. Debrue, M.; Hamilton, E.; Joubert, P.; Lajoie-Kadoch, S.; Lavoie, J.P. Chronic exacerbation of equine heaves is associated with an increased expression of interleukin-17 mRNA in bronchoalveolar lavage cells. *Vet. Immunol. Immunopathol.* **2005**, *105*, 25–31. [CrossRef]

68. Murcia, R.Y.; Vargas, A.; Lavoie, J.P. The interleukin-17 induced activation and increased survival of equine neutrophils is insensitive to glucocorticoids. *PLoS ONE* **2016**, *11*, e0154755. [CrossRef]

69. Kehrli, D.; Jandova, V.; Fey, K.; Jahn, P.; Gerber, V. Multiple hypersensitivities including recurrent airway obstruction, insect bite hypersensitivity, and urticaria in 2 warmblood horse populations. *J. Vet. Intern. Med.* **2015**, *29*, 320–326. [CrossRef]

70. Lanz, S.; Brunner, A.; Graubner, C.; Marti, E.; Gerber, V. Insect Bite Hypersensitivity in Horses is Associated with Airway Hyperreactivity. *J. Vet. Intern. Med.* **2017**, *31*, 1877–1883. [CrossRef]

71. Lo Feudo, C.M.; Stucchi, L.; Alberti, E.; Conturba, B.; Zucca, E.; Ferrucci, F. Intradermal testing results in horses affected by mild-moderate and severe equine asthma. *Animals* **2021**, *11*, 2086. [CrossRef]

72. Klier, J.; Lindner, D.; Reese, S.; Mueller, R.S.; Gehlen, H. Comparison of Four Different Allergy Tests in Equine Asthma Affected Horses and Allergen Inhalation Provocation Test. *J. Equine Vet. Sci.* **2021**, *102*, 103433. [CrossRef] [PubMed]

73. Couëtil, L.L.; Ward, M.P. Analysis of risk factors for recurrent airway obstruction in North American horses: 1,444 Cases (1990-1999). *J. Am. Vet. Med. Assoc.* **2003**, *223*, 1645–1650. [CrossRef] [PubMed]

74. Hansen, S.; Baptiste, K.E.; Fjeldborg, J.; Horohov, D.W. A review of the equine age-related changes in the immune system: Comparisons between human and equine aging, with focus on lung-specific immune-aging. *Ageing Res. Rev.* **2015**, *20*, 11–23. [CrossRef] [PubMed]

75. Bullone, M.; Lavoie, J.P. The contribution of oxidative stress and inflamm-aging in human and equine asthma. *Int. J. Mol. Sci.* **2017**, *18*, 2612. [CrossRef]

76. Franceschi, C.; Bonafè, M.; Valensin, S.; Olivieri, F.; De Luca, M.; Ottaviani, E.; De Benedictis, G. Inflamm-aging. An evolutionary perspective on immunosenescence. *Ann. N. Y. Acad. Sci.* **2000**, *908*, 244–254. [CrossRef]

77. Adams, A.A.; Breathnach, C.C.; Katepalli, M.P.; Kohler, K.; Horohov, D.W. Advanced age in horses affects divisional history of T cells and inflammatory cytokine production. *Mech. Ageing Dev.* **2008**, *129*, 656–664. [CrossRef]

78. Robbin, M.G.; Wagner, B.; Noronha, L.E.; Antczak, D.F.; De Mestre, A.M. Subpopulations of equine blood lymphocytes expressing regulatory T cell markers. *Vet. Immunol. Immunopathol.* **2011**, *140*, 90–101. [CrossRef]

79. Suagee, J.K.; Corl, B.A.; Crisman, M.V.; Pleasant, R.S.; Thatcher, C.D.; Geor, R.J. Relationships between Body Condition Score and Plasma Inflammatory Cytokines, Insulin, and Lipids in a Mixed Population of Light-Breed Horses. *J. Vet. Intern. Med.* **2013**, *27*, 157–163. [CrossRef]

80. Sage, S.E.; Bedenice, D.; McKinney, C.A.; Long, A.E.; Pacheco, A.; Wagner, B.; Mazan, M.R.; Paradis, M.R. Assessment of the impact of age and of blood-derived inflammatory markers in horses with colitis. *J. Vet. Emerg. Crit. Care* **2021**, *31*, 779–787. [CrossRef]

81. McFarlane, D.; Holbrook, T.C. Cytokine dysregulation in aged horses and horses with pituitary pars intermedia dysfunction. *J. Vet. Intern. Med.* **2008**, *22*, 436–442. [CrossRef]

82. Hansen, S.; Sun, L.; Baptiste, K.E.; Fjeldborg, J.; Horohov, D.W. Age-related changes in intracellular expression of IFN-γ and TNF-α in equine lymphocytes measured in bronchoalveolar lavage and peripheral blood. *Dev. Comp. Immunol.* **2013**, *39*, 228–233. [CrossRef]

83. Davis, J.D.; Wypych, T.P. Cellular and functional heterogeneity of the airway epithelium. *Mucosal Immunol.* **2021**, *14*, 978–990. [CrossRef] [PubMed]

84. Rose, M.C.; Voynow, J.A. Respiratory tract mucin genes and mucin glycoproteins in health and disease. *Physiol. Rev.* **2006**, *86*, 245–278. [CrossRef] [PubMed]
85. Bullone, M.; Hélie, P.; Joubert, P.; Lavoie, J.P. Development of a Semiquantitative Histological Score for the Diagnosis of Heaves Using Endobronchial Biopsy Specimens in Horses. *J. Vet. Intern. Med.* **2016**, *30*, 1739–1746. [CrossRef] [PubMed]
86. Lee, G.K.C.; Tessier, L.; Bienzle, D. Salivary Scavenger and Agglutinin (SALSA) Is Expressed in Mucosal Epithelial Cells and Decreased in Bronchial Epithelium of Asthmatic Horses. *Front. Vet. Sci.* **2019**, *6*, 418. [CrossRef]
87. Kaup, F.-J.; Drommer, W.; Deegen, E. Ultrastructural findings in horses with chronic obstructive pulmonary disease (COPD) I: Alterations of the larger conducting airways. *Equine Vet. J.* **1990**, *22*, 343–348. [CrossRef]
88. Dacre, K.J.; McGorum, B.C.; Marlin, D.J.; Bartner, L.R.; Brown, J.K.; Shaw, D.J.; Robinson, N.E.; Deaton, C.; Pemberton, A.D. Organic dust exposure increases mast cell tryptase in bronchoalveolar lavage fluid and airway epithelium of heaves horses. *Clin. Exp. Allergy* **2007**, *37*, 1809–1818. [CrossRef]
89. Abs, V.; Bonicelli, J.; Kacza, J.; Zizzadoro, C.; Abraham, G. Equine bronchial fibroblasts enhance proliferation and differentiation of primary equine bronchial epithelial cells co-cultured under air-liquid interface. *PLoS ONE* **2019**, *14*, e0225025. [CrossRef]
90. Sha, Q.; Truong-Tran, A.Q.; Plitt, J.R.; Beck, L.A.; Schleimer, R.P. Activation of airway epithelial cells by toll-like receptor agonists. *Am. J. Respir. Cell Mol. Biol.* **2004**, *31*, 358–364. [CrossRef]
91. Frellstedt, L.; Gosset, P.; Kervoaze, G.; Hans, A.; Desmet, C.; Pirottin, D.; Bureau, F.; Lekeux, P.; Art, T. The innate immune response of equine bronchial epithelial cells is altered by training. *Vet. Res.* **2015**, *46*, 3. [CrossRef]
92. Ainsworth, D.M.; Matychak, M.B.; Reyner, C.L.; Erb, H.N.; Young, J.C. Effect of in vitro exposure to hay dust on the gene expression of chemokines and cell-surface receptors in primary bronchial epithelial cell cultures established from horses with chronic recurrent airway obstruction. *Am. J. Vet. Res.* **2009**, *70*, 365–372. [CrossRef] [PubMed]
93. Parbhakar, O.P.; Duke, T.; Townsend, H.G.G.; Singh, B. Depletion of pulmonary intravascular macrophages partially inhibits lipopolysaccharide-induced lung inflammation in horses. *Vet. Res.* **2005**, *36*, 557–569. [CrossRef] [PubMed]
94. Laan, T.T.J.M.; Bull, S.; van Nieuwstadt, R.A.; Fink-Gremmels, J. The effect of aerosolized and intravenously administered clenbuterol and aerosolized fluticasone propionate on horses challenged with Aspergillus fumigatus antigen. *Vet. Res. Commun.* **2006**, *30*, 623–635. [CrossRef] [PubMed]
95. Joubert, P.; Cordeau, M.E.; Lavoie, J.P. Cytokine mRNA expression of pulmonary macrophages varies with challenge but not with disease state in horses with heaves or in controls. *Vet. Immunol. Immunopathol.* **2011**, *142*, 236–242. [CrossRef] [PubMed]
96. Aharonson-Raz, K.; Lohmann, K.L.; Townsend, H.G.; Marques, F.; Singh, B. Pulmonary intravascular macrophages as proinflammatory cells in heaves, an asthma-like equine disease. *Am. J. Physiol. -Lung Cell. Mol. Physiol.* **2012**, *303*, 189–198. [CrossRef]
97. Nahrendorf, M.; Swirski, F.K. Abandoning M1/M2 for a network model of macrophage function. *Circ. Res.* **2016**, *119*, 414–417. [CrossRef]
98. Karagianni, A.E.; Kapetanovic, R.; Summers, K.M.; McGorum, B.C.; Hume, D.A.; Pirie, R.S. Comparative transcriptome analysis of equine alveolar macrophages. *Equine Vet. J.* **2017**, *49*, 375–382. [CrossRef]
99. Mantovani, A.; Biswas, S.K.; Galdiero, M.R.; Sica, A.; Locati, M. Macrophage plasticity and polarization in tissue repair and remodelling. *J. Pathol.* **2013**, *229*, 176–185. [CrossRef] [PubMed]
100. Karagianni, A.E.; Kapetanovic, R.; McGorum, B.C.; Hume, D.A.; Pirie, S.R. The equine alveolar macrophage: Functional and phenotypic comparisons with peritoneal macrophages. *Vet. Immunol. Immunopathol.* **2013**, *155*, 219–228. [CrossRef]
101. Brazil, T.J.; Dagleish, M.P.; McGorum, B.C.; Dixon, P.M.; Haslett, C.; Chilvers, E.R. Kinetics of pulmonary neutrophil recruitment and clearance in a natural and spontaneously resolving model of airway inflammation. *Clin. Exp. Allergy* **2005**, *35*, 854–865. [CrossRef] [PubMed]
102. Mosser, D.M.; Edwards, J.P. Exploring the full spectrum of macrophage activation. *Nat. Rev. Immunol.* **2008**, *8*, 958–969. [CrossRef] [PubMed]
103. Jackson, K.A.; Stott, J.L.; Horohov, D.W.; Watson, J.L. IL-4 induced CD23 (FcεRII) up-regulation in equine peripheral blood mononuclear cells and pulmonary alveolar macrophages. *Vet. Immunol. Immunopathol.* **2004**, *101*, 243–250. [CrossRef] [PubMed]
104. Varin, A.; Gordon, S. Alternative activation of macrophages: Immune function and cellular biology. *Immunobiology* **2009**, *214*, 630–641. [CrossRef] [PubMed]
105. Wilson, M.E.; McCandless, E.E.; Olszewski, M.A.; Robinson, N.E. Alveolar macrophage phenotypes in severe equine asthma. *Vet. J.* **2020**, *256*, 105436. [CrossRef]
106. Kang, H.; Bienzle, D.; Lee, G.K.C.; Piché, É.; Viel, L.; Odemuyiwa, S.O.; Beeler-Marfisi, J. Flow cytometric analysis of equine bronchoalveolar lavage fluid cells in horses with and without severe equine asthma. *Vet. Pathol.* **2022**, *59*, 91–99. [CrossRef]
107. Bureau, F.; Delhalle, S.; Bonizzi, G.; Fiévez, L.; Dogné, S.; Kirschvink, N.; Vanderplasschen, A.; Merville, M.-P.; Bours, V.; Lekeux, P. Mechanisms of Persistent NF-κB Activity in the Bronchi of an Animal Model of Asthma. *J. Immunol.* **2000**, *165*, 5822–5830. [CrossRef] [PubMed]
108. Turlej, R.K.; Fiévez, L.; Sandersen, C.F.; Dogné, S.; Kirschvink, N.; Lekeux, P.; Bureau, F. Enhanced survival of lung granulocytes in an animal model of asthma: Evidence for a role of GM-CSF activated STAT5 signalling pathway. *Thorax* **2001**, *56*, 696–702. [CrossRef]
109. Niedzwiedz, A.; Jaworski, Z.; Tykalowski, B.; Smialek, M. Neutrophil and macrophage apoptosis in bronchoalveolar lavage fluid from healthy horses and horses with recurrent airway obstruction (RAO). *BMC Vet. Res.* **2014**, *10*, 29. [CrossRef]
110. Mitsi, E.; Kamng'ona, R.; Rylance, J.; Solórzano, C.; Jesus Reiné, J.; Mwandumba, H.C.; Ferreira, D.M.; Jambo, K.C. Human alveolar macrophages predominately express combined classical M1 and M2 surface markers in steady state. *Respir. Res.* **2018**, *19*, 66. [CrossRef] [PubMed]

111. Evren, E.; Ringqvist, E.; Tripathi, K.P.; Sleiers, N.; Rives, I.C.; Alisjahbana, A.; Gao, Y.; Sarhan, D.; Halle, T.; Sorini, C.; et al. Distinct developmental pathways from blood monocytes generate human lung macrophage diversity. *Immunity* **2021**, *54*, 259–275.e7. [CrossRef]

112. Léguillette, R. Recurrent airway obstruction—Heaves. *Vet. Clin. N. Am. -Equine Pract.* **2003**, *19*, 63–86. [CrossRef]

113. Uberti, B.; Morán, G. Role of neutrophils in equine asthma. *Anim. Health Res. Rev.* **2018**, *19*, 65–73. [CrossRef]

114. Davis, K.U.; Sheats, M.K. The Role of Neutrophils in the Pathophysiology of Asthma in Humans and Horses. *Inflammation* **2021**, *44*, 450–465. [CrossRef] [PubMed]

115. Kolaczkowska, E.; Kubes, P. Neutrophil recruitment and function in health and inflammation. *Nat. Rev. Immunol.* **2013**, *13*, 159–175. [CrossRef]

116. Cheng, O.Z.; Palaniyar, N. NET balancing: A problem in inflammatory lung diseases. *Front. Immunol.* **2013**, *4*, 1. [CrossRef] [PubMed]

117. Fox, S.; Leitch, A.E.; Duffin, R.; Haslett, C.; Rossi, A.G. Neutrophil apoptosis: Relevance to the innate immune response and inflammatory disease. *J. Innate Immun.* **2010**, *2*, 216–227. [CrossRef] [PubMed]

118. Simpson, J.L.; Grissell, T.V.; Douwes, J.; Scott, R.J.; Boyle, M.J.; Gibson, P.G. Innate immune activation in neutrophilic asthma and bronchiectasis. *Thorax* **2007**, *62*, 211–218. [CrossRef] [PubMed]

119. Poole, J.A.; Romberger, D.J. Immunological and inflammatory responses to organic dust in agriculture. *Curr. Opin. Allergy Clin. Immunol.* **2012**, *12*, 126–132. [CrossRef] [PubMed]

120. Nocker, R.E.T.; Schoonbrood, D.F.M.; Van de Graaf, E.A.; Hack, E.; Lutter, R.; Jansen, H.M.; Out, T.A. Lnterleukin-8 in airway inflammation in patients with asthma and chronic obstructive pulmonary disease. *Int. Arch. Allergy Immunol.* **1996**, *109*, 183–191. [CrossRef]

121. Medoff, B.D.; Sauty, A.; Tager, A.M.; Maclean, J.A.; Smith, R.N.; Mathew, A.; Dufour, J.H.; Luster, A.D. IFN-γ-Inducible Protein 10 (CXCL10) Contributes to Airway Hyperreactivity and Airway Inflammation in a Mouse Model of Asthma. *J. Immunol.* **2002**, *168*, 5278–5286. [CrossRef]

122. Singh, S.R.; Sutcliffe, A.; Kaur, D.; Gupta, S.; Desai, D.; Saunders, R.; Brightling, C.E. CCL2 release by airway smooth muscle is increased in asthma and promotes fibrocyte migration. *Allergy Eur. J. Allergy Clin. Immunol.* **2014**, *69*, 1189–1197. [CrossRef] [PubMed]

123. Molet, S.; Hamid, Q.; Davoine, F.; Nutku, E.; Taha, R.; Pagé, N.; Olivenstein, R.; Elias, J.; Chakir, J. IL-17 is increased in asthmatic airways and induces human bronchial fibroblasts to produce cytokines. *J. Allergy Clin. Immunol.* **2001**, *108*, 430–438. [CrossRef] [PubMed]

124. Franchini, M.; Gill, U.; Von Fellenberg, R.; Bracher, V.D. Interleukin-8 concentration and neutrophil chemotactic activity in bronchoalveolar lavage fluid of horses with chronic obstructive pulmonary disease following exposure to hay. *Am. J. Vet. Res.* **2000**, *61*, 1369–1374. [CrossRef] [PubMed]

125. Riihimäki, M.; Raine, A.; Art, T.; Lekeux, P.; Couëtil, L.; Pringle, J. Partial divergence of cytokine mRNA expression in bronchial tissues compared to bronchoalveolar lavage cells in horses with recurrent airway obstruction. *Vet. Immunol. Immunopathol.* **2008**, *122*, 256–264. [CrossRef] [PubMed]

126. Korn, A.; Miller, D.; Dong, L.; Buckles, E.L.; Wagner, B.; Ainsworth, D.M. Differential gene expression profiles and selected cytokine protein analysis of mediastinal lymph nodes of horses with chronic recurrent airway obstruction (RAO) support an interleukin-17 immune response. *PLoS ONE* **2015**, *10*, e0142622. [CrossRef]

127. Ouyang, W.; Kolls, J.K.; Zheng, Y. The Biological Functions of T Helper 17 Cell Effector Cytokines in Inflammation. *Immunity* **2008**, *28*, 454–467. [CrossRef] [PubMed]

128. Ramirez-Carrozzi, V.; Sambandam, A.; Luis, E.; Lin, Z.; Jeet, S.; Lesch, J.; Hackney, J.; Kim, J.; Zhou, M.; Lai, J.; et al. IL-17C regulates the innate immune function of epithelial cells in an autocrine manner. *Nat. Immunol.* **2011**, *12*, 1159–1166. [CrossRef] [PubMed]

129. Wolf, L.; Sapich, S.; Honecker, A.; Jungnickel, C.; Seiler, F.; Bischoff, M.; Wonnenberg, B.; Herr, C.; Schneider-Daum, N.; Lehr, C.M.; et al. IL-17A-mediated expression of epithelial IL-17C promotes inflammation during acute Pseudomonas aeruginosa pneumonia. *Am. J. Physiol. -Lung Cell. Mol. Physiol.* **2016**, *311*, L1015–L1022. [CrossRef] [PubMed]

130. Honda, K.; Wada, H.; Nakamura, M.; Nakamoto, K.; Inui, T.; Sada, M.; Koide, T.; Takata, S.; Yokoyama, T.; Saraya, T.; et al. IL-17A synergistically stimulates TNF-α-induced IL-8 production in human airway epithelial cells: A potential role in amplifying airway inflammation. *Exp. Lung Res.* **2016**, *42*, 205–216. [CrossRef]

131. Berndt, A.; Derksen, F.J.; Venta, P.J.; Ewart, S.; Yuzbasiyan-Gurkan, V.; Robinson, N.E. Elevated amount of Toll-like receptor 4 mRNA in bronchial epithelial cells is associated with airway inflammation in horses with recurrent airway obstruction. *Am. J. Physiol. -Lung Cell. Mol. Physiol.* **2007**, *292*, 936–943. [CrossRef]

132. Porto, B.N.; Stein, R.T. Neutrophil extracellular traps in pulmonary diseases: Too much of a good thing? *Front. Immunol.* **2016**, *7*, 311. [CrossRef] [PubMed]

133. Martinelli, S.; Urosevic, M.; Baryadel, A.; Oberholzer, P.A.; Baumann, C.; Fey, M.F.; Dummer, R.; Simon, H.U.; Yousefi, S. Induction of genes mediating interferon-dependent extracellular trap formation during neutrophil differentiation. *J. Biol. Chem.* **2004**, *279*, 44123–44132. [CrossRef] [PubMed]

134. Papayannopoulos, V.; Metzler, K.D.; Hakkim, A.; Zychlinsky, A. Neutrophil elastase and myeloperoxidase regulate the formation of neutrophil extracellular traps. *J. Cell Biol.* **2010**, *191*, 677–691. [CrossRef] [PubMed]

135. Twaddell, S.H.; Baines, K.J.; Grainge, C.; Gibson, P.G. The Emerging Role of Neutrophil Extracellular Traps in Respiratory Disease. *Chest* **2019**, *156*, 774–782. [CrossRef]
136. Saffarzadeh, M.; Juenemann, C.; Queisser, M.A.; Lochnit, G.; Barreto, G.; Galuska, S.P.; Lohmeyer, J.; Preissner, K.T. Neutrophil extracellular traps directly induce epithelial and endothelial cell death: A predominant role of histones. *PLoS ONE* **2012**, *7*, e32366. [CrossRef]
137. Narasaraju, T.; Yang, E.; Samy, R.P.; Ng, H.H.; Poh, W.P.; Liew, A.A.; Phoon, M.C.; Van Rooijen, N.; Chow, V.T. Excessive neutrophils and neutrophil extracellular traps contribute to acute lung injury of influenza pneumonitis. *Am. J. Pathol.* **2011**, *179*, 199–210. [CrossRef]
138. Pham, D.L.; Ban, G.Y.; Kim, S.H.; Shin, Y.S.; Ye, Y.M.; Chwae, Y.J.; Park, H.S. Neutrophil autophagy and extracellular DNA traps contribute to airway inflammation in severe asthma. *Clin. Exp. Allergy* **2017**, *47*, 57–70. [CrossRef]
139. Côté, O.; Clark, M.E.; Viel, L.; Labbé, G.; Seah, S.Y.K.; Khan, M.A.; Douda, D.N.; Palaniyar, N.; Bienzle, D. Secretoglobin 1A1 and 1A1A differentially regulate neutrophil reactive oxygen species production, phagocytosis and extracellular trap formation. *PLoS ONE* **2014**, *9*, e96217. [CrossRef]
140. Vargas, A.; Boivin, R.; Cano, P.; Murcia, Y.; Bazin, I.; Lavoie, J.P. Neutrophil extracellular traps are downregulated by glucocorticosteroids in lungs in an equine model of asthma. *Respir. Res.* **2017**, *18*, 207. [CrossRef]
141. Fu, J.; Tobin, M.C.; Thomas, L.L. Neutrophil-like low-density granulocytes are elevated in patients with moderate to severe persistent asthma. *Ann. Allergy, Asthma Immunol.* **2014**, *113*, 635–640.e2. [CrossRef]
142. Herteman, N.; Vargas, A.; Lavoie, J.P. Characterization of Circulating Low-Density Neutrophils Intrinsic Properties in Healthy and Asthmatic Horses. *Sci. Rep.* **2017**, *7*, 7743. [CrossRef] [PubMed]
143. Nguyen, G.T.; Green, E.R.; Mecsas, J. Neutrophils to the ROScue: Mechanisms of NADPH oxidase activation and bacterial resistance. *Front. Cell. Infect. Microbiol.* **2017**, *7*, 373. [CrossRef] [PubMed]
144. Chan, T.K.; Tan, W.S.D.; Peh, H.Y.; Wong, W.S.F. Aeroallergens Induce Reactive Oxygen Species Production and DNA Damage and Dampen Antioxidant Responses in Bronchial Epithelial Cells. *J. Immunol.* **2017**, *199*, 39–47. [CrossRef] [PubMed]
145. Bucchieri, F.; Puddicombe, S.M.; Lordan, J.L.; Richter, A.; Buchanan, D.; Wilson, S.J.; Ward, J.; Zummo, G.; Howarth, P.H.; Djukanović, R.; et al. Asthmatic bronchial epithelium is more susceptible to oxidant-induced apoptosis. *Am. J. Respir. Cell Mol. Biol.* **2002**, *27*, 179–185. [CrossRef]
146. Zeng, H.; Wang, Y.; Gu, Y.; Wang, J.; Zhang, H.; Gao, H.; Jin, Q.; Zhao, L. Polydatin attenuates reactive oxygen species-induced airway remodeling by promoting Nrf2-mediated antioxidant signaling in asthma mouse model. *Life Sci.* **2019**, *218*, 25–30. [CrossRef]
147. Csiszar, A.; Wang, M.; Lakatta, E.G.; Ungvari, Z. Inflammation and endothelial dysfunction during aging: Role of NF-κB. *J. Appl. Physiol.* **2008**, *105*, 1333–1341. [CrossRef]
148. Schuliga, M. NF-kappaB signaling in chronic inflammatory airway disease. *Biomolecules* **2015**, *5*, 1266–1283. [CrossRef]
149. Frossi, B.; De Carli, M.; Daniel, K.C.; Rivera, J.; Pucillo, C. Oxidative stress stimulates IL-4 and IL-6 production in mast cells by an APE/Ref-1-dependent pathway. *Eur. J. Immunol.* **2003**, *33*, 2168–2177. [CrossRef]
150. Niedzwiedz, A.; Jaworski, Z. Oxidant-Antioxidant Status in the Blood of Horses with Symptomatic Recurrent Airway Obstruction (RAO). *J. Vet. Intern. Med.* **2014**, *28*, 1845–1852. [CrossRef]
151. Niedzwiedz, A.; Borowicz, H.; Januszewska, L.; Markiewicz-Gorka, I.; Jaworski, Z. Serum 8-hydroxy-2-deoxyguanosine as a marker of DNA oxidative damage in horses with recurrent airway obstruction. *Acta Vet. Scand.* **2016**, *58*, 38. [CrossRef]
152. Deaton, C.M.; Marlin, D.J.; Smith, N.C.; Roberts, C.A.; Harris, P.A.; Schroter, R.C.; Kelly, F.J. Antioxidant and inflammatory responses of healthy horses and horses affected by recurrent airway obstruction to inhaled ozone. *Equine Vet. J.* **2005**, *37*, 243–249. [CrossRef] [PubMed]
153. Deaton, C.M.; Marlan, D.J.; Smith, N.C.; Harris, P.A.; Dagleish, M.P.; Schroter, R.C.; Kelly, F.J. Effect of acute airway inflammation on the pulmonary antioxidant status. *Exp. Lung Res.* **2005**, *31*, 653–670. [CrossRef] [PubMed]
154. Pourali Dogaheh, S.; Boivin, R.; Lavoie, J.P. Studies of molecular pathways associated with blood neutrophil corticosteroid insensitivity in equine asthma. *Vet. Immunol. Immunopathol.* **2021**, *237*, 110265. [CrossRef] [PubMed]
155. Greene, C.M.; McElvaney, N.G. Proteases and antiproteases in chronic neutrophilic lung disease - Relevance to drug discovery. *Br. J. Pharmacol.* **2009**, *158*, 1048–1058. [CrossRef]
156. Katavolos, P.; Ackerley, C.A.; Clark, M.E.; Bienzle, D. Clara cell secretory protein increases phagocytic and decreases oxidative activity of neutrophils. *Vet. Immunol. Immunopathol.* **2011**, *139*, 1–9. [CrossRef]
157. Sabbione, F.; Keitelman, I.A.; Iula, L.; Ferrero, M.; Giordano, M.N.; Baldi, P.; Rumbo, M.; Jancic, C.; Trevani, A.S. Neutrophil Extracellular Traps Stimulate Proinflammatory Responses in Human Airway Epithelial Cells. *J. Innate Immun.* **2017**, *9*, 387–402. [CrossRef]
158. Katavolos, P.; Ackerley, C.A.; Viel, L.; Clark, M.E.; Wen, X.; Bienzle, D. Clara cell secretory protein is reduced in equine recurrent airway obstruction. *Vet. Pathol.* **2009**, *46*, 604–613. [CrossRef]
159. Ortega-Gómez, A.; Perretti, M.; Soehnlein, O. Resolution of inflammation: An integrated view. *EMBO Mol. Med.* **2013**, *5*, 661–674. [CrossRef]
160. Amur, S. Biomarker qualification program in CDER, FDA. Available online: https://www.fda.gov/drugs/drug-development-tool-ddt-qualification-programs/biomarker-qualification-program (accessed on 21 February 2022).
161. Diamant, Z.; Vijverberg, S.; Alving, K.; Bakirtas, A.; Bjermer, L.; Custovic, A.; Dahlen, S.E.; Gaga, M.; Gerth van Wijk, R.; Del Giacco, S.; et al. Toward clinically applicable biomarkers for asthma: An EAACI position paper. *Allergy Eur. J. Allergy Clin. Immunol.* **2019**, *74*, 1835–1851. [CrossRef]

162. Bedenice, D.; Mazan, M.R.; Hoffman, A.M. Association between cough and cytology of bronchoalveolar lavage fluid and pulmonary function in horses diagnosed with inflammatory airway disease. *J. Vet. Intern. Med.* **2008**, *22*, 1022–1028. [CrossRef]
163. Zareba, L.; Szymanski, J.; Homoncik, Z.; Czystowska-Kuzmicz, M. Evs from balf—mediators of inflammation and potential biomarkers in lung diseases. *Int. J. Mol. Sci.* **2021**, *22*, 3651. [CrossRef] [PubMed]
164. Paggiaro, P.L.; Chanez, P.; Holz, O.; Ind, P.W.; Djukanović, R.; Maestrelli, P.; Sterk, P.J. Sputum induction. *Eur. Respir. J. Suppl.* **2002**, *20*, 3s–8s. [CrossRef]
165. Duz, M.; Whittaker, A.G.; Love, S.; Parkin, T.D.H.; Hughes, K.J. Exhaled breath condensate hydrogen peroxide and pH for the assessment of lower airway inflammation in the horse. *Res. Vet. Sci.* **2009**, *87*, 307–312. [CrossRef] [PubMed]
166. Popović-Grle, S.; Štajduhar, A.; Lampalo, M.; Rnjak, D. Biomarkers in different asthma phenotypes. *Genes* **2021**, *12*, 801. [CrossRef]
167. Choy, D.F.; Hart, K.M.; Borthwick, L.A.; Shikotra, A.; Nagarkar, D.R.; Siddiqui, S.; Jia, G.; Ohri, C.M.; Doran, E.; Vannella, K.M.; et al. TH2 and TH17 inflammatory pathways are reciprocally regulated in asthma. *Sci. Transl. Med.* **2015**, *7*, 301ra129. [CrossRef]
168. Tajiri, T.; Matsumoto, H.; Gon, Y.; Ito, R.; Hashimoto, S.; Izuhara, K.; Suzukawa, M.; Ohta, K.; Ono, J.; Ohta, S.; et al. Utility of serum periostin and free IgE levels in evaluating responsiveness to omalizumab in patients with severe asthma. *Allergy Eur. J. Allergy Clin. Immunol.* **2016**, *71*, 1472–1479. [CrossRef]
169. Ibrahim, B.; Basanta, M.; Cadden, P.; Singh, D.; Douce, D.; Woodcock, A.; Fowler, S.J. Non-invasive phenotyping using exhaled volatile organic compounds in asthma. *Thorax* **2011**, *66*, 804–809. [CrossRef]
170. Brightling, C.E.; Chanez, P.; Leigh, R.; O'Byrne, P.M.; Korn, S.; She, D.; May, R.D.; Streicher, K.; Ranade, K.; Piper, E. Efficacy and safety of tralokinumab in patients with severe uncontrolled asthma: A randomised, double-blind, placebo-controlled, phase 2b trial. *Lancet Respir. Med.* **2015**, *3*, 692–701. [CrossRef]
171. Hagan, J.B.; Laidlaw, T.M.; Divekar, R.; O'Brien, E.K.; Kita, H.; Volcheck, G.W.; Hagan, C.R.; Lal, D.; Teaford, H.G.; Erwin, P.J.; et al. Urinary Leukotriene E4 to Determine Aspirin Intolerance in Asthma: A Systematic Review and Meta-Analysis. *J. Allergy Clin. Immunol. Pract.* **2017**, *5*, 990–997.e1. [CrossRef]
172. Woodrow, J.S.; Hines, M.; Sommardahl, C.; Flatland, B.; Davis, K.U.; Lo, Y.; Wang, Z.; Sheats, M.K.; Lennon, E.M. Multidimensional analysis of bronchoalveolar lavage cytokines and mast cell proteases reveals Interferon-γ as a key biomarker in equine asthma syndrome. *bioRxiv* **2020**. [CrossRef]
173. Frigas, E.; Gleich, G.J. The eosinophil and the pathophysiology of asthma. *J. Allergy Clin. Immunol.* **1986**, *77*, 527–537. [CrossRef]
174. Lee, G.K.C.; Beeler-Marfisi, J.; Viel, L.; Piché, É.; Kang, H.; Sears, W.; Bienzle, D. Bronchial brush cytology, endobronchial biopsy, and SALSA immunohistochemistry in severe equine asthma. *Vet. Pathol.* **2022**, *59*, 100–111. [CrossRef] [PubMed]
175. Bright, L.A.; Dittmar, W.; Nanduri, B.; McCarthy, F.M.; Mujahid, N.; Costa, L.R.; Burgess, S.C.; Swiderski, C.E. Modeling the pasture-associated severe equine asthma bronchoalveolar lavage fluid proteome identifies molecular events mediating neutrophilic airway inflammation. *Vet. Med. Res. Rep.* **2019**, *10*, 43–63. [CrossRef]
176. Lavoie-Lamoureux, A.; Leclere, M.; Lemos, K.; Wagner, B.; Lavoie, J.P. Markers of Systemic Inflammation in Horses with Heaves. *J. Vet. Intern. Med.* **2012**, *26*, 1419–1426. [CrossRef] [PubMed]
177. Larsen, K.; Macleod, D.; Nihlberg, K.; Gürcan, E.; Bjermer, L.; Marko-Varga, G.; Westergren-Thorsson, G. Specific haptoglobin expression in bronchoalveolar lavage during differentiation of circulating fibroblast progenitor cells in mild asthma. *J. Proteome Res.* **2006**, *5*, 1479–1483. [CrossRef] [PubMed]
178. Kim, C.K.; Chung, C.Y.; Koh, Y.Y. Changes in serum haptoglobin level after allergen challenge test in asthmatic children. *Allergy Eur. J. Allergy Clin. Immunol.* **1998**, *53*, 184–189. [CrossRef] [PubMed]
179. Ather, J.L.; Ckless, K.; Martin, R.; Foley, K.L.; Suratt, B.T.; Boyson, J.E.; Fitzgerald, K.A.; Flavell, R.A.; Eisenbarth, S.C.; Poynter, M.E. Serum Amyloid A Activates the NLRP3 Inflammasome and Promotes Th17 Allergic Asthma in Mice. *J. Immunol.* **2011**, *187*, 64–73. [CrossRef]
180. Jouoilahti, P.; Salomaa, V.; Hakala, K.; Rasi, V.; Vahtera, E.; Palosuo, T. The association of sensitive systemic inflammation markers with bronchial asthma. *Ann. Allergy, Asthma Immunol.* **2002**, *89*, 381–385. [CrossRef]
181. Barton, A.K.; Shety, T.; Bondzio, A.; Einspanier, R.; Gehlen, H. Metalloproteinases and their tissue inhibitors in comparison between different chronic pneumopathies in the horse. *Mediat. Inflamm.* **2015**, *2015*, 569512. [CrossRef]
182. Nevalainen, M.; Raulo, S.M.; Brazil, T.J.; Pirie, R.S.; Sorsa, T.; McGorum, B.C.; Maisi, P. Inhalation of organic dusts and lipopolysaccharide increases gelatinolytic matrix metalloproteinases (MMPs) in the lungs of heaves horses. *Equine Vet. J.* **2002**, *34*, 150–155. [CrossRef]
183. Niedźwiedź, A.; Jaworski, Z.; Kubiak, K. Circulating immune complexes and markers of systemic inflammation in RAO-affected horses. *Pol. J. Vet. Sci.* **2014**, *17*, 697–702. [CrossRef] [PubMed]
184. Slowikowska, M.; Bajzert, J.; Miller, J.; Stefaniak, T.; Niedzwiedz, A. The dynamics of circulating immune complexes in horses with severe equine asthma. *Animals* **2021**, *11*, 1001. [CrossRef] [PubMed]
185. Bazzano, M.; Laghi, L.; Zhu, C.; Magi, G.E.; Tesei, B.; Laus, F. Respiratory metabolites in bronchoalveolar lavage fluid (BALF) and exhaled breath condensate (EBC) can differentiate horses affected by severe equine asthma from healthy horses. *BMC Vet. Res.* **2020**, *16*, 233. [CrossRef] [PubMed]
186. Barton, A.K.; Shety, T.; Bondzio, A.; Einspanier, R.; Gehlen, H. Metalloproteinases and their inhibitors are influenced by inhalative glucocorticoid therapy in combination with environmental dust reduction in equine recurrent airway obstruction. *BMC Vet. Res.* **2016**, *12*, 282. [CrossRef]

187. Barton, A.K.; Shety, T.; Klier, J.; Geis, S.; Einspanier, R.; Gehlen, H. Metalloproteinases and their inhibitors under the course of immunostimulation by CPG-ODN and specific antigen inhalation in equine asthma. *Mediat. Inflamm.* **2019**, *2019*, 7845623. [CrossRef]
188. Simonen-Jokinen, T.; Pirie, R.S.; McGorum, B.C.; Maisi, P. Effect of composition and different fractions of hay dust suspension on inflammation in lungs of heaves-affected horses: MMP-9 and MMP-2 as indicators of tissue destruction. *Equine Vet. J.* **2005**, *37*, 412–417. [CrossRef]
189. Van Doren, S.R. Matrix metalloproteinase interactions with collagen and elastin. *Matrix Biol.* **2015**, *44–46*, 224–231. [CrossRef]
190. Arpino, V.; Brock, M.; Gill, S.E. The role of TIMPs in regulation of extracellular matrix proteolysis. *Matrix Biol.* **2015**, *44–46*, 247–254. [CrossRef]
191. Gy, C.; Leclere, M.; Vargas, A.; Grimes, C.; Lavoie, J.P. Investigation of blood biomarkers for the diagnosis of mild to moderate asthma in horses. *J. Vet. Intern. Med.* **2019**, *33*, 1789–1795. [CrossRef]
192. du Preez, S.; Raidal, S.L.; Doran, G.S.; Prescott, M.; Hughes, K.J. Exhaled breath condensate hydrogen peroxide, pH and leukotriene B 4 are associated with lower airway inflammation and airway cytology in the horse. *Equine Vet. J.* **2019**, *51*, 24–32. [CrossRef]
193. Beck, J.M. ABCs of the lung microbiome. *Ann. Am. Thorac. Soc.* **2014**, *11*, 3–6. [CrossRef] [PubMed]
194. Murcia, P.R. Clinical insights: The equine microbiome. *Equine Vet. J.* **2019**, *51*, 714–715. [CrossRef] [PubMed]
195. Segal, L.N.; Rom, W.N.; Weiden, M.D. Lung microbiome for clinicians: New discoveries about bugs in healthy and diseased lungs. *Ann. Am. Thorac. Soc.* **2014**, *11*, 108–116. [CrossRef] [PubMed]
196. Iwase, T.; Uehara, Y.; Shinji, H.; Tajima, A.; Seo, H.; Takada, K.; Agata, T.; Mizunoe, Y. Staphylococcus epidermidis Esp inhibits Staphylococcus aureus biofilm formation and nasal colonization. *Nature* **2010**, *465*, 346–349. [CrossRef]
197. de Steenhuijsen Piters, W.A.A.; Jochems, S.P.; Mitsi, E.; Rylance, J.; Pojar, S.; Nikolaou, E.; German, E.L.; Holloway, M.; Carniel, B.F.; Chu, M.L.J.N.; et al. Interaction between the nasal microbiota and S. pneumoniae in the context of live-attenuated influenza vaccine. *Nat. Commun.* **2019**, *10*, 2981. [CrossRef]
198. Barcik, W.; Boutin, R.C.T.; Sokolowska, M.; Finlay, B.B. The Role of Lung and Gut Microbiota in the Pathology of Asthma. *Immunity* **2020**, *52*, 241–255. [CrossRef]
199. Segal, L.N.; Clemente, J.C.; Tsay, J.C.J.; Koralov, S.B.; Keller, B.C.; Wu, B.G.; Li, Y.; Shen, N.; Ghedin, E.; Morris, A.; et al. Enrichment of the lung microbiome with oral taxa is associated with lung inflammation of a Th17 phenotype. *Nat. Microbiol.* **2016**, *1*, 16031. [CrossRef]
200. Dickson, R.P.; Erb-Downward, J.R.; Freeman, C.M.; McCloskey, L.; Falkowski, N.R.; Huffnagle, G.B.; Curtis, J.L. Bacterial topography of the healthy human lower respiratory tract. *MBio* **2017**, *8*, e02287-16. [CrossRef]
201. Jankauskaitė, L.; Misevičienė, V.; Vaidelienė, L.; Kėvalas, R. Lower airway virology in health and disease—From invaders to symbionts. *Medicina* **2018**, *54*, 72. [CrossRef]
202. van Woerden, H.C.; Gregory, C.; Brown, R.; Marchesi, J.R.; Hoogendoorn, B.; Matthews, I.P. Differences in fungi present in induced sputum samples from asthma patients and non-atopic controls: A community based case control study. *BMC Infect. Dis.* **2013**, *13*, 69. [CrossRef]
203. Bond, S.L.; Timsit, E.; Workentine, M.; Alexander, T.; Léguillette, R. Upper and lower respiratory tract microbiota in horses: Bacterial communities associated with health and mild asthma (inflammatory airway disease) and effects of dexamethasone. *BMC Microbiol.* **2017**, *17*, 184. [CrossRef] [PubMed]
204. Fillion-Bertrand, G.; Dickson, R.P.; Boivin, R.; Lavoie, J.P.; Huffnagle, G.B.; Leclere, M. Lung microbiome is influenced by the environment and asthmatic status in an equine model of asthma. *Am. J. Respir. Cell Mol. Biol.* **2019**, *60*, 189–197. [CrossRef] [PubMed]
205. Dickson, R.P.; Erb-Downward, J.R.; Freeman, C.M.; McCloskey, L.; Beck, J.M.; Huffnagle, G.B.; Curtis, J.L. Spatial variation in the healthy human lung microbiome and the adapted island model of lung biogeography. *Ann. Am. Thorac. Soc.* **2015**, *12*, 821–830. [CrossRef] [PubMed]
206. Kaiser-Thom, S.; Hilty, M.; Gerber, V. Effects of hypersensitivity disorders and environmental factors on the equine intestinal microbiota. *Vet. Q.* **2020**, *40*, 97–107. [CrossRef] [PubMed]
207. Leclere, M.; Costa, M.C. Fecal microbiota in horses with asthma. *J. Vet. Intern. Med.* **2020**, *34*, 996–1006. [CrossRef] [PubMed]
208. Tavenner, M.K.; McDonnell, S.M.; Biddle, A.S. Development of the equine hindgut microbiome in semi-feral and domestic conventionally-managed foals. *Anim. Microbiome* **2020**, *2*, 43. [CrossRef] [PubMed]
209. Husso, A.; Jalanka, J.; Alipour, M.J.; Huhti, P.; Kareskoski, M.; Pessa-Morikawa, T.; Iivanainen, A.; Niku, M. The composition of the perinatal intestinal microbiota in horse. *Sci. Rep.* **2020**, *10*, 441. [CrossRef]
210. Mach, N.; Lansade, L.; Bars-Cortina, D.; Dhorne-Pollet, S.; Foury, A.; Moisan, M.P.; Ruet, A. Gut microbiota resilience in horse athletes following holidays out to pasture. *Sci. Rep.* **2021**, *11*, 5007. [CrossRef]
211. Daniels, S.P.; Leng, J.; Swann, J.R.; Proudman, C.J. Bugs and drugs: A systems biology approach to characterising the effect of moxidectin on the horse's faecal microbiome. *Anim. Microbiome* **2020**, *2*, 38. [CrossRef]
212. Di Pietro, R.; Arroyo, L.G.; Leclere, M.; Costa, M.C. Species-level gut microbiota analysis after antibiotic-induced dysbiosis in horses. *Animals* **2021**, *11*, 2859. [CrossRef]
213. Biesbroek, G.; Tsivtsivadze, E.; Sanders, E.A.M.; Montijn, R.; Veenhoven, R.H.; Keijser, B.J.F.; Bogaert, D. Early respiratory microbiota composition determines bacterial succession patterns and respiratory health in children. *Am. J. Respir. Crit. Care Med.* **2014**, *190*, 1283–1292. [CrossRef] [PubMed]

214. Teo, S.M.; Mok, D.; Pham, K.; Kusel, M.; Serralha, M.; Troy, N.; Holt, B.J.; Hales, B.J.; Walker, M.L.; Hollams, E.; et al. The infant nasopharyngeal microbiome impacts severity of lower respiratory infection and risk of asthma development. *Cell Host Microbe* **2015**, *17*, 704–715. [CrossRef] [PubMed]
215. Sokolowska, M.; Frei, R.; Lunjani, N.; Akdis, C.A.; O'Mahony, L. Microbiome and asthma. *Asthma Res. Pract.* **2018**, *4*, 1. [CrossRef] [PubMed]
216. Huang, Y.J.; Nelson, C.E.; Brodie, E.L.; Desantis, T.Z.; Baek, M.S.; Liu, J.; Woyke, T.; Allgaier, M.; Bristow, J.; Wiener-Kronish, J.P.; et al. Airway microbiota and bronchial hyperresponsiveness in patients with suboptimally controlled asthma. *J. Allergy Clin. Immunol.* **2011**, *127*, 372–381.e1–3. [CrossRef]
217. Denner, D.R.; Sangwan, N.; Becker, J.B.; Hogarth, D.K.; Oldham, J.; Castillo, J.; Sperling, A.I.; Solway, J.; Naureckas, E.T.; Gilbert, J.A.; et al. Corticosteroid therapy and airflow obstruction influence the bronchial microbiome, which is distinct from that of bronchoalveolar lavage in asthmatic airways. *J. Allergy Clin. Immunol.* **2016**, *137*, 1398–1405.e3. [CrossRef]
218. Kuhnert, P.; Korczak, B.; Falsen, E.; Straub, R.; Hoops, A.; Boerlin, P.; Frey, J.; Mutters, R. Nicoletella semolina gen. nov., sp. nov., a new member of Pasteurellaceae isolated from horses with airway disease. *J. Clin. Microbiol.* **2004**, *42*, 5542–5548. [CrossRef]
219. Hansson, I.; Johansson, K.E.; Persson, M.; Riihimäki, M. The clinical significance of Nicoletella semolina in horses with respiratory disorders and a screening of the bacterial flora in the airways of horses. *Vet. Microbiol.* **2013**, *162*, 695–699. [CrossRef]
220. Payette, F.; Charlebois, A.; Fairbrother, J.H.; Beauchamp, G.; Leclere, M. Nicoletella semolina in the airways of healthy horses and horses with severe asthma. *J. Vet. Intern. Med.* **2021**, *35*, 1612–1619. [CrossRef]
221. Manguin, E.; Pépin, E.; Boivin, R.; Leclere, M. Tracheal microbial populations in horses with moderate asthma. *J. Vet. Intern. Med.* **2020**, *34*, 986–995. [CrossRef]
222. Wood, J.L.N.; Burrell, M.H.; Roberts, C.A.; Chanter, N.; Shaw, Y. Streptococci and Pasteurella spp. associated with disease of the equine lower respiratory tract. *Equine Vet. J.* **1993**, *25*, 314–318. [CrossRef]
223. Hilty, M.; Burke, C.; Pedro, H.; Cardenas, P.; Bush, A.; Bossley, C.; Davies, J.; Ervine, A.; Poulter, L.; Pachter, L.; et al. Disordered microbial communities in asthmatic airways. *PLoS ONE* **2010**, *5*, e8578. [CrossRef] [PubMed]
224. Alnahas, S.; Hagner, S.; Raifer, H.; Kilic, A.; Gasteiger, G.; Mutters, R.; Hellhund, A.; Prinz, I.; Pinkenburg, O.; Visekruna, A.; et al. IL-17 and TNF-α are key mediators of Moraxella catarrhalis triggered exacerbation of allergic airway inflammation. *Front. Immunol.* **2017**, *8*, 1562. [CrossRef] [PubMed]
225. Charlson, E.S.; Diamond, J.M.; Bittinger, K.; Fitzgerald, A.S.; Yadav, A.; Haas, A.R.; Bushman, F.D.; Collman, R.G. Lung-enriched organisms and aberrant bacterial and fungal respiratory microbiota after lung transplant. *Am. J. Respir. Crit. Care Med.* **2012**, *186*, 536–545. [CrossRef] [PubMed]
226. Bond, S.L.; Workentine, M.; Hundt, J.; Gilkerson, J.R.; Léguillette, R. Effects of nebulized dexamethasone on the respiratory microbiota and mycobiota and relative equine herpesvirus-1, 2, 4, 5 in an equine model of asthma. *J. Vet. Intern. Med.* **2020**, *34*, 307–321. [CrossRef] [PubMed]
227. Goleva, E.; Jackson, L.P.; Harris, J.K.; Robertson, C.E.; Sutherland, E.R.; Hall, C.F.; Good, J.T.; Gelfand, E.W.; Martin, R.J.; Leung, D.Y.M. The effects of airway microbiome on corticosteroid responsiveness in asthma. *Am. J. Respir. Crit. Care Med.* **2013**, *188*, 1193–1201. [CrossRef]
228. Essilfie, A.T.; Simpson, J.L.; Dunkley, M.L.; Morgan, L.C.; Oliver, B.G.; Gibson, P.G.; Foster, P.S.; Hansbro, P.M. Combined Haemophilus influenzae respiratory infection and allergic airways disease drives chronic infection and features of neutrophilic asthma. *Thorax* **2012**, *67*, 588–599. [CrossRef]
229. Chung, K.F. Airway microbial dysbiosis in asthmatic patients: A target for prevention and treatment? *J. Allergy Clin. Immunol.* **2017**, *139*, 1071–1081. [CrossRef]
230. Marsland, B.J.; Trompette, A.; Gollwitzer, E.S. The gut-lung axis in respiratory disease. *Ann. Am. Thorac. Soc.* **2015**, *12*, S150–S156. [CrossRef]
231. Qin, N.; Zheng, B.; Yao, J.; Guo, L.; Zuo, J.; Wu, L.; Zhou, J.; Liu, L.; Guo, J.; Ni, S.; et al. Influence of H7N9 virus infection and associated treatment on human gut microbiota. *Sci. Rep.* **2015**, *5*, 14771. [CrossRef]
232. Enaud, R.; Prevel, R.; Ciarlo, E.; Beaufils, F.; Wieërs, G.; Guery, B.; Delhaes, L. The Gut-Lung Axis in Health and Respiratory Diseases: A Place for Inter-Organ and Inter-Kingdom Crosstalks. *Front. Cell. Infect. Microbiol.* **2020**, *10*, 9. [CrossRef]
233. Le Poul, E.; Loison, C.; Struyf, S.; Springael, J.Y.; Lannoy, V.; Decobecq, M.E.; Brezillon, S.; Dupriez, V.; Vassart, G.; Van Damme, J.; et al. Functional characterization of human receptors for short chain fatty acids and their role in polymorphonuclear cell activation. *J. Biol. Chem.* **2003**, *278*, 25481–25489. [CrossRef] [PubMed]
234. Mirković, B.; Murray, M.A.; Lavelle, G.M.; Molloy, K.; Azim, A.A.; Gunaratnam, C.; Healy, F.; Slattery, D.; McNally, P.; Hatch, J.; et al. The role of short-chain fatty acids, produced by anaerobic bacteria, in the cystic fibrosis airway. *Am. J. Respir. Crit. Care Med.* **2015**, *192*, 1314–1324. [CrossRef] [PubMed]
235. Young, R.P.; Hopkins, R.J.; Marsland, B. The gut-liver-lung axis: Modulation of the innate immune response and its possible role in chronic obstructive pulmonary disease. *Am. J. Respir. Cell Mol. Biol.* **2016**, *54*, 161–169. [CrossRef] [PubMed]
236. Pugin, B.; Barcik, W.; Westermann, P.; Heider, A.; Wawrzyniak, M.; Hellings, P.; Akdis, C.A.; O'Mahony, L. A wide diversity of bacteria from the human gut produces and degrades biogenic amines. *Microb. Ecol. Health Dis.* **2017**, *28*, 1353881. [CrossRef]
237. Roduit, C.; Frei, R.; Ferstl, R.; Loeliger, S.; Westermann, P.; Rhyner, C.; Schiavi, E.; Barcik, W.; Rodriguez-Perez, N.; Wawrzyniak, M.; et al. High levels of butyrate and propionate in early life are associated with protection against atopy. *Allergy Eur. J. Allergy Clin. Immunol.* **2019**, *74*, 799–809. [CrossRef]

238. McLoughlin, R.; Berthon, B.S.; Rogers, G.B.; Baines, K.J.; Leong, L.E.X.; Gibson, P.G.; Williams, E.J.; Wood, L.G. Soluble fibre supplementation with and without a probiotic in adults with asthma: A 7-day randomised, double blind, three way cross-over trial. *EBioMedicine* **2019**, *46*, 473–485. [CrossRef]

239. Arpaia, N.; Campbell, C.; Fan, X.; Dikiy, S.; Van Der Veeken, J.; Deroos, P.; Liu, H.; Cross, J.R.; Pfeffer, K.; Coffer, P.J.; et al. Metabolites produced by commensal bacteria promote peripheral regulatory T-cell generation. *Nature* **2013**, *504*, 451–455. [CrossRef]

240. Trompette, A.; Gollwitzer, E.S.; Yadava, K.; Sichelstiel, A.K.; Sprenger, N.; Ngom-Bru, C.; Blanchard, C.; Junt, T.; Nicod, L.P.; Harris, N.L.; et al. Gut microbiota metabolism of dietary fiber influences allergic airway disease and hematopoiesis. *Nat. Med.* **2014**, *20*, 159–166. [CrossRef]

241. Cait, A.; Hughes, M.R.; Antignano, F.; Cait, J.; Dimitriu, P.A.; Maas, K.R.; Reynolds, L.A.; Hacker, L.; Mohr, J.; Finlay, B.B.; et al. Microbiome-driven allergic lung inflammation is ameliorated by short-chain fatty acids. *Mucosal Immunol.* **2018**, *11*, 785–795. [CrossRef]

242. Hua, X.; Goedert, J.J.; Pu, A.; Yu, G.; Shi, J. Allergy associations with the adult fecal microbiota: Analysis of the American Gut Project. *EBioMedicine* **2016**, *3*, 172–179. [CrossRef]

243. Zhang, Q.; Cox, M.; Liang, Z.; Brinkmann, F.; Cardenas, P.A.; Duff, R.; Bhavsar, P.; Cookson, W.; Moffatt, M.; Chung, K.F. Airway microbiota in severe asthma and relationship to asthma severity and phenotypes. *PLoS ONE* **2016**, *11*, e0152724. [CrossRef]

244. Arrieta, M.C.; Stiemsma, L.T.; Dimitriu, P.A.; Thorson, L.; Russell, S.; Yurist-Doutsch, S.; Kuzeljevic, B.; Gold, M.J.; Britton, H.M.; Lefebvre, D.L.; et al. Early infancy microbial and metabolic alterations affect risk of childhood asthma. *Sci. Transl. Med.* **2015**, *7*, 307ra152. [CrossRef] [PubMed]

245. Zhang, J.; Ma, J.; Li, Q.; Su, H.; Sun, X. Exploration of the effect of mixed probiotics on microbiota of allergic asthma mice. *Cell. Immunol.* **2021**, *367*, 104399. [CrossRef] [PubMed]

246. Yang, X.; Li, H.; Ma, Q.; Zhang, Q.; Wang, C. Neutrophilic Asthma Is Associated with Increased Airway Bacterial Burden and Disordered Community Composition. *Biomed Res. Int.* **2018**, *2018*, 9230234. [CrossRef]

247. Huang, Y.J.; Nariya, S.; Harris, J.M.; Lynch, S.V.; Choy, D.F.; Arron, J.R.; Boushey, H. The airway microbiome in patients with severe asthma: Associations with disease features and severity. *J. Allergy Clin. Immunol.* **2015**, *136*, 874–884. [CrossRef]

248. Panettieri, R.A.; Sjöbring, U.; Péterffy, A.M.; Wessman, P.; Bowen, K.; Piper, E.; Colice, G.; Brightling, C.E. Tralokinumab for severe, uncontrolled asthma (STRATOS 1 and STRATOS 2): Two randomised, double-blind, placebo-controlled, phase 3 clinical trials. *Lancet Respir. Med.* **2018**, *6*, 511–525. [CrossRef]

 *animals*

MDPI

Article

Lung Function Variation during the Estrus Cycle of Mares Affected by Severe Asthma

Sophie Mainguy-Seers, Mouhamadou Diaw and Jean-Pierre Lavoie *

Faculty of Veterinary Medicine, Department of Clinical Sciences, Université de Montréal,
Saint-Hyacinthe, QC J2S 2M2, Canada; sophie.mainguy-seers@umontreal.ca (S.M.-S.);
mouhamadou.diaw@umontreal.ca (M.D.)
* Correspondence: jean-pierre.lavoie@umontreal.ca

Simple Summary: The estrus cycle and sex hormones influence asthma development and severity in humans, but whether the same is occurring in the asthma of horses is unknown. Severe equine asthma (SEA) is characterized by breathing difficulty, even at rest, and although it can be controlled by management and medication, it remains incurable. Stabling and hay feeding are the main contributors to disease exacerbation, but other factors could possibly alter the respiratory compromise. Therefore, the objective of this study was to evaluate the effects of the estrus cycle on airway dysfunction in five mares affected by SEA by assessing the lung function during the follicular and luteal phases of the reproductive cycle. The inspiratory obstruction improved during the luteal phase and the variation in progesterone and the dominant follicle size correlated with lung function parameters, suggesting a role for sex hormones in asthma pathophysiology. This first description of the estrus cycle's modulation of airway obstruction in horses supports further studies to uncover the effects of sex hormones in asthma in horses and humans.

Abstract: While the prevalence of asthma is higher in boys than in girls during childhood, this tendency reverses at puberty, suggesting an effect of sex hormones on the disease pathophysiology. Fluctuations of asthma severity concurring with the estrus cycle are reported in women, but this phenomenon has never been investigated in mares to date. The objective of this exploratory study was to determine whether the estrus cycle modulates airway obstruction in severe equine asthma (SEA). Five mares with SEA during exacerbation of the disease were studied. The whole breath, expiratory and inspiratory resistance, and reactance were compared during the follicular and luteal phases of the estrus cycle. The reproductive tract was evaluated by rectal palpation, ultrasound, and serum progesterone levels. The inspiratory resistance and reactance improved during the luteal phase of the estrus cycle, and variation in progesterone levels and the dominant follicle size correlated with several lung function parameters. The fluctuation of airway dysfunction during the estrus cycle is noteworthy as deterioration of the disease could perhaps be expected and prevented by horse owners and veterinarians. Further studies are required to determine if the equine species could be a suitable model to evaluate the effects of sex hormones on asthma.

Keywords: equine asthma; lung function; estrus cycle; progesterone; sex hormones

Citation: Mainguy-Seers, S.; Diaw, M.; Lavoie, J.-P. Lung Function Variation during the Estrus Cycle of Mares Affected by Severe Asthma. *Animals* **2022**, *12*, 494. https://doi.org/10.3390/ani12040494

Academic Editors: Harold C. McKenzie III and Chris W. Rogers

Received: 10 December 2021
Accepted: 12 February 2022
Published: 17 February 2022

Publisher's Note: MDPI stays neutral with regard to jurisdictional claims in published maps and institutional affiliations.

1. Introduction

Studying the sex disparities in asthma pathophysiology is requisite to comprehending the influence of sex hormones to ultimately provide personalized care to asthmatic patients [1]. The higher prevalence of asthma in boys reverses at puberty, after which the disease becomes more prevalent in women [2]. Importantly, women are over-represented in some corticosteroid-resistant and severe asthma clusters [3,4], and the sex difference in hospitalization admissions are more pronounced during the reproductive years [5]. However, conflicting results obscure the precise influence of sex hormones in asthma. While some studies described a deterioration of respiratory symptoms or lung function during the perimenstrual phase or menses [6–9], when progesterone and estrogen levels are low, others have reported no variation during the estrus cycle [10]. Also, increased emergency visits for asthma have been described both in the pre-ovulatory [11,12] and perimenstrual periods [11].

Severe equine asthma (SEA) is a chronic respiratory disease affecting approximately 14% of adult horses living in a temperate climate [13], and it is mostly influenced by environmental conditions [14] and genetics [15]. A predisposition for females has been reported [16], but this is not a consistent finding [13], and whether the disease is altered by the estrus cycle is currently unknown. Therefore, the objective of this exploratory study was to determine if lung function differed during the luteal and follicular phases of the estrus cycle in mares affected by SEA. A significant improvement in the inspiratory obstruction was observed during the luteal phase of the estrus cycle in this study.

2. Materials and Methods

All experimental procedures were performed in accordance with the Canadian Council for Animal Care guidelines and were approved by the Animal Care Committee of the Faculty of Veterinary Medicine of the Université de Montréal on 27 May 2021 (Protocol # 21-Rech-2128). This manuscript follows the recommendations of the ARRIVE guidelines.

Six mares with SEA donated to a research herd were studied during the 2021 reproductive season in the northern hemisphere (between May and August). The mares were mixed breeds, aged 17.7 (16–21) years old and weighed 573 (509–664) kg. One mare had foaled in the past, 3 were never bred and the information was unknown for 2 mares. A priori power analysis was not performed as no data were available for the equine species for calculations. The diagnosis of SEA was based on a history of repeated periods of labored breathing at rest, abnormal lung function (transpulmonary pressure >15 cm H_2O), and >25% neutrophils on bronchoalveolar lavage fluid (BALF) cytology when horses were stabled and fed hay, as previously recommended [17]. The mares had been part of the research herd for 2 to 6 years. At the beginning of the study, disease exacerbation was triggered by exposure to environmental antigens by stabling (wood shaving bedding) and dry hay feeding until respiratory efforts were visible at rest (median exposure of 36 days before the study). All mares had access to timed daily turnout on a dirt paddock, and the management remained the same for the duration of the experiment. The mares were conditioned to stand in a stock and to wear a mask. Endpoints included anorexia, decreased manure production, colic, hyperthermia, respiratory distress, or any other medical conditions that would have required treatment.

2.1. Pulmonary Function Tests

Lung function was evaluated with the Equine MasterScreen impulse oscillometry system (IOS; Jaeger GmbH, Würzburg, Germany) as previously described [18], in unsedated mares standing in stocks with the head in resting physiological position. Briefly, multi-frequency impulses produced by a loudspeaker were superimposed to the tidal breathing of the horse through an airtight mask. Simultaneously, pressure transducers connected to a pneumotachograph placed in front of the mask acquired the pressure-flow signal response of the respiratory system. The device was calibrated on each test day and accuracy was verified with a resistive test load before experimental measurements. Lung function data were acquired with LabManager (version 4.53, Jaeger, Würzburg, Germany) and analyzed with FAMOS (IMC, Meßsysteme, Berlin, Germany) using the fast Fourier transform method. Three 30 s recordings were averaged for analyses. Inspiratory (insp), expiratory (exp) and whole breath resistance (R), reactance (X), and coherence of the respiratory system from 2 to 7 Hertz (Hz) were analyzed. Briefly, the impulse of higher frequencies travels a shorter distance and thus represents changes occurring in central and upper airways, while lower frequency impulses travel deeper into the lungs where it detects peripheral airway dysfunction in diseases such as asthma [19].

2.2. Reproductive Tract Evaluation

After assessment of the lung function, the mares were sedated with xylazine (0.3 mg/kg IV) if needed, and the reproductive tract was examined by rectal palpation and ultrasound to determine whether the mares were in the follicular or the luteal phase of the estrus cycle by examining the cervix tonus, uterine edema, and the presence of relevant ovarian structures (major follicles and corpus luteum). The mares were scheduled to be evaluated every 7–10 days until an assessment of the lung function was obtained in the follicular and luteal phases, which was later confirmed by serum progesterone levels.

2.3. Serum Progesterone

Blood was collected in glass vacutainer tubes (BD Biosciences, Mississauga, Canada) before any other manipulations. After 20 min sedimentation, serum was obtained after centrifugation at $900 \times g$ for 10 min at room temperature, then stored at $-80\,^{\circ}C$ until batch analysis. Serum progesterone levels were measured by chemiluminescence (IMMULITE, Siemens, Erlangen, Germany) by the Centre de diagnostic vétérinaire de l'Université de Montréal with a detection limit of 0.2 ng/mL. Results below the limit of detection were attributed the arbitrary value of 0.2 ng/mL for statistical comparisons. The values of ≤ 1 ng/mL and ≥ 5 ng/mL were considered representative of the follicular and luteal phases, respectively [20]. In instances where more than two serum progesterone levels were obtained in the same mare (when palpation and ultrasound gave ambiguous results and assessment was repeated at a later date within the same cycle), the lung function values at the time points with the lowest and highest progesterone levels were used for statistical analysis.

2.4. Statistical Analysis

Data were analyzed using GraphPad Prism version 8.4.3 for Windows (GraphPad Software, San Diego, CA, USA). Normality was assessed with Shapiro–Wilk tests. As lung function data were normally distributed, two-way ANOVA for repeated measures was used to evaluate the effects of the phase of the estrus cycle (follicular or luteal) and the IOS impulse frequency (from 2 to 7 Hz) on the lung function parameters. When a significant effect was observed with the ANOVA, Bonferroni multiple comparison tests were used to compare the data between the luteal and follicular phases at each frequency. Associations between the variation in lung function data, progesterone levels and the dominant follicle diameter were explored with Pearson or Spearman correlations as appropriate. Results are reported with mean, or with median if not normally distributed, and range for descriptive data and with 95% confidence interval (CI) for statistical results. Ambient conditions (temperature and humidity) were compared with paired t-tests, or Wilcoxon matched-paired signed rank tests, as appropriate.

3. Results

The temperature and humidity were not different between the evaluations in the luteal and follicular phases (respectively, a mean of 19.6 °C (16.4–22.9) and 19.4 °C (16.4–24.9), and a median of 45% (36–76) and 46% (36–84)).

3.1. Mares

Six mares were initially included in this study. In one mare, no reproductive tract was detected by rectal palpation or ultrasound on three occasions, a hymen persistence was present, and two serum progesterone levels taken at a 12-day interval were under the limit of detection. Combined, these findings suggested a congenital developmental anomaly, and data from that mare were excluded. The other five mares had normal reproductive tracts, based on clinical and ultrasonographic examinations.

The mean progesterone level was 9.4 (5.0–13.2) and 0.5 (<0.2–1.4) ng/mL in the luteal and follicular phases, respectively. The mare with the nadir of progesterone of 1.4 ng/mL was still included based on an eight-fold difference with the progesterone value during the luteal phase [21].

3.2. Lung Function Variation

The exacerbation of the disease was confirmed at the first lung function assessment (ratio of the resistance at 3 and 7 Hz (R3/R7) $\geq$1 and negative reactance from 2–7 Hz). Three mares had their first evaluation during the follicular phase, and two during the luteal phase.

3.2.1. Whole-Breath Analysis

There was no significant effect of the phases of the estrus cycle on the whole-breath R and X (Figures 1a and 2a). The whole-breath R was frequency-dependent ($p = 0.009$), and a similar trend was present for the whole-breath X ($p = 0.06$).

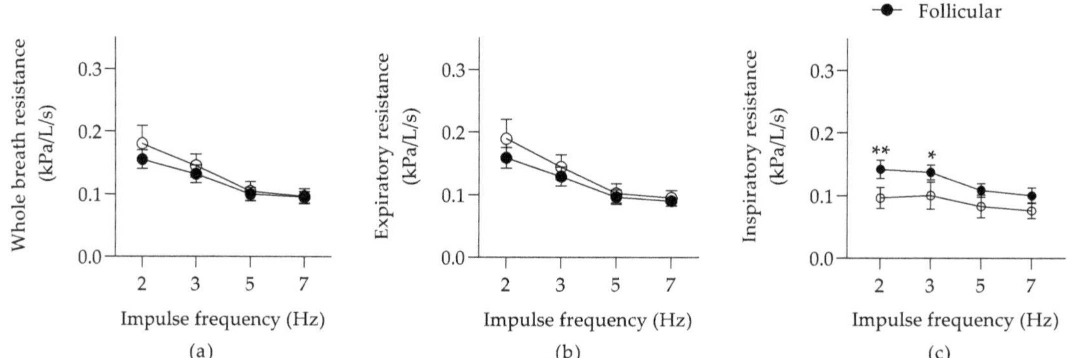

Figure 1. Pulmonary resistance (means $\pm$ SEM) during whole-breath (**a**), expiration (**b**), and inspiration (**c**). * $p < 0.05$ and ** $p < 0.01$ between the estrus phases with Bonferroni's multiple comparison tests.

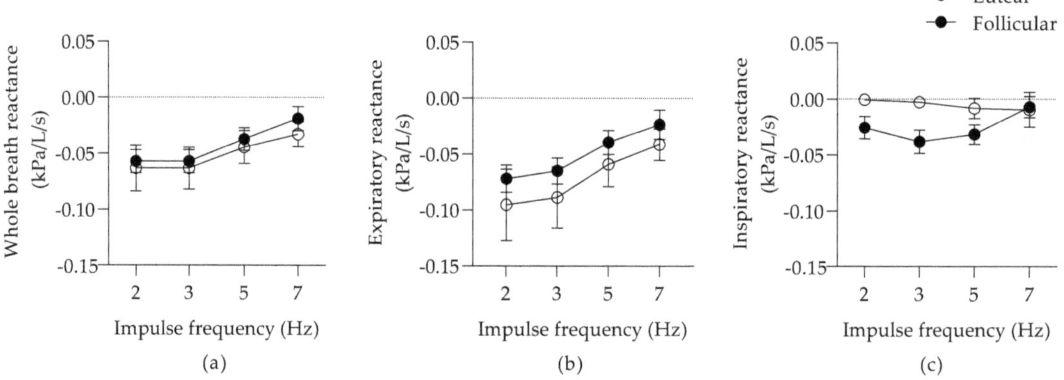

Figure 2. Pulmonary reactance (means $\pm$ SEM) during whole-breath (**a**), expiration (**b**), and inspiration (**c**).

3.2.2. Within-Breath Analysis

There was a significant frequency dependence of Rexp ($p = 0.005$) and Xexp ($p = 0.047$), but no effect of the estrus cycle during expiration (Figures 1b and 2b). At inspiration, there was a significant effect of the estrus cycle on the pulmonary Rinsp ($p < 0.0001$; Figure 1c) and Xinsp ($p = 0.016$; Figure 2c). The Rinsp at 2 and 3 Hz were improved during the luteal phase (a mean decrease of 0.046 (95% CI; 0.012–0.079) kPa/L/s and 0.037 (95% CI; 0.004–0.07) kPa/L/s, respectively).

3.2.3. Coherence

The coherence represents the causality between the flow input and the resultant pressure signals of the respiratory system and is an indicator of the accuracy of the mathematical model to predict the impedance data. Although it may be used as a quality control in oscillometry, coherence is also influenced by the severity of the disease itself [22]. The whole-breath ($p = 0.0006$; Figure 3a) and expiratory ($p = 0.0002$; Figure 3b) coherence were frequency-dependent and were both influenced by the estrus cycle (respectively, $p = 0.002$ and $p = 0.008$). There was a significant decrease of the whole-breath coherence at 2 Hz during the luteal phase (a mean reduction of 0.117 (95% CI; 0.014–0.219)). The inspiratory coherence was influenced by the impulse frequency ($p = 0.0003$; Figure 3c), but not by the estrus cycle.

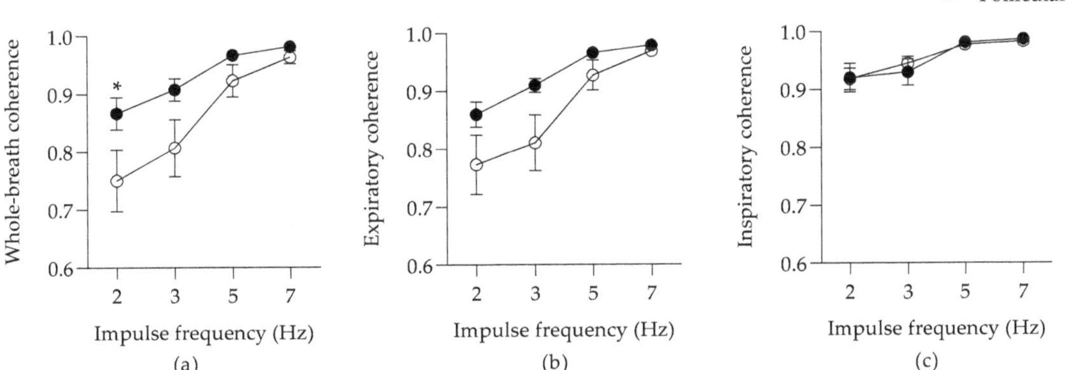

Figure 3. Coherence (means ± SEM) during whole-breath (**a**), expiration (**b**), and inspiration (**c**). * $p < 0.05$ between the estrus phases with Bonferroni's multiple comparison tests.

3.3. Correlations

The variation (difference between the luteal and follicular values) of serum progesterone was positively correlated with the variation of whole-breath R ($r = 0.89$, $p = 0.046$; Figure 4a), expiratory R ($r = 0.94$, $p = 0.017$), and X ($r = 0.98$, $p = 0.004$; Figure 4b) at 7 Hz. These results indicate that increasing progesterone was associated with a worsening of the resistance but an improving reactance. Correlations between the variation in lung function (the difference between the luteal and follicular values) and the diameter of the dominant follicle during the follicular phase were also assessed, as systemic estrogen has been shown to strongly correlate with the size of the dominant follicle in mares [23]. The diameter of the dominant follicle was negatively correlated with the variation of the whole-breath X at 3 Hz ($r = -0.90$, $p = 0.036$), 5 Hz ($r = -0.97$, $p = 0.006$), and 7 Hz ($r = -0.92$, $p = 0.027$; Figure 4c), and with the variation of Xexp at 3 Hz ($r = -0.90$, $p = 0.04$) and 5 Hz ($r = -0.95$, $p = 0.012$). It was correlated positively with the Rexp at 2 Hz ($r = 0.89$, $p = 0.044$). These results show that the size of the dominant follicle was associated with an improvement in the reactance and in the small airway's resistance during the follicular phase.

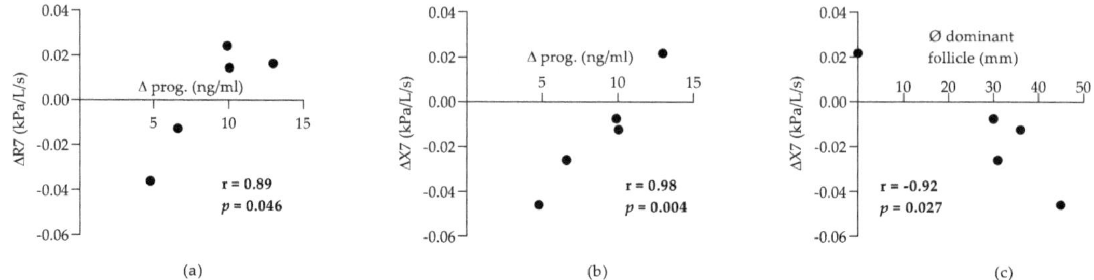

Figure 4. Correlations between the variation in progesterone (prog.) and the variation in pulmonary resistance (R) at 7 Hz (**a**), the variation of pulmonary reactance (X) at 7 Hz (**b**). Correlation between the diameter (Ø) of the dominant follicle in the follicular phase and variation in pulmonary X at 7 Hz (**c**).

4. Discussion

This study shows that the estrus cycle alters the lung function of mares with SEA, with an improvement in the inspiratory resistance and reactance during the luteal phase. These findings, combined with the correlations between variation in progesterone levels and the dominant follicle size with lung function parameters, suggest an influence of sex hormones on asthma pathophysiology in horses.

4.1. Estrus Cycle, Sex Hormones, and Lung Function

Asthma in humans and horses is characterized by expiratory flow limitation [24]. Expiratory parameters are indicative of intra-thoracic, thus smaller airway function, while inspiratory data represents predominantly extra-thoracic airways. Therefore, the improvement of inspiratory resistance during the luteal phase observed in the current study suggests a change of airway caliber occurring in the upper or most central airways [25]. Interestingly, the genioglossus muscle, an important upper airway dilator, has increased activity in women during the luteal phase [21], and synthetic progestins can improve inspiratory dysfunction during sleep apnea in humans, a disorder characterized by upper airway obstruction [26]. As the genioglossus muscle is also deemed important for stabilization of the equine pharynx [27], it is possible that modulation of its activity during the estrus cycle contributed to the inspiratory modifications observed in the current study. Alternatively, variations in lung function might be more easily detected during inspiration as the expiratory phase is much more compromised in asthma. As such, selected inspiratory parameters are better indicators of day-to-day variation in lung function in asthmatic adolescents [28].

The variation in serum progesterone levels between the luteal and follicular phases was strongly positively correlated with the difference in whole-breath and expiratory R and X at 7 Hz, indicating an association between progesterone and worsening resistance but improving reactance. These results could be caused by opposing actions of this hormone on separate airway compartments. Progesterone induces relaxation of the urogenital tract smooth muscle, and a similar effect is expected for airway smooth muscle (ASM) [29], which could explain the improved reactance. However, ASM relaxation would normally also improve pulmonary resistance, at least if it was occurring in the central airways. The tidal volume is higher during the luteal phase in healthy women when progesterone levels are high, compared to the follicular phase [30]. If a similar modification of tidal volume was present in mares, it could contribute to the association between improving reactance and progesterone. A decrease in peribronchial collagen deposition, reported after progesterone administration in mice, could also improve reactance [31]. These beneficial effects of progesterone could be relevant clinically, as suggested by the stabilization of peak flow rate after its administration in women with life-threatening exacerbation of asthma during the premenstrual phase [32]. Other properties of progesterone could explain its association with increasing airflow resistance, such as a decrease in epithelial ciliary beat frequency [33] and central hyperventilation [34]. Indeed, hyperventilation has been shown to increase pulmonary resistance as assessed by oscillometry in healthy horses [35]. Of note, progesterone receptors exist in two isoforms (A and B) and each mediates different biological activities [36]. Determining the proportion and the localization of these receptors in the equine respiratory tract is necessary to uncover the distinctive effects of progesterone on lung function observed in this study.

The size of the dominant follicle was positively correlated with the variation in whole-breath reactance at 5 Hz and the expiratory reactance at 5 and 7 Hz, suggesting positive effects of estrogen during the follicular phase [23]. Estrogen relaxes the ASM [37,38], decreases the proliferation of lung myofibroblasts [39], and enhances the effects of β2-adrenergic agonists in vitro [29]. In rodent experimental models, estrogen reduces airway hyperresponsiveness [40,41] and lung inflammation [41] in ovariectomized animals, and the absence of the estrogen receptor-α (knockout mice) is associated with lung dysfunction [42]. While estrogen has been shown to increase mucus production in vitro [43], the contrary was observed in a mouse model of allergic asthma [41]. Despite these interesting properties in experimental conditions, the clinical effects of estrogen in women are difficult to delineate and often contradictory. Its administration improves asthma symptoms in women in some studies [9], but not in all [44]. Its levels are negatively associated with lung function in adolescents [45], and the use of hormonal replacement therapy is related to increased odds of new-onset asthma after menopause [46]. Furthermore, single nucleotide polymorphisms of the estrogen receptor α are associated with a decline in lung function in asthmatics, particularly in women [47]. Taken together, these results suggest both beneficial and deleterious impacts of estrogen, and its actions might vary depending on the physiological status of the patients. In mares, the biological effects of estrogen on airway cells, the distribution and proportion of each receptor subtypes (α and β) along the respiratory tract, and the influence of endogenous or exogenous estrogen in SEA have not been studied to date. However, tamoxifen, a synthetic selective estrogen receptor modulator, induces a mild reduction of airway resistance in horses with SEA, suggesting a possible effect of this sex hormone [48].

Estrogen and progesterone could act synergistically to modulate lung function. The isometric contraction of mouse tracheal rings is more strongly attenuated by a combination of estrogen and progesterone, in concentrations representative of levels observed during human pregnancy, compared to the sole effect of each hormone [38]. Furthermore, both estrogen and progesterone levels correlated positively with the peak expiratory flow in a woman with perimenstrual asthma [49]. These findings would fit well with the data from the current study, which suggests that both hormones are associated with an improvement in reactance. This would also be consistent with anecdotal reports of improvement of respiratory signs during the gestation of mares with SEA (personal communication) and during pregnancy in some women [50].

4.2. Limitations and Areas of Future Research

The main limitation of this study was the assessment of a low number of mares during only one estrus cycle. Ideally, future investigations should follow larger cohorts during multiple cycles to obtain a more precise understanding of the phenomenon. The mares included in this study were aged from 16 to 21 years old, which is expected as aging is associated with an increased risk of SEA [13,16]. However, the estrus cycle is also influenced by aging in mares [51,52], with hormonal modifications and subfertility starting in the teens and culminating in the cessation of ovarian activity around 25 years old [52]. Therefore, future studies should ideally include mares within a wider age range. Additionally, as the effects of the estrus cycle on the lung function in healthy mares are unknown, a control group of healthy mares and of males with SEA should be included in future studies.

The effects of other hormones that could influence asthma physiology, such as testosterone [53,54] and sex hormone-binding globulin [55], were not investigated in this study, and estrogen levels were not directly measured. Interestingly, endogenous cortisol was shown to vary during the estrus cycle, with a higher value during the mid-late luteal phase in pony mares [56]; however, this was not observed by others [57]. Perhaps a variation of endogenous glucocorticoids could have contributed to the modulation of lung function in the current study, and this should be explored in future experiments. Furthermore, other factors that mediate sex differences in asthma, such as the smaller airway caliber in women [2], have never been investigated in horses to our knowledge.

Finally, the data from 2–7 Hz were reported in this study because the variation between recordings was low at these frequencies (mostly <15% [22]), lower frequencies better described the dysfunction in SEA [58], and coherence values were acceptable. The frequency dependence of the expiratory resistance and reactance is typical of small airway disease and has been previously reported by a forced oscillometry technique in SEA [58]. Additionally, the frequency-dependence of the coherence was not surprising as more severe lung dysfunction in smaller airways is expected during exacerbation. However, the cause of the poorer coherence at lower frequencies during the luteal phase for whole-breath and expiratory parameters in the current study is unknown. It might be related to increased heterogeneity of the respiratory system, perhaps suggesting a worsening of expiratory lung dysfunction during the luteal phase. Artifacts during measurements, such as leaks, swallowing, and coughing, could also result in lower coherence, but care was taken to repeat recording when such an event occurred. As measurements were performed concurrently for luteal and follicular assessment in different mares, the difference in coherence related to the estrus cycle is unlikely to be caused by a technical variation. Importantly, the differences in R and X during the estrus cycle occurred during the inspiratory phase, which had high coherence values that were not influenced by the phase of the estrus cycle and removing the few data with low coherence values (0.6–0.7) did not modify the results. Confirming the variation of lung function through the estrus cycle with standard lung function would be relevant, but given the current results showing differences only with within-breath analysis, it is unlikely that pleural pressure measurements would have the sensitivity required to detect these changes.

5. Conclusions

This study describes for the first time the influence of the estrus cycle on the natural course of SEA and supports further investigations to determine if horses could be a relevant model to explore the roles of sex hormones in asthma. Indeed, the reproductive physiology of mares and women is similar in many aspects, including the prolonged follicular phase and the monovolution [52]. The ease of reproductive tract evaluation and the long life span of horses, compared to rodent models, are additional valuable features [52]. Even the seasonality of reproduction in mares represents an opportunity to assess the effects of exogenous sex hormones when estrus activity is null. To further delineate the significance of the current results, future investigations should examine the localization and proportion of progesterone and estrogen receptors in the equine respiratory tract and the effects of these sex hormones on the biology of airway cells.

Author Contributions: Conceptualization, S.M.-S. and J.-P.L.; methodology, S.M.-S., M.D. and J.-P.L.; formal analysis, S.M.-S. and J.-P.L.; investigation, S.M.-S. and M.D.; resources, J.-P.L.; data curation, S.M.-S.; writing—original draft preparation, S.M.-S.; writing—review and editing, S.M.-S., M.D. and J.-P.L.; visualization, S.M.-S.; funding acquisition, J.-P.L. All authors have read and agreed to the published version of the manuscript.

Funding: This research was funded by the Canadian Institutes of Health Research (grant #PJT-148807) and a doctoral grant from the Fonds de Recherche du Québec en Santé (grant # 272483). The APC were waved for the publication in this special issue.

Institutional Review Board Statement: All experimental procedures were performed in accordance with the Canadian Council for Animal Care guidelines and were approved by the Animal Care Committee of the Faculty of Veterinary Medicine of the Université de Montréal on 27 May 2021 (Protocol # 21-Rech-2128).

Data Availability Statement: The datasets generated in the current study are available in the Dataverse UdeM repository (https://doi.org/10.5683/SP3/XGYUKU).

Conflicts of Interest: The authors declare no conflict of interest. The funders had no role in the design of the study; in the collection, analyses, or interpretation of data; in the writing of the manuscript, or in the decision to publish the results.

References

1. Shah, R.; Newcomb, D.C. Sex Bias in Asthma Prevalence and Pathogenesis. *Front. Immunol.* **2018**, *9*, 2997. [CrossRef] [PubMed]
2. de Marco, R.; Locatelli, F.; Sunyer, J.; Burney, P. Differences in incidence of reported asthma related to age in men and women. A retrospective analysis of the data of the European Respiratory Health Survey. *Am. J. Respir. Crit. Care Med.* **2000**, *162*, 68–74. [CrossRef] [PubMed]
3. Moore, W.C.; Meyers, D.A.; Wenzel, S.E.; Teague, W.G.; Li, H.; Li, X.; D'Agostino, R., Jr.; Castro, M.; Curran-Everett, D.; Fitzpatrick, A.M.; et al. Identification of asthma phenotypes using cluster analysis in the Severe Asthma Research Program. *Am. J. Respir. Crit. Care Med.* **2010**, *181*, 315–323. [CrossRef] [PubMed]
4. Wu, W.; Bang, S.; Bleecker, E.R.; Castro, M.; Denlinger, L.; Erzurum, S.C.; Fahy, J.V.; Fitzpatrick, A.M.; Gaston, B.M.; Hastie, A.T.; et al. Multiview Cluster Analysis Identifies Variable Corticosteroid Response Phenotypes in Severe Asthma. *Am. J. Respir. Crit. Care Med.* **2019**, *199*, 1358–1367. [CrossRef]
5. Chen, Y.; Stewart, P.; Johansen, H.; McRae, L.; Taylor, G. Sex difference in hospitalization due to asthma in relation to age. *J. Clin. Epidemiol.* **2003**, *56*, 180–187. [CrossRef]
6. Juniper, E.F.; Kline, P.A.; Roberts, R.S.; Hargreave, F.E.; Daniel, E.E. Airway responsiveness to methacholine during the natural menstrual cycle and the effect of oral contraceptives. *Am. Rev. Respir. Dis.* **1987**, *135*, 1039–1042. [CrossRef]
7. Gibbs, C.J.; Coutts, I.I.; Lock, R.; Finnegan, O.C.; White, R.J. Premenstrual exacerbation of asthma. *Thorax* **1984**, *39*, 833–836. [CrossRef]
8. Hanley, S.P. Asthma variation with menstruation. *Br. J. Dis. Chest* **1981**, *75*, 306–308. [CrossRef]
9. Chandler, M.H.; Schuldheisz, S.; Phillips, B.A.; Muse, K.N. Premenstrual asthma: The effect of estrogen on symptoms, pulmonary function, and beta 2-receptors. *Pharmacotherapy* **1997**, *17*, 224–234.
10. Weinmann, G.G.; Zacur, H.; Fish, J.E. Absence of changes in airway responsiveness during the menstrual cycle. *J. Allergy Clin. Immunol.* **1987**, *79*, 634–638. [CrossRef]
11. Brenner, B.E.; Holmes, T.M.; Mazal, B.; Camargo, C.A., Jr. Relation between phase of the menstrual cycle and asthma presentations in the emergency department. *Thorax* **2005**, *60*, 806–809. [CrossRef]
12. Zimmerman, J.L.; Woodruff, P.G.; Clark, S.; Camargo, C.A. Relation between phase of menstrual cycle and emergency department visits for acute asthma. *Am. J. Respir. Crit. Care Med.* **2000**, *162*, 512–515. [CrossRef] [PubMed]
13. Hotchkiss, J.W.; Reid, S.W.; Christley, R.M. A survey of horse owners in Great Britain regarding horses in their care. Part 2: Risk factors for recurrent airway obstruction. *Equine Vet. J.* **2007**, *39*, 301–308. [CrossRef] [PubMed]
14. Bullone, M.; Murcia, R.Y.; Lavoie, J.P. Environmental heat and airborne pollen concentration are associated with increased asthma severity in horses. *Equine Vet. J.* **2016**, *48*, 479–484. [CrossRef] [PubMed]
15. Gerber, V.; Baleri, D.; Klukowska-Rotzler, J.; Swinburne, J.E.; Dolf, G. Mixed inheritance of equine recurrent airway obstruction. *J. Vet. Intern. Med.* **2009**, *23*, 626–630. [CrossRef]
16. Couetil, L.L.; Ward, M.P. Analysis of risk factors for recurrent airway obstruction in North American horses: 1,444 cases (1990–1999). *J. Am. Vet. Med. Assoc.* **2003**, *223*, 1645–1650. [CrossRef]
17. Couetil, L.L.; Cardwell, J.M.; Gerber, V.; Lavoie, J.P.; Leguillette, R.; Richard, E.A. Inflammatory Airway Disease of Horses–Revised Consensus Statement. *J. Vet. Intern. Med.* **2016**, *30*, 503–515. [CrossRef]
18. van Erck, E.; Votion, D.; Art, T.; Lekeux, P. Measurement of respiratory function by impulse oscillometry in horses. *Equine Vet. J.* **2004**, *36*, 21–28. [CrossRef]
19. Desai, U.; Joshi, J.M. Impulse oscillometry. *Adv. Respir. Med.* **2019**, *87*, 235–238. [CrossRef]
20. Noden, P.A.; Oxender, W.D.; Hafs, H.D. The cycle of oestrus, ovulation and plasma levels of hormones in the mare. *J. Reprod. Fertil. Suppl.* **1975**, *23*, 189–192.
21. Popovic, R.M.; White, D.P. Upper airway muscle activity in normal women: Influence of hormonal status. *J. Appl. Physiol.* **1998**, *84*, 1055–1062. [CrossRef]
22. King, G.G.; Bates, J.; Berger, K.I.; Calverley, P.; de Melo, P.L.; Dellaca, R.L.; Farre, R.; Hall, G.L.; Ioan, I.; Irvin, C.G.; et al. Technical standards for respiratory oscillometry. *Eur. Respir. J.* **2020**, *55*, 1900753. [CrossRef]
23. Satue, K.; Fazio, E.; Ferlazzo, A.; Medica, P. Intrafollicular and systemic serotonin, oestradiol and progesterone concentrations in cycling mares. *Reprod. Domest. Anim.* **2019**, *54*, 1411–1418. [CrossRef] [PubMed]
24. Bond, S.; Leguillette, R.; Richard, E.A.; Couetil, L.; Lavoie, J.P.; Martin, J.G.; Pirie, R.S. Equine asthma: Integrative biologic relevance of a recently proposed nomenclature. *J. Vet. Intern. Med.* **2018**, *32*, 2088–2098. [CrossRef] [PubMed]
25. Klein, C.; Smith, H.J.; Reinhold, P. The use of impulse oscillometry for separate analysis of inspiratory and expiratory impedance parameters in horses: Effects of sedation with xylazine. *Res. Vet. Sci.* **2006**, *80*, 201–208. [CrossRef] [PubMed]

26. Saaresranta, T.; Aittokallio, T.; Polo-Kantola, P.; Helenius, H.; Polo, O. Effect of medroxyprogesterone on inspiratory flow shapes during sleep in postmenopausal women. *Respir. Physiol. Neurobiol.* **2003**, *134*, 131–143. [CrossRef]

27. Morello, S.L.; Ducharme, N.G.; Hackett, R.P.; Warnick, L.D.; Mitchell, L.M.; Soderholm, L.V. Activity of selected rostral and caudal hyoid muscles in clinically normal horses during strenuous exercise. *Am. J. Vet. Res.* **2008**, *69*, 682–689. [CrossRef] [PubMed]

28. Goldman, M.D.; Carter, R.; Klein, R.; Fritz, G.; Carter, B.; Pachucki, P. Within- and between-day variability of respiratory impedance, using impulse oscillometry in adolescent asthmatics. *Pediatr. Pulmonol.* **2002**, *34*, 312–319. [CrossRef]

29. Foster, P.S.; Goldie, R.G.; Paterson, J.W. Effect of steroids on beta-adrenoceptor-mediated relaxation of pig bronchus. *Br. J. Pharmacol.* **1983**, *78*, 441–445. [CrossRef]

30. White, D.P.; Douglas, N.J.; Pickett, C.K.; Weil, J.V.; Zwillich, C.W. Sexual influence on the control of breathing. *J. Appl. Physiol. Respir. Environ. Exerc. Physiol.* **1983**, *54*, 874–879. [CrossRef]

31. Zhang, X.; Bao, W.; Fei, X.; Zhang, Y.; Zhang, G.; Zhou, X.; Zhang, M. Progesterone attenuates airway remodeling and glucocorticoid resistance in a murine model of exposing to ozone. *Mol. Immunol.* **2018**, *96*, 69–77. [CrossRef] [PubMed]

32. Beynon, H.L.; Garbett, N.D.; Barnes, P.J. Severe premenstrual exacerbations of asthma: Effect of intramuscular progesterone. *Lancet* **1988**, *2*, 370–372. [CrossRef]

33. Jain, R.; Ray, J.M.; Pan, J.H.; Brody, S.L. Sex hormone-dependent regulation of cilia beat frequency in airway epithelium. *Am. J. Respir. Cell Mol. Biol.* **2012**, *46*, 446–453. [CrossRef] [PubMed]

34. Bayliss, D.A.; Millhorn, D.E. Central neural mechanisms of progesterone action: Application to the respiratory system. *J. Appl. Physiol.* **1992**, *73*, 393–404. [CrossRef] [PubMed]

35. Van Erck, E.; Votion, D.; Kirschvink, N.; Genicot, B.; Lindsey, J.; Art, T.; Lekeux, P. Influence of breathing pattern and lung inflation on impulse oscillometry measurements in horses. *Vet. J.* **2004**, *168*, 259–269. [CrossRef]

36. Asavasupreechar, T.; Saito, R.; Miki, Y.; Edwards, D.P.; Boonyaratanakornkit, V.; Sasano, H. Systemic distribution of progesterone receptor subtypes in human tissues. *J. Steroid. Biochem. Mol. Biol.* **2020**, *199*, 105599. [CrossRef]

37. Intapad, S.; Dimitropoulou, C.; Snead, C.; Piyachaturawat, P.; Catravas, J.D. Regulation of asthmatic airway relaxation by estrogen and heat shock protein 90. *J. Cell Physiol.* **2012**, *227*, 3036–3043. [CrossRef]

38. Dimitropoulou, C.; White, R.E.; Ownby, D.R.; Catravas, J.D. Estrogen reduces carbachol-induced constriction of asthmatic airways by stimulating large-conductance voltage and calcium-dependent potassium channels. *Am. J. Respir. Cell Mol. Biol.* **2005**, *32*, 239–247. [CrossRef]

39. Flores-Delgado, G.; Bringas, P.; Buckley, S.; Anderson, K.D.; Warburton, D. Nongenomic Estrogen Action in Human Lung Myofibroblasts. *Biochem. Biophys. Res. Commun.* **2001**, *283*, 661–667. [CrossRef]

40. Degano, B.; Mourlanette, P.; Valmary, S.; Pontier, S.; Prevost, M.C.; Escamilla, R. Differential effects of low and high-dose estradiol on airway reactivity in ovariectomized rats. *Respir. Physiol. Neurobiol.* **2003**, *138*, 265–274. [CrossRef]

41. Dimitropoulou, C.; Drakopanagiotakis, F.; Chatterjee, A.; Snead, C.; Catravas, J.D. Estrogen replacement therapy prevents airway dysfunction in a murine model of allergen-induced asthma. *Lung* **2009**, *187*, 116–127. [CrossRef] [PubMed]

42. Carey, M.A.; Card, J.W.; Bradbury, J.A.; Moorman, M.P.; Haykal-Coates, N.; Gavett, S.H.; Graves, J.P.; Walker, V.R.; Flake, G.P.; Voltz, J.W.; et al. Spontaneous airway hyperresponsiveness in estrogen receptor-alpha-deficient mice. *Am. J. Respir. Crit. Care Med.* **2007**, *175*, 126–135. [CrossRef] [PubMed]

43. Tam, A.; Wadsworth, S.; Dorscheid, D.; Man, S.F.; Sin, D.D. Estradiol increases mucus synthesis in bronchial epithelial cells. *PLoS ONE* **2014**, *9*, e100633. [CrossRef] [PubMed]

44. Ensom, M.H.; Chong, G.; Zhou, D.; Beaudin, B.; Shalansky, S.; Bai, T.R. Estradiol in premenstrual asthma: A double-blind, randomized, placebo-controlled, crossover study. *Pharmacotherapy* **2003**, *23*, 561–571. [CrossRef]

45. DeBoer, M.D.; Phillips, B.R.; Mauger, D.T.; Zein, J.; Erzurum, S.C.; Fitzpatrick, A.M.; Gaston, B.M.; Myers, R.; Ross, K.R.; Chmiel, J.; et al. Effects of endogenous sex hormones on lung function and symptom control in adolescents with asthma. *BMC Pulm. Med.* **2018**, *18*, 58. [CrossRef]

46. McCleary, N.; Nwaru, B.I.; Nurmatov, U.B.; Critchley, H.; Sheikh, A. Endogenous and exogenous sex steroid hormones in asthma and allergy in females: A systematic review and meta-analysis. *J. Allergy Clin. Immunol.* **2018**, *141*, 1510–1513.e1518. [CrossRef]

47. Dijkstra, A.; Howard, T.D.; Vonk, J.M.; Ampleford, E.J.; Lange, L.A.; Bleecker, E.R.; Meyers, D.A.; Postma, D.S. Estrogen receptor 1 polymorphisms are associated with airway hyperresponsiveness and lung function decline, particularly in female subjects with asthma. *J. Allergy Clin. Immunol.* **2006**, *117*, 604–611. [CrossRef]

48. Mainguy-Seers, S.; Picotte, K.; Lavoie, J.P. Efficacy of tamoxifen for the treatment of severe equine asthma. *J. Vet. Intern. Med.* **2018**, *32*, 1748–1753. [CrossRef]

49. Lam, S.M.; Huang, S.C. Premenstrual asthma: Report of a case with hormonal studies. *J. Microbiol. Immunol. Infect.* **1998**, *31*, 197–199.

50. Schatz, M.; Harden, K.; Forsythe, A.; Chilingar, L.; Hoffman, C.; Sperling, W.; Zeiger, R.S. The course of asthma during pregnancy, post partum, and with successive pregnancies: A prospective analysis. *J. Allergy Clin. Immunol.* **1988**, *81*, 509–517. [CrossRef]

51. Claes, A.; Ball, B.A.; Scoggin, K.E.; Roser, J.F.; Woodward, E.M.; Davolli, G.M.; Squires, E.L.; Troedsson, M.H.T. The influence of age, antral follicle count and diestrous ovulations on estrous cycle characteristics of mares. *Theriogenology* **2017**, *97*, 34–40. [CrossRef] [PubMed]

52. Carnevale, E.M. The mare model for follicular maturation and reproductive aging in the woman. *Theriogenology* **2008**, *69*, 23–30. [CrossRef]

53. Matteis, M.; Polverino, F.; Spaziano, G.; Roviezzo, F.; Santoriello, C.; Sullo, N.; Bucci, M.R.; Rossi, F.; Polverino, M.; Owen, C.A.; et al. Effects of sex hormones on bronchial reactivity during the menstrual cycle. *BMC Pulm. Med.* **2014**, *14*, 108. [CrossRef] [PubMed]

54. Scott, H.A.; Gibson, P.G.; Garg, M.L.; Upham, J.W.; Wood, L.G. Sex hormones and systemic inflammation are modulators of the obese-asthma phenotype. *Allergy* **2016**, *71*, 1037–1047. [CrossRef] [PubMed]

55. Arathimos, R.; Granell, R.; Haycock, P.; Richmond, R.C.; Yarmolinsky, J.; Relton, C.L.; Tilling, K. Genetic and observational evidence supports a causal role of sex hormones on the development of asthma. *Thorax* **2019**, *74*, 633–642. [CrossRef] [PubMed]

56. Asa, C.S.; Robinson, J.A.; Ginther, O.J. Changes in plasma cortisol concentrations during the ovulatory cycle of the mare. *J. Endocrinol.* **1983**, *99*, 329–334. [CrossRef]

57. Satue, K.; Montesinos, P.; Munoz, A. Modulation of the renin-angiotensin-aldosterone system by steroid hormones during the oestrous cycle in mares. *Acta Vet. Hung* **2020**, *68*, 79–84. [CrossRef]

58. Young, S.S.; Tesarowski, D.; Viel, L. Frequency dependence of forced oscillatory respiratory mechanics in horses with heaves. *J. Appl. Physiol.* **1997**, *82*, 983–987. [CrossRef]

Article

Associations between Exercise-Induced Pulmonary Hemorrhage (EIPH) and Fitness Parameters Measured by Incremental Treadmill Test in Standardbred Racehorses

Chiara Maria Lo Feudo [1,†], Luca Stucchi [2,†], Giovanni Stancari [2], Elena Alberti [1], Bianca Conturba [2], Enrica Zucca [1] and Francesco Ferrucci [1,*]

[1] Equine Sports Medicine Laboratory "Franco Tradati", Department of Veterinary Medicine and Animal Sciences, Università Degli Studi di Milano, 26900 Lodi, Italy; chiara.lofeudo@unimi.it (C.M.L.F.); elena.alberti@unimi.it (E.A.); enrica.zucca@unimi.it (E.Z.)

[2] Veterinary Teaching Hospital, Department of Veterinary Medicine and Animal Sciences, Università Degli Studi di Milano, 26900 Lodi, Italy; luca.stucchi@unimi.it (L.S.); giovanni.stancari@unimi.it (G.S.); bianca.conturba@unimi.it (B.C.)

[*] Correspondence: francesco.ferrucci@unimi.it; Tel.: +39-0250334146

[†] These authors contributed equally to this paper.

Simple Summary: Exercise-induced pulmonary hemorrhage (EIPH) frequently affects racehorses worldwide and has been widely associated with poor performance; however, scientific evidence supporting this observation is low. The present retrospective study aims to evaluate objectively whether the presence and grade of EIPH could affect some fitness parameters, measured during an incremental treadmill test, in poorly performing Standardbred racehorses. For this purpose, the association between EIPH and the results of a treadmill metabolic test (including blood lactate analysis and venous blood gas analysis) were evaluated in 81 Standardbred racehorses. No relationship between EIPH and aerobic/anaerobic capacity was observed, suggesting that EIPH may affect performance in a different manner. However, EIPH-affected horses were shown to reach higher hematocrit values during exercise compared to EIPH-negative horses; therefore, it may be hypothesized that hemoconcentration may take part in the pathogenesis of EIPH by increasing the pulmonary capillary pressure.

Abstract: Exercise-induced pulmonary hemorrhage (EIPH) is a condition affecting up to 95% of racehorses, diagnosed by detecting blood in the trachea after exercise and/or the presence of hemosiderophages in the bronchoalveolar lavage fluid (BALf). Although EIPH is commonly associated with poor performance, scientific evidence is scarce. The athletic capacity of racehorses can be quantified through some parameters obtained during an incremental treadmill test; in particular, the speed at a heart rate of 200 bpm (V200), and the speed (VLa4) and the heart rate (HRLa4) at which the blood lactate concentration reaches 4 mmol/L are considered good fitness indicators. The present retrospective study aims to evaluate whether EIPH could influence fitness parameters in poorly performing Standardbreds. For this purpose, data from 81 patients regarding their V200, VLa4, HRLa4, peak lactate, maximum speed, minimum pH, and maximum hematocrit were reviewed; EIPH scores were assigned based on tracheobronchoscopy and BALf cytology. The association between the fitness parameters and EIPH was evaluated through Spearman's correlation analysis. No relationship between EIPH and V200, VLa4, and HRLa4 was observed. Interestingly, EIPH-positive horses showed higher hematocrit values ($p = 0.0072$, r = 0.47), suggesting the possible influence of the hemoconcentration on the increase of pulmonary capillary pressure as a part of the pathogenesis of EIPH.

Keywords: EIPH; horses; racehorses; standardbred; treadmill test; lactate; poor performance; sports medicine

Citation: Lo Feudo, C.M.; Stucchi, L.; Stancari, G.; Alberti, E.; Conturba, B.; Zucca, E.; Ferrucci, F. Associations between Exercise-Induced Pulmonary Hemorrhage (EIPH) and Fitness Parameters Measured by Incremental Treadmill Test in Standardbred Racehorses. *Animals* **2022**, *12*, 449. https://doi.org/10.3390/ani12040449

Academic Editor: Harold C. McKenzie III

Received: 31 December 2021
Accepted: 10 February 2022
Published: 12 February 2022

Publisher's Note: MDPI stays neutral with regard to jurisdictional claims in published maps and institutional affiliations.

1. Introduction

Exercise-induced pulmonary hemorrhage (EIPH), defined as bleeding occurring from the lungs during exercise, is a highly frequent condition among racehorses worldwide [1], affecting from 43% to 75% of racehorses when diagnosed on a single examination, and up to 95% in the case of repeated examinations [2]. Its diagnosis is based on the detection of blood in the trachea from 30 to 120 min after strenuous exercise [3], and/or the presence in the bronchoalveolar lavage fluid (BALf) of red blood cells, acutely, or hemosiderophages in subacute and chronic cases [4]. EIPH has frequently been associated with poor performance in Thoroughbreds and Standardbreds [5–9]; however, scientific evidence supports this hypothesis only partially. Different authors reported an association between the presence of EIPH and the likelihood of a lower position in races [7,10,11]. In particular, in a study, horses with EIPH of grade $\geq$ 2 showed to have lower odds of finishing in the first three positions, and horses with EIPH of grade $\geq$ 1 finished at a longer distance behind the winner compared to EIPH-negative horses [7]; also, in another study, EIPH-positive horses tended to finish unplaced [11]. Moreover, severe forms of EIPH (grade 4) have been associated with a shorter duration of the racing career [9], and EIPH of grade $\geq$ 2 has been associated with a lower likelihood to be in the 90th higher percentile for race earnings [7]. In contrast, other authors failed to detect any influence of EIPH on the finishing position of racehorses [12–16], nor on the finishing time in a race [15]. However, to date, no studies have managed to quantify objectively the lower performance of EIPH-affected horses compared to EIPH-negative horses. To assess in a quantitative way the athletic capacity of the horses, a standardized incremental exercise test on a treadmill may be performed. The maximal oxygen consumption (VO_{2max}) reflects the maximal aerobic capacity and is considered as a good predictor of athletic performance [17]. Moreover, the measurement of the blood lactate concentration can provide useful information on both the aerobic and anaerobic capacities of the horse; the velocity at which the horse's blood lactate concentration reaches 4 mmol/L (VLa4) represents the aerobic–anaerobic threshold and is a good indicator of aerobic capacity [18]. High values of VLa4 have been associated with superior performance in different studies [19–21]. An indication of anaerobic capacity could be provided by the measurement of the post-exercise peak lactate levels [22], and higher concentrations are reported to be found in faster horses [23,24] with better performance [25]; however, the lactate concentration is affected by many variables, including the rates of flux between fluid compartments. Therefore the use of the peak lactate concentrations as an indicator of anaerobic capacity appears limited [26]. The present study aims to evaluate whether the presence and grade of EIPH could have an influence on selected performance parameters measured during a standardized incremental treadmill test in poorly performing Standardbred horses.

2. Materials and Methods

2.1. Horses

The clinical records from a population of Standardbred racehorses referred to the Equine Medicine Unit of the Veterinary Teaching Hospital of the University of Milan (Italy) for poor performance evaluation between 2002 and 2020 were retrospectively reviewed. All horses were in full training and underwent a complete diagnostic protocol, including the collection of history, clinical physical examination, laboratory analysis (complete blood count, blood chemistry, and arterial blood gas analysis), lameness evaluation, electrocardiogram, thoracic ultrasonography, upper airways endoscopy at rest, incremental high-speed treadmill metabolic test (plasma lactate analysis and Holter registration during exercise), dynamic endoscopy of the upper airways on a high-speed treadmill, tracheobronchoscopy 30 min after dynamic endoscopy, and BALf collection, including its cytological examination, performed 24 h after the dynamic endoscopy on the treadmill.

Horses showing signs of systemic illness, lameness, cardiac murmurs of grade > 3/6 [27], clinically significant valvular regurgitations [28], clinically significant cardiac arrhythmias [29,30], dynamic upper-airway obstructions (DUAOs), or rhabdomyolysis (stiffness, reluctance to move, sweating, tachypnea, tachycardia, myoglobinuria, and/or creatin-kinase > 735 U/l at 6 h after exercise [31]) were excluded from the study, since these afflictions could influence athletic performance.

2.2. Incremental Treadmill Test

Before performing the incremental treadmill test, the horses were conditioned to the high-speed treadmill (Sato I, Uppsala, Sweden) with two daily sessions. On the third day, the test was performed: the belt was inclined with a 5% slope, the protocol started with a warm-up of 4 min walk (1.5 m/s) and 3 min trot (6 m/s), followed by 1 min phases increasing the speed by 1 m/s until the horse was no longer able to maintain the treadmill speed; at the end of the test, the horses were walked for 30 min with a 0% slope to cool down [31].

Blood samples were collected during the test with the aid of a 14 G Teflon venous catheter placed in the left jugular vein and connected to an extension tube; blood samples were taken at rest, after the warm-up phase, at the end of each speed phase, and at 1, 5, 15, and 30 min during the cool down. To perform plasma lactate analysis, blood was transferred into tubes containing 10 mg sodium fluoride and 2 mg potassium oxalate for 1 mL of blood. The samples were centrifugated within 15 min and refrigerated, and the plasma lactate was measured with the enzymatic colorimetric method using a lactate dry-fast kit for the automatic system (Uni Fast System II Analyzer, Sclavo, Italy) and reagents supplied by the manufacturer [32]. In some horses, an aliquot of blood collected in heparinized syringes at each phase was used to measure the blood pH (48 horses) and hematocrit (31 horses) by means of a blood gas analyzer (Opti CCA, Opti Medical System, Roswell, NM, USA).

Throughout the duration of the treadmill test, the heart rate was monitored in real-time using a heart rate monitor (Polar, Equine Inzone FT1, Steinhausen, Switzerland); moreover, an ECG was obtained continuously before, during. and after exercise by means of a Holter recorder (Cardioline® Click Holter, Trento, Italy) [30].

2.3. Fitness Parameters

Data about the values of plasma lactate, heart rate, and eventually pH and hematocrit at each speed phase and during the cool down were obtained at the end of the test and collected on an electronic sheet (Microsoft Excel, Redmond, WA, USA). The fitness parameters obtained from the treadmill test were:

- VLa4: speed at a plasma lactate concentration of 4 mmol/L;
- HRLa4: heart rate at a plasma lactate concentration of 4 mmol/L;
- V200: speed at a heart rate of 200 bpm;
- Peak lactate: maximum plasma lactate concentration reached during the treadmill test or cool down;
- V max: maximum speed reached during the test until fatigue;
- pH min: minimum pH reached during the test;
- Ht max: maximum hematocrit reached during the test.

The values of VLA4 and HRLa4 were calculated by means of specific software (Lactate-E 1.0) [33]; after entering the data about the plasma lactate concentration and heart rate collected during treadmill exercise at each speed, this software used inverse prediction to provide precise lactate threshold markers.

2.4. Airway Endoscopy

The day after the incremental test, the horses underwent a dynamic upper airway treadmill endoscopy at racing speed. After a warm-up of a 4 min walk (1.5 m/s) and 5 min trot (6 m/s) with a 5% slope, the treadmill was temporarily stopped. A videoendoscope (ETM PVG-325, Storz, Tuttlingen, Germany) was passed into the nasopharynx of the horse and held in position with straps. Then, the treadmill was rapidly accelerated up to maximal speed (speed at maximum heart rate) for a 1600–2100 m distance or until fatigue. Thirty minutes after the end of the exercise, tracheobronchoscopy was performed. The horses were restrained in a stock and contained with a twitch. A flexible videoendoscope (EC-530WL-P, Fujifilm, Tokyo, Japan) was passed through the left nostril, and the upper and lower tracts of the respiratory system were visualized. The eventual presence of blood in the trachea and the mainstem bronchi was graded from 0 to 4 based on a reported scoring system (tracheal blood score, TBS) [3].

2.5. BALf Collection and Cytological Examination

Twenty-four hours after the treadmill exercise, BALf was collected; with this aim, horses were restrained in a stock and sedated with detomidine hydrochloride (0.01 mg/kg IV; Domosedan; Vetoquinol, Italy). Airway endoscopy was performed as described above. To perform the BAL, 60 mL of a 0.5% lidocaine hydrochloride solution was sprayed at the level of the tracheal bifurcation in order to inhibit the coughing reflex; then, the endoscope was passed into the bronchial tree until it was wedged firmly within a segmental bronchus. Here, a 300 mL pre-warmed sterile saline 0.9% was instilled, and the fluid was immediately aspirated. The BALf sample was stored in sterile ethylenediaminetetraacetic acid (EDTA) tubes and processed within 90 min. To perform the cytological examination, a few drops of pooled BALf were cytocentrifugated (Rotofix 32, Hettich Cyto System, Tuttlingen, Germany) at 500 rpm for 5 min. The slides were air dried, stained with May-Grünwald Giemsa and Perl's Prussian blue, and observed under a light microscope at $400\times$ and $1000\times$ for 400-cell leukocyte differential counting [34]. To evaluate EIPH, 100 macrophages were assessed. The percentage of hemosiderophages on the total of macrophages was calculated and hemosiderin was scored from 0 to 4 based on the grading of blue coloration in the cytoplasm of the macrophages [35]; then, the percentage of hemosiderophages was multiplied by the median hemosiderin score to obtain a simplified total hemosiderin score (sTHS), with a maximum score of 400 [36].

2.6. Statistical Analysis

All data were analyzed using descriptive statistics and evaluated for normality by means of the Shapiro–Wilk test. The influence of age and weight on every parameter was evaluated using the Spearman correlation, while the influence of sex was evaluated by means of the Kruskal–Wallis test (when considering stallions, geldings, and mares) and the Mann–Whitney test (when considering males and females). The association between the TBS and all the fitness parameters, the BALf leukocyte differential cell count, and the sTHS was evaluated using Spearman's correlation. The same test was used to analyze the relationship between the sTHS with the fitness parameters and the BALf leukocyte differential cell count. The fitness parameters, the BALf leukocyte differential cell count, and the sTHS in the EIPH positive (TBS $\geq$ 1) and negative groups were compared by means of the Mann–Whitney test and the unpaired t test. The relationship between the BALf leukocyte differential cell count and the fitness parameters was evaluated by means of Spearman's correlation. Since the percentage of neutrophils is related to some fitness parameters [32], a group of horses presenting a neutrophils percentage of < 5% [37] in the BALf was identified (Neu5 group); among them, the association of the different fitness parameters with age, weight, TBS, and sTHS was evaluated with Spearman's correlation. The data are presented as the mean $\pm$ standard deviation (SD) if normally distributed, and as the median and interquartile range (IQR) if not normally distributed. Statistical significance was set at $p < 0.05$. The data were analyzed using a commercially available statistical software package (GraphPad Prism 9.1.0 for MacOS; GraphPad Software, San Diego, CA, USA).

3. Results

3.1. Study Population

Among the 230 Standardbred racehorses that underwent a complete diagnostic protocol for poor performance, 81 met the inclusion criteria for the study, while 149 were excluded for the presence of DUAOs, clinically significant cardiac arrhythmias or valvular regurgitations, rhabdomyolysis, or lameness detected either during clinical examination or the treadmill tests. The study population consisted of 27 mares, 44 stallions, and 10 geldings aged from 2 to 8 years old (median 3, IQR 3–4) and weighing from 372 to 530 kg (mean 453 ± 34.41 kg).

Twenty-six horses were EIPH-negative at tracheoscopy (32.1%), while the remaining 55 (67.9%) were EIPH-positive; 19 horses had a TBS of 1 (23.46%), 23 had a TBS of 2 (28.39%), 11 had a TBS of 3 (13.58%), and 2 had a TBS of 4 (2.47%). The sTHS ranged from 0 to 272 (median 34, IQR 12–68). The results of the BALf leukocyte differential count in the enrolled horses and in the horses included in the Neu5 group (n = 28) are displayed in Table 1.

Table 1. Leukocyte differential counts of the bronchoalveolar lavage fluids of all horses and those included in the Neu5 Group. Data are expressed as mean ± standard deviation if normally distributed, or median (IQR) if not normally distributed.

Cell Population	All Horses	Neu5 Group
Macrophages	43.33% ± 10.66%	39.46% ± 10.59%
Lymphocytes	36.43% ± 13.71%	46% (38.25–52.75%)
Neutrophils	9% (5–17%)	3% (2.25–5%)
Eosinophils	1% (0–3%)	1.5% (0–6%)
Mast cells	4% (3–6%)	5% (4–7%)

Concerning the fitness parameters, the median VLa4 value was 8.6 m/s (IQR 7.6–9.3 m/s), the median HRLa4 was 208 bpm (IQR 199.6–214 bpm), and the median V200 was 8 m/s (IQR 7–8.5 m/s). The horses reached a mean peak lactate concentration of 20.5 ± 7.28 mmol/L and a median V max of 11 m/s (IQR 11-11). The values of pH min were measured in 48 horses, while those of Ht max in 31 horses; the average pH min was 7.152 ± 0.099 and the mean Ht max was 64.13 ± 4.13%.

3.2. Age, Sex, and Weight

Age was positively correlated with TBS (all horses: p = 0.0038, r = 0.32; Neu5: p = 0.0002, r = 0.65), sTHS (p = 0.0269, r = 0.25), and V max (all horses: p = 0.0013, r = 0.35; Neu5: p = 0.0017, r = 0.58); in contrast, it was inversely correlated with the BALf eosinophils count (p = 0.0104, r = −0.28). When considering only the Neu5 group, no association was found between age and sTHS.

Regarding the sex, geldings were older than mares (p = 0.0113), while no differences concerning age were observed when considering males vs. females. Females had significantly higher BALf macrophage counts compared to males (p = 0.0106) and lower BALf lymphocytes (p = 0.0285) and mast cells (p = 0.0308). The BALf eosinophils percentages were higher in stallions compared to geldings (p = 0.0191). Males showed significantly higher values of VLa4 (p = 0.0037), HRLa4 (p = 0.0013) and V200 (p = 0.047) compared to females. No association between sex and the EIPH parameters was observed.

Weight was positively correlated with V200 (p = 0.0126, r = 0.28; Neu5: p = 0.0098, r = 0.48); when considering only the Neu5 group, weight was also correlated with sTHS (p = 0.0472, r = 0.38) and VLa4 (p = 0.0024, r = 0.55).

3.3. EIPH-Related Parameters

The TBS was positively correlated with the sTHS (p = 0.0430, r = 0.23); moreover, a positive correlation was observed between the TBS and the Ht max (p = 0.0072, r = 0.47). When considering only the Neu5 group, a statistical correlation was found between the TBS and the V max (p = 0.0323, r = 0.41). Furthermore, horses showing to be EIPH-positive at the tracheobronchoscopy had slightly higher BALf mast cell counts compared to EIPH-negative horses (p = 0.0489) and reached higher values of Ht max (p = 0.0276) (Figure 1). The sTHS was positively correlated with the BALf mast cells count (p = 0.0392, r = 0.23; Neu5: p = 0.0024, r = 0.55) and, only in the Neu5 group, with the pH min value (p = 0.0488, r = 0.65).

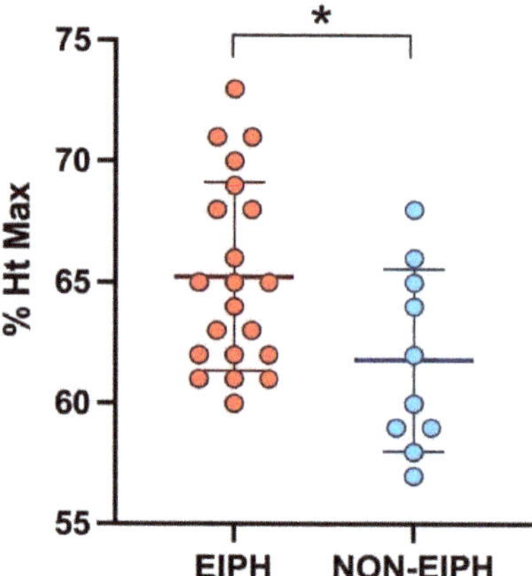

Figure 1. Scatter plot showing the mean and standard deviation of the maximum hematocrit reached during exercise in EIPH and non-EIPH horses. The statistical significance is shown as * ($p < 0.05$).

3.4. BALf Leukocyte Differential Cell Count

The percentage of BALf macrophages was inversely correlated with the values of VLa4 ($p = 0.0026$, r = −0.33) and HRLa4 ($p = 0.0006$, r = −0.37). The lymphocytes count was positively correlated with VLa4 ($p = 0.0017$, r = 0.34), HRLa4 ($p = 0.0242$, r = 0.25) and V200 ($p = 0.0342$, r = 0.24). The count of neutrophils was inversely correlated with VLa4 ($p = 0.0035$, r = −0.32) and V200 ($p = 0.0145$, r = −0.27), and positively correlated with peak lactate ($p = 0.0027$, r = 0.33). No associations were observed between the eosinophils and mast cells counts and any fitness parameters.

4. Discussion

Since the influence of EIPH on performance in racehorses has been widely discussed without reaching univocal results [5–16], the present study aimed to objectively define whether EIPH affected some fitness parameters in Standardbreds based on an incremental exercise test on a treadmill. Although our study found no association between EIPH and the main treadmill parameters, the causative role of EIPH in poor racing performance cannot be excluded, as it may be related to different mechanisms. Moreover, higher values of maximum hematocrit reached during exercise were observed in EIPH-affected horses, hinting that the hemoconcentration may contribute to the increase in pulmonary capillary pressure.

Among the population included in the current study, post-exercise tracheobron-choscopy revealed the presence of EIPH in 68% of the horses, which reflects the previously reported prevalence in racehorses [2]. However, the prevalence detected in the present work may be underestimated, as exercise on a treadmill does not accurately resemble what happens during a race. In our study, a significant correlation was observed between the tracheal blood score and the simplified total hemosiderin score, although a single tracheo-bronchoscopic evaluation was performed. This finding suggests that the EIPH episodes detected at endoscopy did not represent single and isolated events, but EIPH-positive horses were prone to develop repeated episodes of EIPH.

In the present study, both the tracheal blood score and total hemosiderin score were associated with increasing age: in fact, EIPH is a progressive condition [1], which has been associated with the number of racings starts [38] and, consequently, with age [11,13,14]. Moreover, a correlation between age and the maximum speed reached during the treadmill test was observed; it has been reported that trotters have their fastest racing times at about 6 years of age [39,40]. This could be explained by the fact that adult horses are more trained, fit, and physically mature compared to two-year-old horses; furthermore, it is reasonable that only the best-performing horses are selected to keep racing at older ages, while, among novice young racehorses, there may be a wider range of performance quality. Age was also inversely correlated with BALf eosinophilia, in accordance with previous studies [31,41,42]. Males had higher values of VLa4, HRLa4, and V200 compared to females, suggesting that sex had an influence on fitness; an association between sex and VLa4 has already been reported [43], while another study failed to detect any relationship, probably due to the small number of horses included [32]. Moreover, it has been reported that male racehorses are one second faster over one mile [36] and are 1.6 times more likely to win or place than females [44]. In our study, weight was also correlated with VLa4 and V200; it could be hypothesized that bigger horses have a longer stride, making less effort to maintain higher speeds by influencing the economy of locomotion. Moreover, weight was correlated with the simplified total hemosiderin score, suggesting that heavier horses are more prone to suffer from EIPH; to the authors' knowledge, an association between weight and EIPH has not been previously reported. The hypothesis that EIPH may result from locomotory impact-induced trauma [45,46] could explain this finding; in heavier horses, the impact pressure would be higher, and the resulting shear waves may damage the lung parenchyma more easily. However, the exact pathogenetic mechanisms of EIPH have not been identified yet, and this theory has not been widely accepted [47].

In the current study, all horses presented a BALf cytological profile typical of mild–moderate equine asthma (MEA) [37]; this could be due to the fact that the population included in the study consisted mainly of young racehorses in training, among which MEA can have a prevalence higher than 80% [48]. Since MEA, and, in particular, BALf neutrophilia, has been associated with poor performance [48–51], the lack of non-asthmatic horses may represent a limitation to this study. When more than one poor-performance-associated disease is observed in the same subject, the real cause of the lower athletic capacity can be debatable, as it may result from the combination of different factors. Also, in the present study, higher percentages of neutrophils were correlated with lower VLa4 and V200, and with higher peak lactate concentration, confirming the relationship between neutrophilic lung inflammation and lower athletic capacity observed in a previous study [32]. Therefore, among our study population, we selected a group of horses showing a neutrophils percentage of < 5% in order to evaluate the influence of EIPH alone on the fitness parameters, in particular, those associated with neutrophilia (VLa4, V200, and peak lactate). Nevertheless, no association between EIPH and V200, VLa4, HRLa4, and peak lactate was detected, even in the Neu5 group. Similarly, previous studies including a small number of horses failed to observe any relationship between the presence of EIPH and blood lactate values [52,53]; only one study identified higher peak lactate after exercise in EIPH-affected horses [54]. Moreover, no difference in the neutrophil percentage between the EIPH-positive and negative groups was detected, and no association between neutrophils count and sTHS was observed. This result suggests that EIPH does not directly influence the fitness capacity of racehorses, which depends on multiple variables. However, in our study, the maximal oxygen consumption (VO_{2max}), expressing the maximal aerobic capacity and reflecting athletic performance [17], was not measured; future studies may be conducted to investigate a possible association between EIPH and VO_{2max}. Finally, a possible role of EIPH in the racing performance cannot be ruled out; this should not be sought in decreased fitness, but it should be reasonably investigated in a different way. Interestingly, horses with higher percentages of mast cells had higher tracheal blood scores and simplified total hemosiderin scores, suggesting an association between BALf mastocytosis and EIPH; similar results had been reported by a previous study, where the hemosiderophage counts were higher in horses affected by mastocytic MEA compared to those with neutrophilic MEA [55].

In our study, the tracheal blood score, observed at endoscopy after treadmill exercise, was correlated with higher speeds reached during the treadmill test; it has been reported that the risk of EIPH is higher for sprinter horses racing 1000–1200 m compared with horses racing longer distances [37], and that epistaxis is more common after races < 1600 m than in longer races [56]. However, different studies detected no direct association between race speed and EIPH [5,37], while a faster average early/mid-race speed has been associated with EIPH scores of ≥ 3 [5]; this could be explained by the fact that rapid acceleration triggers higher pulmonary vascular pressures than a gradual incremental increase to the same speed [57]. Also, a study reported that barrel racing horses with the most severe grade of EIPH were faster than the EIPH-negative ones [58], confirming rapid acceleration as a risk factor for EIPH.

Moreover, EIPH-positive horses reached, in the present study, higher values of Ht max during exercise—the more the hematocrit raises, the more the oxygen-carrying capacity of the horse's blood during exercise increases; however, no association has been reported between post-exercise Ht and performance [26]. To explain the association between EIPH and Ht, it should be considered that higher values of Ht have been related to a higher mean arterial pressure during treadmill exercise in horses [59] and, in human medicine, red blood cell aggregation seems to participate in the increase of pulmonary capillary pressure [60]; therefore, a hypothetical explanation may be that hemoconcentration might play a role in the pathogenesis of pulmonary hemorrhage in horses. Moreover, it could be hypothesized that higher values of Ht max could provide a more efficient buffering capacity of the acid–base disturbance during exercise [61] in EIPH-positive horses; interestingly, in our study, the simplified total hemosiderin score was associated with higher values of pH min reached during the treadmill test, and, therefore, a lower grade of exercise-induced acidosis, supporting this hypothesis. However, as hematocrit variations during exercise are influenced by different factors, such as the splenic reserve and the intercompartmental fluid shifts, further studies are needed to investigate the relationship and the possible pathogenetic role of Ht in EIPH.

5. Conclusions

The present study showed no influence of EIPH on treadmill parameters, such as VLA4, HRLa4, and V200, suggesting that EIPH does not impair fitness in Standardbred racehorses; however, this finding does not rule out the causative role of EIPH in decreased racing performance, which should reasonably be further investigated. Interestingly, horses affected by EIPH reached higher values of hematocrit, suggesting a possible role of hemoconcentration in the increase of pulmonary capillary pressure.

Author Contributions: Conceptualization, C.M.L.F., L.S. and F.F.; methodology, C.M.L.F. and L.S.; formal analysis, C.M.L.F.; investigation, C.M.L.F., L.S., E.A., G.S., B.C., E.Z. and F.F.; writing—original draft preparation, C.M.L.F., L.S. and F.F.; writing—review and editing, C.M.L.F., L.S., E.A., B.C., G.S., E.Z. and F.F.; visualization, C.M.L.F.; supervision, L.S. and F.F. All authors have read and agreed to the published version of the manuscript.

Funding: This research received no external funding.

Institutional Review Board Statement: For the present study, no ethical approval was required, as all the procedures were performed on clinical horses for diagnostic purposes and included informed owner consent for the use of clinical data.

Informed Consent Statement: Not applicable.

Data Availability Statement: The data presented in this study are available on request from the corresponding author.

Acknowledgments: The authors wish to acknowledge the University of Milan for the support through the APC initiative.

Conflicts of Interest: The authors declare no conflict of interest.

References

1. Hinchcliff, K.W.; Couetil, L.L.; Knight, P.K.; Morley, P.S.; Robinson, N.E.; Sweeney, C.R.; Van Erck, E. Exercise Induced Pulmonary Hemorrhage in horses: American College of Veterinary Internal Medicine Consensus Statement. *J. Vet. Intern. Med.* **2015**, *29*, 743–758. [CrossRef] [PubMed]
2. McGilvray, T.A.; Cardwell, J.M. Training related risk factors for exercise induced pulmonary haemorrhage in British National Hunt racehorses. *Equine Vet. J.* **2021**, *54*, 283–289. [CrossRef] [PubMed]
3. Hinchcliff, K.W.; Jackson, M.A.; Brown, J.A.; Dredge, A.F.; O'Callaghan, P.A.; McCaffrey, J.P.; Morley, P.S.; Slocombe, R.E.; Clarke, A.F. Tracheobronchoscopic assessment of exercise-induced pulmonary hemorrhage in horses. *Am. J. Vet. Res.* **2005**, *66*, 596–598. [CrossRef] [PubMed]
4. Hoffman, A.M. Bronchoalveolar Lavage: Sampling Technique and Guidelines for Cytologic Preparation and Interpretation. *Vet. Clin. North Am. Equine Pract.* **2008**, *24*, 423–435. [CrossRef] [PubMed]
5. Crispe, E.J.; Lester, G.D.; Secombe, C.J.; Perera, D.I. The association between exercise-induced pulmonary haemorrhage and race-day performance in Thoroughbred racehorses. *Equine Vet. J.* **2017**, *49*, 584–589. [CrossRef] [PubMed]
6. Crispe, E.J.; Secombe, C.J.; Perera, D.I.; Manderson, A.A.; Turlach, B.A.; Lester, G.D. Exercise-induced pulmonary haemorrhage in Thoroughbred racehorses: A longitudinal study. *Equine Vet. J.* **2019**, *51*, 45–51. [CrossRef]
7. Hinchcliff, K.W.; Jackson, M.A.; Morley, P.S.; Brown, J.A.; Dredge, A.F.; O'Callaghan, P.A.; McCaffrey, J.P.; Slocombe, R.E.; Clarke, A.E. Association between exercise-induced pulmonary hemorrhage and performance in Thoroughbred racehorses. *J. Am. Vet. Med. Assoc.* **2005**, *227*, 768–774. [CrossRef]
8. Morley, P.S.; Bromberek, J.L.; Saulez, M.N.; Hinchcliff, K.W.; Guthrie, A.J. Exercise-induced pulmonary haemorrhage impairs racing performance in Thoroughbred racehorses. *Equine Vet. J.* **2015**, *47*, 358–365. [CrossRef]
9. Sullivan, S.L.; Anderson, G.A.; Morley, P.S.; Hinchcliff, K.W. Prospective study of the association between exercise-induced pulmonary haemorrhage and long-term performance in Thoroughbred racehorses. *Equine Vet. J.* **2015**, *47*, 350–357. [CrossRef]
10. De Mello Costa, M.F.; Thomassian, A.; Gomes, T.S. Study of exercise induced pulmonary haemorrhage (EIPH) in flat racing Thoroughbred horses through 1889 respiratory endoscopies after races. *Rev. Bras. Cienc. Vet.* **2005**, *12*, 89–91.
11. Costa, M.F.M.; Thomassian, A. Evaluation of race distance, track surface and season of the year on exercise-induced pulmonary haemorrhage in flat racing Thoroughbreds in Brazil. *Equine Vet. J. Suppl.* **2006**, *36*, 487–489. [CrossRef] [PubMed]
12. Birks, E.K.; Shuler, K.M.; Soma, L.R.; Martin, B.B.; Marconato, L.; Del Piero, F.; Teleis, D.C.; Schar, D.; Hessinger, A.E.; Uboh, C.E. EIPH: Postrace endoscopic evaluation of Standardbreds and Thoroughbreds. *Equine Vet. J. Suppl.* **2002**, *34*, 375–378. [CrossRef] [PubMed]
13. Pascoe, J.R.; Ferraro, G.L.; Cannon, J.H.; Arthur, R.M.; Wheat, J.D. Exercise-induced pulmonary hemorrhage in racing thoroughbreds: A preliminary study. *Am. J. Vet. Res.* **1981**, *42*, 703–707. [PubMed]
14. Raphel, C.F.; Soma, L.R. Exercise-induced pulmonary hemorrhage in Thoroughbred after racing and breezing. *Am. J. Vet. Res.* **1982**, *43*, 1123–1127.
15. Lapointe, J.M.; Vrins, A.; McCarvill, E. A survey of exercise-induced pulmonary haemorrhage in Quebec standardbred racehorses. *Equine Vet. J.* **1994**, *26*, 482–485. [CrossRef]
16. MacNamara, B.; Bauer, S.; Iafe, J. Endoscopic evaluation of exercise-induced pulmonary hemorrhage and chronic obstructive pulmonary disease in association with poor performance in racing Standardbreds. *J. Am. Vet. Med. Assoc.* **1990**, *196*, 443–445.
17. Franklin, S.; Allen, K. Laboratory exercise testing. In *Equine Sports Medicine and Surgery*, 2nd ed.; Hinchcliff, K.W., Kaneps, A.J., Geor, R.J., Eds.; Saunders Elsevier: Philadelphia, PA, USA, 2014; pp. 11–24.
18. Persson, S.G.B. Evaluation of exercise tolerance and fitness in the performance horse. In *Equine Exercise Physiology*; Snow, D.H., Persson, S.G.B., Rose, R.J., Eds.; Granta Editions: Cambridge, UK, 1983; Volume 1, pp. 441–457.
19. Leleu, C.; Cotrel, C.; Courouce-Malblanc, A. Relationships between physiological variables and race performance in French standardbred trotters. *Vet. Rec.* **2005**, *156*, 339–342. [CrossRef]
20. Lindner, A.E. Relationships between racing times of standardbreds and v4 and v200. *J. Anim. Sci.* **2010**, *88*, 950–954. [CrossRef]
21. Couroucé, A.; Chatard, J.C.; Auvinet, B. Estimation of performance potential of Standardbred trotters from blood lactate concentrations measured in field conditions. *Equine Vet. J.* **1997**, *29*, 365–369. [CrossRef]
22. McGowan, C. Clinical pathology in the racing horse: The role of clinical pathology in assessing fitness and performance in the racehorse. *Vet. Clin. North Am. Equine Pract.* **2008**, *24*, 405–421. [CrossRef]
23. Saibene, F.; Cortili, G.; Gavazzi, P.; Sala, A.; Faina, M.; Sardella, F. Maximal anaerobic (lactic) capacity and power of the horse. *Equine Vet. J.* **1985**, *17*, 130–132. [CrossRef] [PubMed]
24. Bayly, W.M.; Grant, B.D.; Pearson, R.C. Lactate concentrations in Thoroughbred horses following maximal exercise under field conditions. In *Equine Exercise Physiology*; Gillespie, J.R., Robinson, N.E., Eds.; ICEEP publications: Davis, CA, USA, 1987; Volume 2, pp. 426–437.
25. Rasanen, L.A.; Lampinen, K.J.; Poso, A.R. Responses of blood and plasma lactate and plasma purine concentrations to maximal exercise and their relation to performance in Standardbred trotters. *Am. J. Vet. Res.* **1995**, *56*, 1651–1656. [PubMed]
26. Evans, D.L.; Harris, R.C.; Snow, D.H. Correlation of racing performance with blood lactate and heart rate after exercise in Thoroughbred horses. *Equine Vet. J.* **1993**, *25*, 441–445. [CrossRef]
27. Evans, D.L.; Young, L.E. Cardiac responses to exercise and training. In *Cardiology of the Horse*, 2nd ed.; Marr, C.M., Bowen, I.M., Eds.; Saunders Elsevier: Edinburgh, UK, 2010.

28. Reef, V.B. Advances in echocardiography. *Vet. Clin. North. Am. Equine Pract.* **1991**, *7*, 435–450. [CrossRef]
29. Reef, V.B.; Bonagura, J.; Buhl, R.; McGurrin, M.K.J.; Schwarzwald, C.C.; van Loon, G.; Young, L. Recommendations for Management of Equine Athletes with Cardiovascular Abnormalities. *J. Vet. Intern. Med.* **2014**, *28*, 749–761. [CrossRef] [PubMed]
30. Alberti, E.; Stucchi, L.; Lo Feudo, C.M.; Stancari, G.; Conturba, B.; Ferrucci, F.; Zucca, E. Evaluation of cardiac arrhythmias before, during, and after treadmill exercise testing in poorly performing standardbred racehorses. *Animals* **2021**, *11*, 2413. [CrossRef]
31. Stucchi, L.; Valli, C.; Stancari, G.; Zucca, E.; Ferrucci, F. Creatine-kinase reference intervals at rest and after maximal exercise in Standardbred racehorses. *Comp. Exerc. Physiol.* **2019**, *15*, 319–325. [CrossRef]
32. Stucchi, L.; Alberti, E.; Stancari, G.; Conturba, B.; Zucca, E.; Ferrucci, F. The relationship between lung inflammation and aerobic threshold in standardbred racehorses with mild-moderate equine asthma. *Animals* **2020**, *10*, 1278. [CrossRef]
33. Newell, J.; Higgins, D.; Madden, N.; Cruickshank, J.; Einbeck, J.; McMillan, K.; McDonald, R. Software for calculating blood lactate endurance markers. *J. Sports Sci.* **2007**, *25*, 1403–1409. [CrossRef]
34. Lo Feudo, C.M.; Stucchi, L.; Alberti, E.; Stancari, G.; Conturba, B.; Zucca, E.; Ferrucci, F. The role of thoracic ultrasonography and airway endoscopy in the diagnosis of equine asthma and exercise-induced pulmonary hemorrhage. *Vet. Sci.* **2021**, *8*, 276. [CrossRef]
35. Golde, D.W.; Drew, W.L.; Klein, H.Z.; Finley, T.N.; Cline, M.J. Occult pulmonary haemorrhage in leukaemia. *Br. Med. J.* **1975**, *2*, 166–168. [CrossRef] [PubMed]
36. Doucet, Y.; Viel, L. Alveolar Macrophage Graded Hemosiderin Score from Bronchoalveolar Lavage in Horses with Exercise-Induced Pulmonary Hemorrhage and Controls. *J. Vet. Intern. Med.* **2002**, *16*, 281–286. [CrossRef] [PubMed]
37. Couetil, L.L. Inflammatory diseases of the lower airway of athletic horses. In *Equine Sports Medicine and Surgery*, 2nd ed.; Hinchcliff, K.W., Kaneps, A.J., Geor, R.J., Eds.; Saunders Elsevier: Philadelphia, PA, USA, 2014; pp. 605–632.
38. Hinchcliff, K.W.; Morley, P.S.; Jackson, M.A.; Brown, J.A.; Dredge, A.F.; O'Callaghan, P.A.; McCaffrey, J.P.; Slocombe, R.F.; Clarke, A.F. Risk factors for exercise-induced pulmonary haemorrhage in Thoroughbred racehorses. *Equine Vet. J.* **2010**, *42*, 228–234. [CrossRef] [PubMed]
39. Physick-Sheard, P.W. Career profile of the Canadian Standardbred. II. Influence of age, gait and sex upon number of races, money won and race times. *Can. J. Vet. Res.* **1986**, *50*, 457–470. [PubMed]
40. Thiruvenkadan, A.K.; Kandasamy, N.; Panneerselvam, S. Inheritance of racing performance of trotter horses: An overview. *Livest. Sci.* **2009**, *124*, 163. [CrossRef]
41. Ivester, K.M.; Couetil, L.L.; Moore, G.E.; Zimmerman, N.J.; Raskin, R.E. Environmental exposures and airway inflammation in young thoroughbred horses. *J. Vet. Intern. Med.* **2014**, *28*, 918–924. [CrossRef]
42. Couetil, L.L.; Cardwell, J.M.; Gerber, V.; Lavoie, J.P.; Leguillette, R.; Richard, E.A. Inflammatory Airway Disease of Horses—Revised Consensus Statement. *J. Vet. Intern. Med.* **2016**, *30*, 503–515. [CrossRef]
43. Persson, S.G.B. Heart Rate and Blood Lactate Responses to Submaximal Treadmill Exercise in the Normally Performing Standard-bred Trotter—Age and Sex Variations and Predictability from the Total Red Blood Cell Volume. *Zentralb. Veterinarmed. A* **1997**, *44*, 125–132. [CrossRef]
44. Tanner, J.C.; Rogers, C.W.; Firth, E.C. The relationship of training milestones with racing success in a population of Standardbred horses in New Zealand. *New Zealand Vet. J.* **2012**, *59*, 323–327. [CrossRef]
45. Schroter, R.C.; Marlin, D.J.; Denny, E. Exercise-induced pulmonary haemorrhage (EIPH) in horses results from locomotory impact induced trauma—A novel unifying concept. *Equine Vet. J.* **1998**, *30*, 186–192. [CrossRef]
46. Newton, J.R.; Rogers, K.; Marlin, D.J.; Wood, J.L.N.; Williams, R.B. Risk factors for epistaxis on British racecourses: Evidence for locomotory impact-induced trauma contributing to the aetiology of exercise-induced pulmonary haemorrhage. *Equine Vet. J.* **2005**, *37*, 402–411. [CrossRef] [PubMed]
47. Hinchcliff, K.W. Exercise-induced pulmonary hemorrhage (EIPH). In *Equine Sports Medicine and Surgery*, 2nd ed.; Hinchcliff, K.W., Kaneps, A.J., Geor, R.J., Eds.; Saunders Elsevier: Philadelphia, PA, USA, 2014; pp. 633–647.
48. Nolen-Walston, R.D.; Harris, M.; Agnew, M.E.; Martin, B.B.; Reef, V.M.; Boston, R.C.; Davidson, E.J. Clinical and diagnostic features of inflammatory airway disease subtypes in horses examined because of poor performance: 98 cases (2004–2010). *J. Am. Vet. Med. Assoc.* **2013**, *242*, 1134–1138. [CrossRef] [PubMed]
49. Richard, E.A.; Fortier, G.D.; Lekeux, P.M.; van Erck, E. Laboratory findings in respiratory fluids of the poorly-performing horse. *Vet. J.* **2010**, *185*, 115–122. [CrossRef] [PubMed]
50. Holcombe, S.J.; Robinson, N.E.; Derksen, F.J.; Bertold, B.; Genovese, R.; Miller, R.; de Feiter Rupp, H.; Carr, E.A.; Eberhart, S.W.; Boruta, D.; et al. Effect of tracheal mucus and tracheal cytology on racing performance in Thoroughbred racehorses. *Equine Vet. J.* **2006**, *38*, 300–304. [CrossRef]
51. Lavoie, J.P.; Cesarini, C.; Lavoie-Lamoureux, A.; Moran, K.; Lutz, S.; Picandet, V.; Jean, D.; Marcoux, M. Bronchoalveolar lavage fluid cytology and cytokine messenger ribonucleic Acid expression of racehorses with exercise intolerance and lower airway inflammation. *J. Vet. Intern. Med.* **2011**, *25*, 322–329. [CrossRef]
52. Sanchez, A.; Couetil, L.L.; Ward, M.P.; Clark, S.P. Effect of Airway Disease on Blood Gas Exchange in Racehorses. *J. Vet. Intern. Med.* **2005**, *19*, 87–92. [CrossRef]
53. Couroucè-Malblanc, A.; Pronost, S.; Fortier, G.; Corde, R.; Rossignol, F. Physiological measurements and upper and lower respiratory tract evaluation in French Standardbred Trotters during a standardised exercise test on the treadmill. *Equine Vet. J. Suppl.* **2002**, *34*, 402–407. [CrossRef]

54. Couetil, L.L.; Denicola, D.B. Blood gas, plasma lactate and bronchoalveolar lavage cytology analyses in racehorses with respiratory disease. *Equine Vet. J. Suppl.* **1999**, *30*, 77–82. [CrossRef]

55. Boshuizen, B.; De Bruijin, M.; Dewulf, J.; Delasalle, C. Sport Horses with Inflammatory Airway Disease (IAD) with Predominantly Eosinophils or Mast Cells Are More Predisposed to Poor Performance, Exercise Induced Pulmonary Haemorrhage (EIPH) and Have Shorter Careers When Compared to Horses with Neutrophilic IAD. *Equine Vet. J.* **2014**, *46*, 7. [CrossRef]

56. Takahashi, T.; Hiraga, A.; Ohmura, H. Frequency of and risk factors for epistaxis associated with exercise-induced pulmonary hemorrhage in horses: 251,609 race starts (1992–1997). *J. Am. Vet. Med. Ass.* **2001**, *218*, 1462–1464. [CrossRef]

57. Manohar, M. Pulmonary vascular pressures of strenuously exercising thoroughbreds after administration of flunixin meglumine and furosemide. *Am. J. Vet. Res.* **1994**, *55*, 1308–1312. [PubMed]

58. Leguillette, R.; Steinmann, M.; Bond, S.L.; Stanton, B. Tracheobronchoscopic assessment of exercise-induced pulmonary hemorrhage and airway inflammation in barrel racing horses. *J. Vet. Intern. Med.* **2016**, *30*, 1327–1332. [CrossRef] [PubMed]

59. McKeever, K.H.; Hinchcliff, K.W.; Reed, S.M.; Hamlin, R.L. Splenectomy alters blood pressure response to incremental treadmill exercise in horses. *Am. J. Physiol.* **1993**, *265*, R409–R413. [CrossRef]

60. Raberin, A.; Nader, E.; Ayerbe, J.L.; Mucchi, P.; Connes, P.; Durand, F. Evolution of blood rheology and its relationship to pulmonary hemodynamic during the first days of exposure to moderate altitude. *Clin. Hemorheol. Microcirc.* **2020**, *74*, 201–208. [CrossRef] [PubMed]

61. Lindinger, M.I.; Heigenhauser, G.J.F. Effects of gas exchange on acid-base balance. *Compar. Physiol.* **2012**, *2*, 2203–2254. [CrossRef]

Review

Advances in the Diagnosis of Equine Respiratory Diseases: A Review of Novel Imaging and Functional Techniques

Natalia Kozłowska, Małgorzata Wierzbicka *, Tomasz Jasiński and Małgorzata Domino *

Department of Large Animal Diseases and Clinic, Institute of Veterinary Medicine, Warsaw University of Life Sciences, 02-787 Warsaw, Poland; natalia_kozlowska@sggw.edu.pl (N.K.); tomasz_jasinski@sggw.edu.pl (T.J.)

* Correspondence: malgorzata_wierzbicka@sggw.edu.pl (M.W.); malgorzata_domino@sggw.edu.pl (M.D.); Tel.: +48-22-59-36-191 (M.W. & M.D.)

Citation: Kozłowska, N.; Wierzbicka, M.; Jasiński, T.; Domino, M. Advances in the Diagnosis of Equine Respiratory Diseases: A Review of Novel Imaging and Functional Techniques. *Animals* **2022**, *12*, 381. https://doi.org/10.3390/ani12030381

Academic Editors: Francesco Ferrucci, Chiara Maria Lo Feudo and Luca Stucchi

Received: 3 January 2022
Accepted: 3 February 2022
Published: 4 February 2022

Publisher's Note: MDPI stays neutral with regard to jurisdictional claims in published maps and institutional affiliations.

Simple Summary: Respiratory problems are common in horses and are often diagnosed as a cause of poor athletic performance. The basic diagnostic techniques of the equine respiratory tract examination are not always sufficient for a complete diagnosis of the disease, its exacerbation, remission, or response to treatment. Therefore, advances have been introduced in the diagnosis of equine respiratory diseases. Among them, we can distinguish the high-resolution imaging modalities like computed tomography (CT) and magnetic resonance (MR) imaging. These techniques have revolutionized the capability of visualizing detailed anatomy of the upper respiratory tract, offering the practitioners an advanced view of airway pathology and allowing for appropriate management planning. On the other hand, the pulmonary function tests (PFTs), which provide sensitive assessment of small functional changes in the lungs, are able to comprehensively characterize the mechanics of the respiratory system. Spirometry and impulse oscillation system (IOS) analyze intra-breath respiratory mechanics, while electrical impedance tomography (EIT) measures changes in lung conductivity. These methods may be successfully applied to detect airway obstruction and mechanical inhomogeneity in breathing patterns. Presented advanced diagnostic techniques comply with owners' and trainers' requirements for accurate and early diagnosis of respiratory tract disorders. This paper reviews advantages, disadvantages, and clinical applications of the advanced diagnostic techniques of the equine respiratory tract.

Abstract: The horse, as a flight animal with a survival strategy involving rapid escape from predators, is a natural-born athlete with enormous functional plasticity of the respiratory system. Any respiratory dysfunction can cause a decline in ventilation and gas exchange. Therefore, respiratory diseases often lead to exercise intolerance and poor performance. This is one of the most frequent problems encountered by equine internists. Routine techniques used to evaluate respiratory tract diseases include clinical examination, endoscopic examination, radiographic and ultrasonographic imaging, cytological evaluation, and bacterial culture of respiratory secretions. New diagnostic challenges and the growing development of equine medicine has led to the implementation of advanced diagnostic techniques successfully used in human medicine. Among them, the use of computed tomography (CT) and magnetic resonance (MR) imaging significantly broadened the possibilities of anatomical imaging, especially in the diagnosis of upper respiratory tract diseases. Moreover, the implementation of spirometry, electrical impedance tomography (EIT), and impulse oscillation system (IOS) sheds new light on functional diagnostics of respiratory tract diseases, especially those affecting the lower part. Therefore, this review aimed to familiarize the clinicians with the advantages and disadvantages of the advanced diagnostic techniques of the equine respiratory tract and introduce their recent clinical applications in equine medicine.

Keywords: respiratory system; imaging; spirometry; electrical impedance tomography; impulse oscillation system; horse; poor performance

1. Introduction

Taking a detailed history and performing a good clinical examination is critical to the diagnostic process of equine respiratory diseases. They often provide crucial data which narrows down the diagnostic workup plan and supports the choice of the most appropriate and informative additional examinations. In the case of the equine respiratory tract, observation of the horse from a distance is important to assess the breath rate or any symptoms of respiratory distress. Attention should also be paid to environmental conditions like bedding, ventilation, or access to the pasture. Besides the basic clinical examination, a detailed assessment of the respiratory tract needs to be performed. The respiratory examination is routinely performed in a sequence of inspection, palpation, auscultation, and—when needed—percussion [1,2]. When the basic diagnosis is established, the basic diagnostic techniques of additional examination are routinely used as the initial imaging modality in the evaluation of most common respiratory diseases. Among the basic diagnostic techniques, radiography, ultrasonography, and endoscopy are most utilized in the field of equine practice [1,3–7]. They are often adequate for diagnosis and monitoring of equine respiratory disorders, although advanced imaging techniques are still often required for more detailed assessment of anatomical structures and functional evaluation (Figure 1).

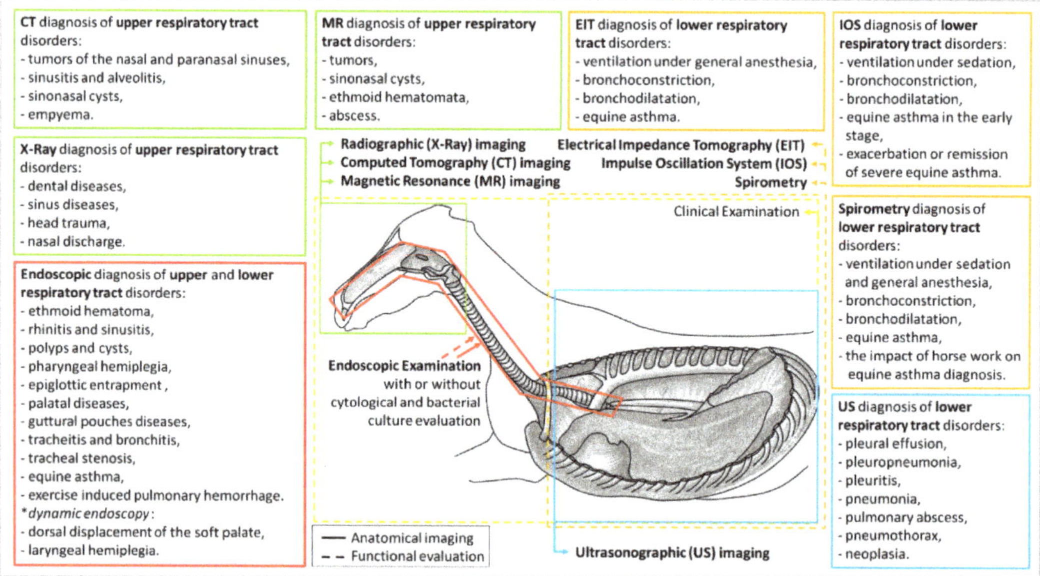

Figure 1. Application of the basic and advanced diagnostic techniques for the anatomical imaging and/or functional evaluation of the areas of the equine respiratory tract. CT—computed tomography; MR—magnetic resonance imaging; EIT—electrical impedance tomography; IOS—impulse oscillation system; X-ray—radiographic imaging; US—ultrasonographic imaging.

Radiography (X-ray) plays an important role in diagnosing upper respiratory tract diseases affecting the fascial, nasal, and paranasal sinus regions of the horse skull [4]. X-ray is a widely available, portable, inexpensive, and well-tolerated diagnostic technique that can be rapidly and easily obtained in both the field and hospital practices. X-ray of the horse's head is commonly used for the diagnosis of dental or sinus diseases, head trauma, or nasal discharge [5]. They allow visualization of sinuses, guttural pouches, and the upper trachea, although obtaining diagnostic quality images and interpreting these findings can be an intimidating task [4]. X-ray has inherent limitations when evaluating complex regions such as the skull or thorax due to the complexity and overlap of the anatomic structures. X-ray is of limited value in the evaluation of soft tissue due to reduced tissue resolution. Disease processes are detected radiographically by identifying disrupted or altered contours, size or shape changes, or abnormal radiopacity, thus an adequate coverage of the area of interest is necessary for proper evaluation. These alterations can be difficult to see when there is complex anatomy [8]. To detect changes in radiopacity, mineral loss needs to be advanced at 30% to 50% to detect osteolysis or bone resorption radiographically. There are few pathognomonic findings for many of the disease processes, which can be easily established when carefully assessing the radiographic findings in conjunction with the clinical presentation.

Ultrasonography (US) is another diagnostic tool that allows inexpensive, accessible, real-time, and radiation-free imaging. This technique is implemented in the diagnosis of lower respiratory tract disorders like pleural effusion, pleuropneumonia, pleuritis, pneumonia, pulmonary abscess, pneumothorax, or neoplasia [3]. In general, US provides an assessment of pathological lesions extending mainly to the peripheral lung surface. An accurate characterization of the amount, location, and characteristics of pleural fluid or pleural thickening is easily obtained. Parenchymal lesions can be localized with high accuracy but often appear very similar and cannot be differentiated via an ultrasound examination [9]. During lung inflammation, the most common occurring artifact is comet tail, which can be detected in 91% of horses with equine asthma or exercise-induced pulmonary hemorrhage [10]. However, the comet-tail artifacts are not specific as they usually suggest the presence of a small amount of fluid, without allowing to distinguish its nature (inflammatory fluid, edema, blood, etc.) [10]. The US complements X-ray as it is easier to use, is more sensitive in the detection of smaller amounts of fluid, and provides information about the fluid character, however, it is limited to the lung surface and has a low specificity [11]. For further evaluation of the pulmonary pathology, an ultrasound-guided biopsy can be performed. Additional uses of US include doppler imaging that offers the characterization of the vascularity of masses of the thoracic wall, neck, or pleural space [3].

Rhinoscopy is of great use in the case of an ethmoid hematoma, rhinitis due to foreign body, polyps, and masses causing restricted airflow [12]. It also may be useful in the case of sinusitis when there is a discharge from the nasomaxillary opening. Endoscopy of the pharynx may lead to the diagnosis of pharyngeal hemiplegia, cysts, epiglottic entrapment, or palatal disorders [2]. Visualization of guttural pouches benefits the definition of problems like mycosis, empyema, or tympany [13]. The trachea is examined to ascertain the discharge presence and appearance or to diagnose anatomic defects like tracheal stenosis [14]. Examination of large bronchi provides information about discharge presence and occurring inflammatory process in the lower airway. Under endoscope guidance or separately, broncho-alveolar lavage fluid (BALF) is collected to retrieve fluid and cells lining the distal airways and alveoli. Microscopic evaluation of BALF detects histological abnormalities in horses with pulmonary disease and is commonly used to stage equine asthma (EA) based on a percentage of neutrophils [15]. In clinically healthy young athletic horses, the distribution of nucleated cells in the BALF varies around 60% of macrophages, 34% of lymphocytes, less than 5% of neutrophils, and less than 2% of mast cells or eosinophils [16]. Horses with mild-to-moderate EA usually reveal a mild-to-moderate increase in the percentage of neutrophils, while severe EA is characterized by severe neutrophilia (>20%) [15]. Although the diagnosis of severe cases of EA is relatively easy, it is difficult to diagnose cases in remission or horses with a mild form of the disease [15].

Dynamic endoscopy, in some cases, may be the only method for diagnosis of dorsal displacement of the soft palate, especially when the symptoms occur mainly during exercise. It then also provides an assessment of laryngeal hemiplegia during horse movement [7]. Both stationary and dynamic endoscopies are available, inexpensive, and easy to perform under field conditions.

New diagnostic challenges and the growing development of equine medicine has led to the implementation of advanced diagnostic techniques successfully used in human medicine. Among them, the use of computed tomography (CT) and magnetic resonance (MR) imaging significantly broadened the possibilities of anatomical imaging, especially in the diagnosis of upper respiratory tract diseases. Moreover, the implementation of spirometry, electrical impedance tomography (EIT), and impulse oscillation system (IOS) sheds new light on functional diagnostics of respiratory tract diseases, especially the lower part. Therefore, this review aimed to familiarize the clinicians with the advantages and disadvantages of the advanced diagnostic techniques of the equine respiratory tract and introduce their recent clinical applications in equine medicine.

2. Advances in the Anatomical Imaging

2.1. Computed Tomography (CT) Imaging of Upper Respiratory Tract

Computed tomography (CT) is an advanced technique established for the detailed imaging of the anatomical structures of the horse's head and limbs, which nowadays is becoming more available in equine clinics. The CT refers to a computerized X-ray imaging procedure where the patient is radiographed slice by slice, using a rotating, highly collimated, X-ray beam that generates cross-sectional images throughout the area of interest. The CT image is represented by a grayscale map of the tissue's ability to attenuate X-ray radiation [17]. The CT is useful in imaging primarily bones and structures containing air due to the highest and lowest ability to X-ray radiation attenuation, respectively, however, the soft tissue evaluation by contrasting imaging is also available. The CT provides high quality and resolution of the image and the possibility of three-dimension rotation of each anatomical structure [18]. In the equine respiratory tract, the applicability of CT is limited to imaging of the horse's head due to the limited diameter of the gantry, in which the examined area must be located during imaging [19]. Smaller horses and foals may fit within the gantry, allowing CT examination of the thoracic cavity and lower airways [20].

2.1.1. Advantages

The CT provides diagnostically important data in situations where X-ray or US has been unrewarding. However, because of costs and the restrictions presented below, it is usually second in line after X-ray or US. The main advantage of CT is the ability to produce high-resolution, three-dimensional, detailed cross-section images [18]. The CT allows more accurate anatomic and morphologic characterization of the anatomic complexity without superimpositions of other anatomical structures within the horse's body. The CT produces images of the area of interest in various planes. It provides information about the involvement of bones and surrounding structures and may indirectly help the surgery by defining the precise surgical margins of removal [17]. As the images are objective and easy to display and share, using the services of a qualified radiologist from another part of the world is possible [21]. The CT operating systems also provide the possibility to manipulate obtained images to highlight particular structures using grey-level mapping, contrast stretching, histogram modification or contrast enhancement [19]. To evaluate lesions characterized by increased vascular permeability in the equine head, intra-arterial, and intravenous contrast enhancement can be applied [22].

2.1.2. Disadvantages

Several problems related to the CT technique that have limited the availability or usefulness in equine practice can be divided into relative and absolute groups. Within the relative group, the high cost of CT equipment and facilities necessary for this imaging is still the main limitation. The monthly maintenance costs of CT scanners are very high and equipment to move anesthetized horses is also expensive [23]. As the cost of CT scanners has come down over the last years, and together with the high demand of horse owners to improve diagnostics, CT imaging is presently not restricted to a limited number of veterinary institutions and large referral centers, and slowly CT examination availability in smaller equine practices has increased [24]. The second relative problem is a need to cooperate with a qualified radiologist. Although performing a CT scan itself is demanding but not very complicated, the viewing and assessment of a large set of highly detailed CT images requires a qualified staff, as for the untrained eye CT imaging may be incomprehensible [24]. Concerning the third relative problem of the imaging artifacts, it should be kept in mind that CT imaging is based on X-ray, ionizing, and radiation with the potential to cause biological effects in living tissues, therefore CT imaging should be done carefully to avoid artifacts and the need for repeated imaging. Within the artifacts which may limit the diagnostic value of CT by masking a lesion or by mimicking a pathologic condition, the partial volume effect and beam hardening should be included [25].

Within the much more important absolute group, the size of the CT scanner is a serious limitation in the use of CT imaging in equine practices. The CT scanners were designed for people, thus in equine practice the diameter of the gantry of the CT scanner, in which the examined area must be located during imaging, ranges from 50–85 cm and allows for insertion of the head, distal limbs, and upper neck of adult full-sized horses [19]. The entire body scan is available only in ponies or foals. Moreover, most available CT scanners require general anesthesia. The horse needs to lie on a table that slides into a gantry of CT scanners. Therefore, the necessity for general anesthesia of the horse during examination is another drawback of CT techniques. When general anesthesia is contraindicated, CT examination will also be contraindicated [18]. The option is the utilization of the modified multidetector systems to perform standing, sedated CT of the equine head. However, the CT examination in the standing position to avoid general anesthesia offers further value, but this technique is still limited to several institutions [18].

2.1.3. Clinical Applications

A wide range of diseases affecting the upper respiratory tract have been diagnosed successfully using CT imaging. The intravascular contrast enhancement can be used to differentiate normal soft tissue from lesions based on an alteration in vascular permeability and perfusion which significantly improves the diagnostic value of equine head CT imaging [22]. The CT examination has been proven to be valuable for surgical and treatment therapy planning, especially in progressing diseases where accurate diagnosis and surgical margins play an important role in future prognosis [26]. Precise diagnosis via CT has been emphasized in order to avoid unviable treatment approaches [27]. Malignant and benign tumors of the nasal and paranasal sinuses including squamous cell carcinoma, undifferentiated carcinoma, hemangiosarcoma, nasal adenocarcinoma, myxoma, chondroblastic osteosarcoma, anaplastic sarcoma, and fibro-osseous lesions have been described [28–31]. Henninger et al. provided detailed CT descriptions regarding the most common features of sinusitis and alveolitis [32]. Diagnosis and surgical treatment of sinonasal cysts have been described [33–35]. Solano et al. depicted CT images of empyema, even though endoscopy is the preferred method for the diagnosis process of guttural pouches [24]. Computed tomography has been reported as a tool that overcomes endoscopy and radiography in the evaluation of temporohyoid osteoarthropathy [36–38]. Both narrow [39] and larger clinical studies [32,40–42] using CT on the equine head have been performed, showing the advantages of this technique in diagnosis and therapy procedures, among which the selected clinical applications of equine CT imaging are summarized in Table 1.

Table 1. The selected diseases of the equine head and neck area and their main findings diagnosed based on computed tomography imaging.

Disease	Area [1]	Main Findings	Authors
Sinusitis	Paranasal sinuses	Thickening of the respiratory epithelium, teeth involvement. The inhomogeneous appearance of the thickened bone, sclerosis of the facial crest, deformed shape of the maxilla, irregularly defined periostitis, bone loss or perforation, soft tissue swelling of the face.	Henninger et al. (2003) [32] Tucker et al. (2001) [19]

Table 1. *Cont.*

Disease	Area [1]	Main Findings	Authors
Laryngeal dysplasia	Larynx	Thyroid cartilage abnormalities: lack of a cricothyroid articulation, a dorsal extension of the thyroid cartilage, absence of the caudal cornu of the thyroid cartilage, absence of the articular process of the cricoid cartilage, and hypoplasia or absence of the cricopharyngeus muscle.	Garrett et al. (2010) [43]
Cysts and cyst-like lesions	Larynx and cranial cervical trachea	Thickening, heterogenous signal intensity of thyroid cartilage laminae, the ventral and lateral aspects of the cricoid cartilage, and the ventral aspect of the first tracheal ring. The thyroid, arytenoid, and cricoid cartilages and the first tracheal ring, presence of focal areas of hyperintense signal consistent with fluid.	Garrett et al. (2010) [43]
Tumors	Paranasal sinuses	Thickening of the nasal mucosa. Narrowed nasal meati. Homogeneous hyperintense signal, consistent with fluid of interior of the nasal septum. Focal mineralization of the soft tissue mass. Fluid lines in one or more paranasal sinuses, dental apex flattening. Bulging and thinning of maxillary bone, partial destruction of the osseous orbit, infraorbital canal changes. Displacement and distortion of the osseous infraorbital or lacrimal canal.	Fenner et al. (2019) [33]
	Paranasal sinuses	Homogenous soft tissue/fluid filling the entire maxillary sinus. Expansion of right maxillary sinus with the erosion of the first molar.	Tucker et al. (2001) [19]
	Paranasal sinuses	Large, clearly demarcated mass within the left caudal maxillary and left conchofrontal sinuses. Lysis of the sphenoid and palatine bones of the medial left orbit and left infraorbital canal. Extension into the left retrobulbar space, with rostral and lateral displacement of the left globe.	Annear et al. (2008) [34]
	Paranasal sinuses	Squamosus cell carcinoma: irregularly surfaced heterogeneous soft tissue mass filling the maxillary sinus and ventral conchal sinus.	Kowalczyk et al. (2011) [39]
	Paranasal sinuses, nasal cavity, tongue, mandible	Squamosus cell carcinoma: soft tissue attenuation filling maxillary sinus, dorsal conchal sinus, ventral conchal sinus, while the conchofrontal and sphenopalatine sinus showed different amount of filling. Nodular masses involved a third of the ipsilateral rostral maxillary sinus and less than a third of the conchofrontal sinus. Involvement of stylohyoid bone. Small nodular soft tissue lesions along the nasal septum.	Strohmayer et al. (2020) [28]
	Neck	Dystrophic mineralized mass at the right side of the vertebral bodies of C3 and C4, associated with bone resorption that caused the thinning of the right transverse process and a widening of the angle between the transverse process and the arch of C3.	De Zani et al. (2011) [30]
	Nasal cavity, Paranasal sinuses	Hemangiosarcoma, nasal adenocarcinoma, myxoma, myxosarcoma, chondroblastic osteosarcoma, anaplastic sarcoma characterized by a homogeneous, poorly defined mass that was iso- or mildly hypoattenuating compared to masseter muscle.	Cissel et al. (2012) [31]
	Paranasal sinuses	Osseous fibroma: well-marginated mass in right nasal passage with destruction of caudal aspect of nasal septum and extension of the mass into the choanae. Rostrocaudal extent of the soft tissue density with loss of bone density in the vicinity of the cribriform plate	Cilliers et al. (2008) [29]
Temporohyoid osteoarthropathy	Temporohyoid articulation	Osseous proliferation of the stylohyoid bone and temoporohyoid articulation, thickening of ceratohyoid bone. Lytic osseous changes of the petrous temporal and stylohyoid bones.	Hilton et al. (2009) [36] Divers et al. (2006) [37] Bras et al. (2014) [38]

[1] Area of the respiratory tract; CT—computed tomography.

2.2. Magnetic Resonance (MR) Imaging of Upper Respiratory Tract

In horses, magnetic resonance (MR), similarly to CT, may support the diagnostic process by providing cross-sectional images of extremities and head. The MR imaging is based on the magnetic dipole nature of the abundant hydrogen protons within tissues and uses the magnetic field and radio waves to create the image of hydrated tissues [19]. The MR images excel in the evaluation of soft tissues, providing anatomic and physiologic information that exceeds tomographic images. The standard image can be acquired in desired orientation, however the commonly obtained planes are sagittal, transverse, and dorsal [44]. Routine MR imaging examinations of the head include three types of pulse-echo sequences: the T1-weighted (T1W) sequence, which is performed before and after contrast administration; proton density (PD); and T2-weighted (T2W) imaging protocols. However, additional imaging sequences may be used [44]. The T1W sequences are based on the longitudinal relaxation properties of tissue and are useful for anatomic detail. T1W images acquired immediately after intravenous contrast administration are compared with identical slices obtained before contrast to detect neovascularisation or dilatation of vessels. Contrast enhancement results in a hyperintense signal in tissues where the contrast has extravasated. PD images are based on the relative concentration of hydrogen protons in different tissues and have the best anatomic detail. T2W images are based on the relaxation interactions between protons [19]. In the equine respiratory tract, the applicability of the high-field closed MR is much more limited than in CT due to the both small gantry diameter and the need to place the center of the imaged object in the center of the gantry [19]. Therefore, the high-field MR imaging of the horse's head, but no further, is available under general anesthesia only for ponies, foals, or the rostral part of larger horses. The equine low-field MR scanners provide an examination in standing sedated horses; however, the scanner should include a specific system, which is not the same as the small system designed for the equine limb scanning [45,46].

2.2.1. Advantages

The MR provides completely different diagnostically important data than CT, X-ray, or US. The main advantage of MR is the type of acquired data, providing not only excellent anatomic features of soft tissue structures but also functional alterations elusive in other imaging techniques. The MR does not use ionizing radiation, thus no negative biological effects in living tissues have so far been demonstrated [44]. In the equine respiratory tract, MR imaging is particularly useful for identifying space-occupying lesions (tumors, sinonasal cysts, or ethmoid hematomata), where the high soft-tissue contrast of the images is needed to allow differentiation of tissue types and establish an accurate relationship with surrounding structures [42]. The second advantage of MR is the ability to produce high-resolution, three-dimensional, detailed cross-section images which can be accomplished in any plane, without a loss of resolution or quality regardless of the orientation of the horse's head in the magnetic field [19].

2.2.2. Disadvantages

The main disadvantage of the MR technique in equine practice is similar to CT imaging limitations, and also could be divided into relative and absolute groups. Within the relative group, as it was noted for CT that the high cost is still the main limitation, it has to be realized that the costs of MR equipment and facilities are extremely high. The MR magnet, especially in the high-field MR, requires a constant high voltage power supply and continuous cooling with the use of liquid helium, even when no imaging is being performed.

Therefore, the monthly maintenance costs of MR scanners are much higher than those of CT [8]. Moreover, in the case of working in a strong magnetic field, specialized non-metallic equipment for anesthesia, monitoring, and the procedure are needed with special precautions [8], therefore MR imaging is still restricted to a limited number of veterinary institutions and large referral centers. The second relative problem is a need to cooperate not only with a highly qualified radiologist, but also close cooperation with a biophysicist or specialized technical support from the MR service company [47]. In the case of MR, both performing an MR scan and viewing and assessing MR images requires narrow, high-quality MR specialist competencies. For the novice, interpretation of MR images can be more challenging than interpretation of CT scans because of the influence of the many patients- and machine-related factors on the gray-scale display. The principles of image contrast are unfamiliar and follow rules that are not as evident or straightforward [47].

Although three routine sequences are mainly used for MR imaging of horses' heads, standardization of protocols and specialist technique knowledge is required to achieve high-quality images [19].

As it was mentioned above, the size of the MR scanner is a serious limitation in the use of high-field MR imaging in equine practice, which should be mentioned as an important absolute disadvantage. The gantry diameter allows for the distal limb and rostral part of the head to be introduced in the center for image acquisition [8]. Moreover, the center of the imaged object has to be positioned in the center of the gantry [8], therefore the high-field MR is available only for ponies or foals. The other option is the utilization of low-field, semi-open MR scanners to perform standing, sedated MR of the equine head. However, the quality, resolution, and number of available imaging sequences are much poorer than in high-field MR imaging [47]. Concerning the second absolute problem, the target tissue, it should be kept in mind that MR imaging displays limited clinical use in the lungs and bones evaluation. MR imaging of the foal thorax may be limited due to the sparse soft tissue structures and low proton density for signal production [48]. Moreover, multiple interfaces between air and soft tissue generate susceptibility artifacts and fast signal decay, as in high-field images the field inhomogeneity susceptibility increases with the increase of air it contains. Therefore, the low proton density-dependent artifacts are prevalent, especially on the boundary of air-containing sinuses as well as bone–soft tissue interfaces. The motion of respiratory, cardiac, and vascular systems may also cause artifacts except when the respiratory/ECG gating is available [8]. Finally, in the case of selected horse head disorders, small lesions affecting flat bones of the skull occur which are not detectable on MR images. Thus, for bone imaging CT images are more detailed [19]. However, the decision between CT and MR depends not only on the target tissue, but also, and perhaps most of all, on the time that can be spent on imaging. The MR imaging takes much more time than performing a CT scan, which is not without significance in the case of unstable horses during general anesthesia.

2.2.3. Clinical Applications

Due to the above limitations, a variety of clinical MR applications in equine practice have been reported regarding limbs [49–51], while head MR imaging has a limited number of cases. Ferrell et al. presented the clinical application of diagnosis of neurologic diseases in twelve horses, wherein eight of them were successfully diagnosed using MR imaging [52]. Spoormakers et al. assessed the application of MR in equine brain abscesses imaging [53]. A larger study was performed by Manso-Diaz et al., where in 84 clinical cases MR imaging showed the exact location, size of the lesions, and relation to surrounding structures due to neurological, sinonasal, and soft tissue disorders. The disorders of the respiratory tract diagnosed there by MR imaging included sinusitis, dental issues, nasal tumors, nasal septum deviation, and ethmoid hematoma [42]. In the report by Tessier et al., the MR imaging technique was used to diagnose sinusitis, paranasal sinus cysts, ethmoid hematoma, and neoplasia [54]. Garrett et al. reported MR as an advanced imaging method of larynx and pharynx, presenting MR features of laryngeal dysplasia and laryngeal cyst-like malformation [43,55,56]. Potentially, the MR might be a superior imaging technique in the diagnosis of foal lung diseases, following the experiences from human medicine. In human medicine, there are reports where magnetic resonance has been compared with high resolution computed tomography (HRCT) in pneumonia diagnosing. In three reports, the MR imaging had 91% [57], 94% [58], and 95% [59] accuracy, respectively, in the diagnosis of pneumonia compared to HRCT with 100% accuracy. The selected current clinical applications of equine MR imaging of the respiratory tract are summarized in Table 2.

Table 2. The selected diseases of the equine head and neck area and their main findings diagnosed based on magnetic resonance imaging.

Disease	Area [1]	Main Findings	Authors
Cyst	Paranasal sinus	Homogeneous On T1W sequences, the cyst was hypointense compared to temporal muscles, no contrast enhancement within the cystic fluid. rim enhancement On T2W sequences, the contents were hyperintense to surrounding muscle, and a wall could be observed consistently surrounding the lesion. This rim was observed consistently around the lesion and could be differentiated from the adjacent mucosa.	Tessier et al. (2013) [54]
Abscess	Ventral Conchal sinus	Well-defined capsule with heterogeneous signal intensity in T2W images, deviation of the dorsal conchal sinus wall, and infraorbital canal.	Manso Diaz et al. (2015) [42]
Tumors	Nasal septum	Chondrosarcoma: heterogeneous intensities on all sequences and no defined borders of the lesion.	Tessier et al. (2013) [54]
	Middle nasal meatus	Osteoma: irregularly shaped mass that was hypointense on both T1W and T2W images, containing small foci isointense to muscle on T2W images, maxillary bone atrophy.	Manso Diaz et al. (2015) [42]
	Nasal cavity	Lymphoma, squamous cell carcinoma: expansile, heterogeneously, and moderate contrast-enhancing mass with complete occlusion of the nasal cavity.	
Laryngeal dysplasia	Larynx	Lack of a cricothyroid articulation, dorsal extension of the thyroid cartilage, absence of the caudal cornu of the thyroid cartilage, absence of the articular process of the cricoid cartilage and hypoplasia, or absence of the cricopharyngeus muscle.	Garrett et al. (2009) [55]

[1] Area of the respiratory tract; MR—magnetic resonance, T1W—T1 weighted image, T2W—T2 weighted image.

3. Advances in the Functional Evaluation

The above basic and advanced imaging techniques predominantly provide data of normality or alterations of the anatomical structures of the respiratory tract. The adaptation of the pulmonary function tests (PFTs) to equine medicine significantly expands the possibilities of the functional evaluation. In contrast to static imaging techniques that allow the direct visualization of the area of interest to diagnose changes like mucus, fluid accumulation, neoplastic changes, or other anatomical abnormalities, dynamic PFTs provide information about the respiratory system in motion. Examples of dynamic measurement are pulmonary resistance and dynamic compliance, which are reliable indicators for airflow obstruction changes [60]. Respiratory abnormalities generally increase respiratory impedance in breathing, and a reduced level of ventilation can be detected objectively by deterioration in breathing mechanics [61]. The PFTs are valuable, noninvasive tools in the investigation and monitoring of breathing mechanics of patients with respiratory diseases. These techniques aid diagnosis, help monitor response to treatment, and can guide decisions regarding further treatment and intervention [62]. However, the PFTs alone cannot be expected to lead to a clinical diagnosis. Further studies on the normal values and appearance of flow-volume curves in equine medicine are required to improve the interpretation of the PETs in horses [63]. Therefore, the PET results should be evaluated in the light of history, physical examination, and diagnostic imaging results.

Among PFS available in human medicine [62], spirometry, electrical impedance tomography, and impulse oscillation systems have been applied to the horses and have enabled tremendous advances in the clinical performance evaluation of the equine athlete [64–66]. The selected clinical applications are summarized in Table 3.

Table 3. The selected diseases of the equine lung area and their main findings diagnosed based on the consecutive pulmonary function tests.

Function	Technique	Area [1]	Main Findings	Authors
Monitor ventilator volumes and respiratory mechanics, control the depth of the anesthesia	Spirometry with pilot-based flow meter	Lungs	Measurement of tidal volume and minute volume, dynamic compliance (Cdyn) of the respiratory system. Visual presentation of pressure-volume (PV) and flow-volume (FV) loop of each breath, representing the compliance (PV) and resistance (FV) of the respiratory system.	Moens et al. (2010) [67]
Correlation of spirometry results and percentage of neutrophils (N%) in tracheal aspirates	Spirometry	Lungs	Wide variation in N% in tracheal aspirates of clinically normal horses with poor racing performance, spirometry results significantly correlated with measurements of N% in tracheal aspirates.	Evans et al. (2011) [68]
Comparison of tidal breathing flow-volume loop (TBFVL) of healthy horses and horses suffering from mild and to severe asthma	Spirometry	Lungs	Disease-related differences in TBFVL indices are affected by the type of work undertaken by a horse.	Herholz et al. (2003) [69]
Measurement respiratory rate, tidal volume, peak inspiratory and expiratory flows, time to peak flow in healthy horses	Spirometry	Lungs	Measurements were repeatable and reproducible, however variable breathing patterns within the same day and on a breath-to-breath basis were present.	Burnheim et al. (2016) [64]
Effect of sedation and salbutamol administration on tidal breathing	Spirometry	Lungs	After sedation, minute ventilation was reduced in association with reduced respiratory rate and decreased expiratory and inspiratory flows. Relative expiratory time was reduced after xylazine, and peak expiratory flow occurred later in the respiratory cycle. Salbutamol administration has a significant effect on most parameters except the increase in peak inspiratory flow during tidal breathing.	Raidal et al. (2017) [70]
Monitoring of ventilation during anesthesia	EIT	Lungs	Inspiratory breath-holding and the redistribution of gas from ventral to dorsal regions of the lung after recovery from general anesthesia.	Mosing et al. (2016) [71]
Monitoring of recruitment maneuvers (RM) during anesthesia	EIT	Lungs	During recruitment maneuvers, ventilation in independent ventral region.	Ambrisco et al. (2015) [72]
Detection of bronchoconstriction and bronchodilatation	EIT	Lungs	EIT-derived flow indices for ventilation significantly changed after histamine administration and returned to control values with subsequent albuterol administration.	Secombe et al. (2021) [73]
Effect of sedation and salbutamol administration on tidal breathing Diagnosis and monitoring of equine asthma	EIT	Lungs	Healthy horses have lower peak expiratory and inspiratory flow compare to horses with mild or severe asthma after exercise.	Herteman et al. (2021) [65]
Diagnosis of bronchoconstriction	IOS	Lungs	IOS parameters in the low-frequency range were sensitive indicators of early methacholine-induced bronchoconstriction.	Van Erck et al. (2003) [74]
Standardization of IOS measurements	IOS	Lungs	IOS measurements were reliable and repeatable. Age, sex, and bodyweight did not influence IOS measurements. Measurements from 5 to 15 Hz were found to be most relevant.	Van Erck et al. (2004) [75]
Effects of sedation on lung airflow	IOS	Lungs	Inspiratory parameters were found to be significantly dependent on the time course of sedation, whereas expiratory parameters were not influenced.	Klein et al. (2006) [76]
Diagnosis and staging of equine asthma	IOS	Lungs	Significant changes were present between horses in exacerbation of EA and control horses within inspiratory and expiratory parameters. The delta reactance (ΔX) shows the presence of tidal expiratory flow limitation (EFLt) and dynamic airway compression in SEA horses in exacerbation of the clinical signs.	Stucchi et al. (2022) [77]

[1] Area of the respiratory tract; EIT—electrical impedance tomography; IOS—impulse oscillometry system.

3.1. Spirometry Evaluation of Lower Respiratory Tract

Spirometry is designed to identify and quantify functional abnormalities of the respiratory tract. In humans, the Global Initiative for Chronic Obstructive Lung Disease recommends spirometry as a diagnostic technique in earlier diagnosis and treatment monitoring in chronic obstructive diseases [78]. In medicine, spirometry provides absolute measures of respiratory function in a simple, reliable, and economical manner. Operating principles are based on three bidirectional pilot flow sensors connected to the face mask that measure breath-by-breath airflow with high resolution [79].

Spirometry begins with a full inspiration, followed by a forced expiration that rapidly empties the lungs. Expiration is continued for as long as possible or until a plateau in exhaled volume is reached. Both efforts during inspiration and expiration are recorded and graphed, demonstrating respiratory frequency, tidal volume, peak inspiratory and expiratory flows, time to peak flow, and forced vital capacity, which is an important spirometric maneuver. The forced vital capacity measurement requires the maximal inspiration followed by the rapid expiration, which should be as complete as possible [79]. Such maneuvers have been performed in horses; however, the use of general anesthesia was necessary to avoid interference of conscious respiratory movements with emptying of the lungs [60,80].

In human medicine, spirometry is the gold standard in the diagnosis of lower obstructive respiratory diseases [78], therefore the adaptation of measurement to the horse practice seems to be a promising advancement in the functional diagnosis of equine respiratory diseases. However, in equine medicine spirometry is still restricted to the university centers, and there is a lack of standardized protocol for horse lung investigation [60,64,67–70,81–86].

3.1.1. Advantages

Spirometry provides repeatable and reproducible data of the respiratory function in horses without the need of very expensive and advanced equipment, contrary to CT and MR imaging. Spirometry is a non-invasive technique that does not require the use of ionizing radiation; moreover, it is well tolerated by horses [82,83]. Therefore, spirometry may be conducted on non-sedated horses in standing position [83], however, the application for a horse's respiratory function monitoring under sedation [70] and general anesthesia [68] is also available. Cooperation with a qualified specialist is required at the beginning for gaining experience with spirometry examination, as the obtained data are simple in interpretation by the practitioners both at rest and during exercise [82,83].

3.1.2. Disadvantages

The main disadvantage of the introduction of spirometry to the equine practice is the necessity of the horse's cooperation to perform voluntary breathing maneuvers [81]. Because getting a non-sedated horse to maximal inspiration followed by the rapid, deep expiration [79] is almost impossible, in equine medicine spirometry cannot be used in the same manner as in human medicine [78]. Therefore, in horses the spirometry-based pulmonary function tests are dependent on involuntary breathing which can be done with much less cooperation from the horse [67–70,82–86]. However, the horse still needs to be cooperative with wearing a mask which might be challenging or time-consuming depending on the horse's activity and temperament [69]. The need to accurately fit the mask to the horse's head also requires possession of numerous masks so that they can be used for foals, ponies, and full-size horses [86]. Moreover, a certain period of training and acclimatizing horses to spirometric procedures is required to achieve informative data [64]. It should be kept in mind that the horses exhibit a significant respiratory reserve, ventilation, and are subject to rapid change in response to excitement, fear, and other emotional states. One should mention the individual variations and age, sex, and usability-related differences in inflow parameters as the spirometry limitations [64,69,82,83]. Therefore, larger studies to evaluate protocols for equine spirometry are required to move spirometry to the next stage of clinical development in equine medicine.

3.1.3. Clinical Applications

Spirometry has been primarily used to characterize the normal equine tidal breathing flow-volume loop in healthy horses and ponies [82,83]. Afterward, Connally et al. measured the maximal expiratory flow-volume loops in horses exercised on a treadmill, finding no marked difference between clinically normal horses and those with airway obstruction [84]. Further reports revealed changes in breathing strategy and disappearance of biphasic airflow pattern in horses with asthma [69,85]. Herholtz et al. provided strong evidence of the impact of horse work on the differentiation ability in diagnosing different degrees of asthma, as the disease-related differences in spirometry-based measures may consequently be obscured by the type of work undertaken by a horse [69]. Moreover, Burnheim et al. indicated high variability of results across days rather than within traces obtained on a single day with preserved high repeatability and reproducibility [64]. These reports suggest the need for further bigger studies to evaluate protocols for equine spirometry.

Raidal et al. used spirometry for evaluation of the effect of xylazine, acepromazine, and salbutamol on lung function in horses where respiration was significantly reduced by the sedative agents in comparison to salbutamol, where no significant changes were noticed except increased peak inspiratory flow [70], whereas previous studies have suggested that bronchodilation therapy has little effect on healthy horses [86]. Moens et al. investigated the continuous measurement of tidal and minute volume on a breath-to-breath basis in anesthetized horses and concluded that spirometry was useful in the detection of changes like depth of sedation or non-fitted tracheal cuff [68].

Among the direct clinical applications of spirometry in the diagnosis of equine respiratory diseases, the spirometry-based early diagnosis of equine asthma was investigated [60,67]. Evans et al. correlated the percentage of neutrophil from tracheal aspirates with spirometry results obtained after exercise and reported that horses with a higher percentage of neutrophils in tracheal aspirates consistently had lower flowtime curves during the second half of both inspiration and expiration [68]. These lower values may be attributed to narrowed airways due to inflammatory exudate, dynamic collapse, and/or airway hyperreactivity associated with asthma. Therefore, the need for pulmonary testing in combination with cytology for compressive diagnosis of equine asthma was strongly suggested [67].

3.2. Electrical Impedance Tomography (EIT) Evaluation of Lower Respiratory Tract

Electrical impedance tomography (EIT) is a non-invasive, radiation-free, real-time imaging modality which allows the assessment of lung ventilation and perfusion [87]. The EIT reconstructs a cross-sectional image of the lung's regional conductivity using electrodes placed circumferentially around the thorax. During EIT examination, a weak alternating current of high frequency and low amplitude is applied between an adjacent pair of electrodes and resulting surface potentials are measured by the remaining electrodes. The measured potentials depend on the tissue bioimpedance and are used to create the functional image of the lower respiratory tract [88]. The tissue bioimpedance changes depending on the fluid content, ion concentration, fat accumulation, or amount of air. Therefore, pathological changes of the tissue composition such as pleural effusion, lung fibrosis, or alveolar fluid accumulation can be easily detected by EIT [87].

In human medicine, EIT is frequently used for functional chest examinations, especially for monitoring regional lung ventilation in mechanically ventilated patients, and for regional PFT in people with chronic lung diseases [87]. The EIT is well situated for adults, neonates, and pediatric patients [89]. In the veterinary field, EIT is at an early stage of clinical development, however, in equine medicine the global and regional peak respiratory flows have recently been investigated [73,90–93].

3.2.1. Advantages

The EIT provides functionally unique clinical data of the lower respiratory tract ventilation which is difficult to obtain by other diagnostic techniques. The data are real-time measured and thus reflect continuous ventilation [94]. The EIT allows for the analysis of the individual region of interest in the lung and comparison, for example, of the dorsal with ventral parts of lung regions, the right and left lung, or the same regions that underwent different conditions [65,93]. The electrical currents used by EIT are imperceptible and safe for body surface application with no ionizing radiation exuded [83]. In contrast to CT and MR imaging and similar to spirometry, the equipment and facilities are affordable; easy to implement; well-tolerated by standing, non-sedated horses; and portable, which makes the EIT system suitable for the field conditions [90]. Electrodes are mainly integrated into one electrode belt which makes the application more user-friendly. The EIT working principle does not limit its use to any size of animal, thus the EIT belt can be adjusted to small and full-sized horses [65,95]. Cooperation with a qualified specialist is required during the first period of operation with the EIT belt, but quick and friendly training can also be done online [87]. As reconstruction algorithms have already been adapted to the horse anatomy [65,90,96], the possibility of the EIT application in equine clinical practice has significantly increased.

3.2.2. Disadvantages

Despite many advantages, the EIT imaging modality shows several limitations. EIT is characterized by very low spatial resolution compared with other imaging techniques, such as CT or MR imaging [93]. The EIT images represent a single cross-section of the thorax [89], whereas CT and MR provide three-dimensional images constructed based on the numerous detailed cross-sections [19]. Other disadvantages are the complexity of EIT data and susceptibility to artifacts. For the understanding of obtained data, specialized training is needed, however, compared to other advanced techniques described here, getting started with the EIT is relatively easy [87]. Concerning the alterations in obtaining data, any patient movement or touching of electrodes during EIT data acquisition leads to the production of the artifacts and thus anatomical and functional distortion. The horse needs to stand quietly, which sometimes may be challenging, and artifacts may still be produced due to gross movement of muscles from muscle fasciculations, interactions with the investigator, sniffing, scratching, or fat tissue accumulation. However, contrary to CT and MR imaging and similar to spirometry, neither general anesthesia or sedation are required [65]. Finally, it should be kept in mind that the EIT in equine medicine is still at an early stage of clinical development, therefore some references and protocols still need to be established [73].

3.2.3. Clinical Applications

The first studies on the EIT application in equine practice have focused on the assessment of ventilation in healthy standing horses [65,93]. Regional distribution of ventilation, the left-to-right lung region impedance ratios, and ventral-to-dorsal lung region impedance ratios were calculated. In healthy physiologically breathing horses, the right lung received a larger fraction of the tidal volume than the left, and the ventral-dependent lung region was more ventilated than the dorsal nondependent one [72]. Schramel et al. described the shift of regional ventilation towards dorsal nondependent regions in progressing pregnancy ponies that reversed seven days after foaling [95].

Application of EIT in anesthetized horses has been well described by Mosing et al. [71]. The EIT-based evaluation of the breathing pattern, distribution of ventilation, and gas exchange during anesthesia revealed the phenomenon of inspiratory breath-holding and the redistribution of gas from ventral to dorsal regions of the lung after recovery from general anesthesia [97]. The phenomenon of auto-recruitment by breath-holding has not been described previously and thus shed new light on horse anesthetized lung function. Mosing et al. also compared the lung function in spontaneously breathing and controlled mechanical ventilation anesthetized horses. In the spontaneously breathing horses, ventilation was essentially centered within dorsal regions of the lungs, while during controlled mechanical ventilation it shifted towards ventral regions [71]. Auer et al. repeated a similar protocol on the lateral recumbent ponies. The ventral shift of the ventilation region was explained by the loss of dorsal movement of the diaphragm when switching from spontaneous ventilation to controlled mechanical ventilation [98].

Moens et al., Wettstein et al., and Mosing et al. made an effort for a better understanding of horse ventilation by applying EIT measurements for continuous monitoring of the dynamic changes in the distribution of ventilation [71,93,99]. The EIT has been proposed as a monitoring tool in alveolar recruitment maneuvers in horses [93,97]. In alveolar recruitment maneuvers, lung opening is based on the implementation of sufficient peak inspiratory pressure and the immediate application of partial end-expiratory pressure [99]. These high airway pressures inevitably induce cardiovascular and pulmonary side effects such as decreases in cardiac output and blood pressure, as well as overdistension of lung parenchyma which results in augmented dead space fractions [93]. Therefore, positive end-expiratory pressure should be adjusted to the lowest pressure that prevents alveolar decruitment [93]. It is worth noting that the EIT-measured effects of the peak inspiratory pressure and partial end-expiratory pressure on the distribution of ventilation were in line with spirometry results in horses [73]. Moreover, the application of EIT in dorsally recumbent anesthetized horses allows for establishing the level of continuous positive airway pressure at which the number of silent spaces in the dependent parts of the lungs decreases [100]. However, to implement this ventilation strategy in clinical practice, the widespread awareness of the possible use of EIT or other appropriate monitoring tools in equine medicine needs to be greatly increased [101,102].

Among the direct clinical applications of the EIT in equine practice, recent research described the global and regional peak respiratory flows in the horses that underwent histamine challenges and drug-induced bronchodilatation [73,90], as well as those suffering from equine asthma [96]. The main component of equine asthma, a chronic disease that greatly affects the horse's physical capacity, is airway inflammation that causes bronchoconstriction with recurring obstruction of air passages, excessive mucus production, and bronchial and pulmonary hyperresponsiveness [103]. It is worth noting that the EIT measurements proved to be effective in the evaluation of histamine-provoked bronchoconstriction in horses [73,90]. Horses were nebulized using histamine saline, and the total impedance change during inspiration and expiration, peak global inspiratory, and peak expiratory global flow were evaluated by calculating the first derivative of the EIT volume signal. In both studies, inspiratory and expiratory global EIT flow variables incrementally increased with bronchoconstriction. A reversal of airflow changes induced by the administration of albuterol after histamine challenge was also EIT detectable [73]. In subsequent studies, EIT was implemented to compare effort-dependent ventilation between horses with asthma and the healthy group. The global expiratory flow was significantly higher in horses affected with mild and severe asthma after 15 min of exercise of the lunge. In horses with airway obstruction, the breathing strategy changes, and the biphasic airflow pattern disappears [90]. In healthy horses, the normal breathing strategy is reflected by a biphasic inspiratory and expiratory airflow pattern [82]. In asthmatic horses, the increase in global flow is more pronounced during expiration, with an increase of 94% compared to inspiration during which an increase of 83% was observed. The EIT measurements proved that asthma affects expiration more than inspiration [65].

3.3. Impulse Oscillation System (IOS) Evaluation of Lower Respiratory Tract

The impulse oscillometry system (IOS) is a non-invasive effort-independent functional modality that allows for measuring both airway resistance (R) and airway reactance (X) [95]. The IOS gains a huge interest in pediatrics, wherein younger children's ability to follow instructions is not required [104]. For a similar reason, the IOS successfully has been introduced into the veterinary field. In the equine application, the harmonic sound waves generated by a loudspeaker flow through the horses' respiratory tract [66]. The harmonic sound waves may flow through the tube attached to the face mask during IOS examination on a standing, non-sedated horse or to the endotracheal tube during anesthesia. Therefore, the IOS may be applied in both clinical applications [75]. A loudspeaker generates the single or multiple frequency harmonic sound waves, which usually range from between 1 and 5 Hz to between 10 and 25 Hz. The impulses generated by the loudspeaker travel superimposed during normal tidal breathing through the large and small airways. Higher frequency harmonic sound waves penetrate out to the lung periphery, thus reflecting the large airways, whereas lower frequency waves travel deeper into the lung reflecting the lower airways. Finally, the inspiratory and expiratory flow and pressure are measured by the pressure and flow transducers [104]. The pressure and flow signals are separated from the breathing pattern by signal filtering. The signal coming back from the airways carries the data representing respiratory impedance which is the sum of all the resistance and reactance opposing the IOS-produced oscillations. The air resistance (R) is a force proportional to energy required to propagate the pressure wave through the airways, whereas the air reactance (X) is another force proportional to the amount of recoil generated against that pressure wave [104].

3.3.1. Advantages

The IOS provides reliable, repeatable, and informative pulmonary functional data of the lower respiratory tract [75], completely different than the data obtained with other diagnostic techniques described. Therefore, it seems to be a very good complement, not a replacement for the basic and advanced diagnostic techniques of the equine respiratory tract. The IOS is non-invasive, fast, and easy to calibrate, and does not require the use of either ionizing radiation or electrical currents. Similar to spirometry and EIT, the equipment and facilities are affordable, portable, easy to implement, and well-tolerated by standing, non-sedated horses, thus the application of the IOS in the field conditions is promising [77]. Contrary to classic spirometry, the IOS requires only passive horse cooperation which is a great advantage in veterinary practice [77] and is the second indicator of applicability for field measurements in equine medicine. Cooperation with a qualified specialist is required during the first measurements conducted with the IOS, and online support of acquired data interpretation is also available [66].

3.3.2. Disadvantages

Although the IOS has many useful promising clinical applications, the examination protocol is subject to some limitations. Despite the IOS protocol being effort-independent, the horse still needs to be cooperative, especially regarding wearing a mask that sometimes might be challenging and time-consuming [77]. Moreover, similar to spirometry, the horse masks need to be adjusted for each horse to avoid air leaks that may interfere with results. Thus, buying a wide spectrum of facemasks is needed to fit them properly to foals, small, and full-sized horses [66]. Maybe for this reason, despite the many advantages, the IOS is currently restricted to research institutions and referrals [66]. Moreover, similar to both spirometry and EIT, the IOS in equine medicine is still at an early stage of clinical development, and research remains ongoing regarding the protocols, interpretation, and clinical application in horses, although more reports are still needed [66].

3.3.3. Clinical Applications

Van Erc et al. made a cornerstone in equine IOS application, providing the normal reference values for adult horses and standardizing the examination protocol concerning the effect of unfitted masks or position of the head. The horse's gender, age, and general physical morphology determined by individual biometrics did not significantly affect the results of IOS measurements [75]. In other studies, van Erck et al. reported that bronchoconstriction resulted in an increase in R at 5 Hz and a decrease in X at 5, 10, 15, 20 Hz frequencies. Young et al. confirmed those results, however, at frequencies between 1 and 3 Hz [74]. Klein et al. investigated the effect of xylazine sedation on the IOS parameters and revealed significant alterations mainly during inspiration [76]. Richard et al. revealed the significant increase in R and decrease in X at lower frequencies 1–10 Hz in a horse group with mild equine asthma [105]. These promising results direct further research towards the assessment of the suitability of the IOS in the diagnosis of equine asthma at an early stage. Stucci et al. initiated the measurement of the Delta X (ΔX) in a horse's respiratory tract, and defined ΔX as the difference between the inspiratory and expiratory reactance at each frequency [77]. In human medicine, ΔX is used in the detection of tidal expiratory flow limitation in chronic obstructive pulmonary disease and human asthma [106]. Based on the Stucci et al. results, ΔX could be used in monitoring the exacerbation or remission of the clinical signs of severe equine asthma and healthy controls, indicating the important clinical application of the IOS in the diagnostic and treatment of diseases of the lower equine respiratory tract [77].

4. Future Development and Practical Applications

In human medicine, specialized centers for respiratory diseases are present and the future may lead to development of similar facilities for horses, as respiratory problems highly impact the equids population. In many cases, the complexity of respiratory disorders makes the full diagnosis under field conditions inapplicable. Thus, specialized centers fully equipped to provide comprehensive diagnoses are needed. Except for the endoscopes and US machines, the remainder of the specialized equipment is not portable. Ambulatory X-ray systems are not suitable for thoracic radiography in adult horses, this area can only be achieved using high powered X-rays which are mainly stationary. Lung function tests are a very promising tool, especially regarding the worldwide occurrence of EA. Early asthmatic horses may not show recognizable signs, but their airway hyperactivity can be detected using a pulmonary function test. PFTs may be especially helpful in early diagnosis of high-performance horses where the smallest amount of respiratory disease may affect the future outcome. PFTs may be also helpful in the evaluation of response to treatment protocol and are more sensitive than BALF retake. Hopefully high demand from the owners and trainers will result in the wide implementation of those techniques in equine respiratory diseases diagnosis, as it would support effective management of affected horses, improve the outcomes, and improve the horses' welfare.

5. Conclusions

Huge improvements have been observed in the last decades and continuous technical progress is changing the capability of the diagnostic techniques, allowing for more accurate anatomical and functional studies of the equine respiratory tract. Within the advanced techniques, the sensitive diagnostic modalities like CT or MR have provided the detailed anatomical features of upper respiratory tract diseases, allowing for the exact visualization of changes and providing vital information for surgical staff in the case of a planned procedure. The further implementation of pulmonary function tests is promising as a non-invasive tool to facilitate an earlier diagnosis of equine lower respiratory tract diseases, especially in the case of equine asthma. However, more studies in the presented field are needed to create protocols that may be widely implemented by equine practitioners.

Author Contributions: Conceptualization, N.K. and M.D.; methodology, N.K. and M.D.; resources, M.W. and T.J.; writing—original draft preparation, N.K. and M.D.; writing—review and editing, M.W. and T.J.; visualization, N.K. and M.D.; supervision, M.W. and M.D. All authors have read and agreed to the published version of the manuscript.

Funding: This research received no external funding.

Informed Consent Statement: Not applicable.

Data Availability Statement: Not applicable.

Conflicts of Interest: The authors declare no conflict of interest.

References

1. Roy, M.-F.; Lavoie, J.-P. Tools for the Diagnosis of Equine Respiratory Disorders. *Vet. Clin. N. Am. Equine Pract.* **2003**, *19*, 1–17. [CrossRef]
2. Savage, C.J. Evaluation of the Equine Respiratory System Using Physical Examination and Endoscopy. *Vet. Clin. N. Am. Equine Pract.* **1997**, *13*, 443–462. [CrossRef]
3. Marr, C. Thoracic Ultrasonography. *Equine Vet. Educ.* **1993**, *5*, 41–46. [CrossRef]
4. Barrett, M.; Park, R. Review of Radiographic Technique and Interpretation of the Equine Skull. *AAEP Proc.* **2011**, *57*, 431–437.
5. Gibbs, C.; Lane, J.G. Radiographic Examination of the Facial, Nasal and Paranasal Sinus Regions of the Horse. II. Radiological Findings. *Equine Vet. J.* **1987**, *19*, 474–482. [CrossRef] [PubMed]
6. Koblinger, K.; Nicol, J.; McDonald, K.; Wasko, A.; Logie, N.; Weiss, M.; Léguillette, R. Endoscopic Assessment of Airway Inflammation in Horses. *J. Vet. Intern. Med.* **2011**, *25*, 1118–1126. [CrossRef] [PubMed]
7. Barakzai, S.Z.; Dixon, P.M. Correlation of Resting and Exercising Endoscopic Findings for Horses with Dynamic Laryngeal Collapse and Palatal Dysfunction. *Equine Vet. J.* **2011**, *43*, 18–23. [CrossRef] [PubMed]

8. Kraft, S.L.; Gavin, P. Physical Principles and Technical Considerations for Equine Computed Tomography and Magnetic Resonance Imaging. *Vet. Clin. N. Am. Equine Pract.* **2001**, *17*, 115–130. [CrossRef]

9. Kutasi, O.; Balogh, N.; Lajos, Z.; Nagy, K.; Szenci, O. Diagnostic Approaches for the Assessment of Equine Chronic Pulmonary Disorders. *J. Equine Vet. Sci.* **2011**, *31*, 400–410. [CrossRef]

10. Lo Feudo, C.M.; Stucchi, L.; Alberti, E.; Stancari, G.; Conturba, B.; Zucca, E.; Ferrucci, F. The Role of Thoracic Ultrasonography and Airway Endoscopy in the Diagnosis of Equine Asthma and Exercise-Induced Pulmonary Hemorrhage. *Vet. Sci.* **2021**, *8*, 276. [CrossRef]

11. Reef, V.B.; Whittier, M.; Allam, L.G. Thoracic Ultrasonography. *Clin. Tech. Equine Pract.* **2004**, *3*, 284–293. [CrossRef]

12. Raphel, C.F. Endoscopic Findings in the Upper Respiratory Tract of 479 Horses. *J. Am. Vet. Med. Assoc.* **1982**, *181*, 470–473.

13. Hardy, J.; Léveillé, R. Diseases of the Guttural Pouches. *Vet. Clin. Equine Pract.* **2003**, *19*, 123–158. [CrossRef]

14. Couëtil, L.L.; Gallatin, L.L.; Blevins, W.; Khadra, I. Treatment of Tracheal Collapse with an Intraluminal Stent in a Miniature Horse. *J. Am. Vet. Med. Assoc.* **2004**, *225*, 1727–1732. [CrossRef]

15. Couetil, L.; Cardwell, J.M.; Leguillette, R.; Mazan, M.; Richard, E.; Bienzle, D.; Bullone, M.; Gerber, V.; Ivester, K.; Lavoie, J.-P.; et al. Equine Asthma: Current Understanding and Future Directions. *Front. Vet. Sci.* **2020**, *7*, 450. [CrossRef]

16. Perkins, G.A.; Viel, L.; Wagner, B.; Hoffman, A.; Erb, H.N.; Ainsworth, D.M. Histamine Bronchoprovocation Does Not Affect Bronchoalveolar Lavage Fluid Cytology, Gene Expression and Protein Concentrations of IL-4, IL-8 and IFN-Gamma. *Vet. Immunol. Immunopathol.* **2008**, *126*, 230–235. [CrossRef]

17. Hathcock, J.T.; Stickle, R.L. Principles and Concepts of Computed Tomography. *Vet. Clin. N. Am. Small Anim. Pract.* **1993**, *23*, 399–415. [CrossRef]

18. Dakin, S.G.; Lam, R.; Rees, E.; Mumby, C.; West, C.; Weller, R. Technical Set-up and Radiation Exposure for Standing Computed Tomography of the Equine Head: Standing CT of the Equine Head. *Equine Vet. Educ.* **2014**, *26*, 208–215. [CrossRef]

19. Tucker, R.L.; Farrell, E. Computed Tomography and Magnetic Resonance Imaging of the Equine Head. *Vet. Clin. N. Am. Equine Pract.* **2001**, *17*, 131–144. [CrossRef]

20. Schliewert, E.-C.; Lascola, K.M.; O'Brien, R.T.; Clark-Price, S.C.; Wilkins, P.A.; Foreman, J.H.; Mitchell, M.A.; Hartman, S.K.; Kline, K.H. Comparison of Radiographic and Computed Tomographic Images of the Lungs in Healthy Neonatal Foals. *Am. J. Vet. Res.* **2015**, *76*, 42–52. [CrossRef]

21. Farrow, C.S. The Equine Skull: Dealing Successfully with Radiographic Complexity. In *Veterinary Diagnostic Imaging: The Horse*; Farrow, C.S., Ed.; Mosby: Saint Louis, MO, USA, 2006; Chapter 18; pp. 329–332. ISBN 978-0-323-01206-5.

22. Crijns, C.P.; Baeumlin, Y.; De Rycke, L.; Broeckx, B.J.G.; Vlaminck, L.; Bergman, E.H.J.; van Bree, H.; Gielen, I. Intra-Arterial versus Intra Venous Contrast-Enhanced Computed Tomography of the Equine Head. *BMC Vet. Res.* **2016**, *12*, 6. [CrossRef] [PubMed]

23. Barbee, D.D.; Allen, J.R.; Gavin, P.R. Computed tomography in horses: Technique. *Vet. Radiol.* **1987**, *28*, 144–151. [CrossRef]

24. Solano, M.; Brawer, R.S. CT of the Equine Head: Technical Considerations, Anatomical Guide, and Selected Diseases. *Clin. Tech. Equine Pract.* **2004**, *3*, 374–388. [CrossRef]

25. Esmaeili, F.; Johari, M.; Haddadi, P.; Vatankhah, M. Beam Hardening Artifacts: Comparison between Two Cone Beam Computed Tomography Scanners. *J. Dent. Res. Dent. Clin. Dent. Prospect.* **2012**, *6*, 49–53. [CrossRef]

26. Witte, T.H.; Perkins, J.D. Early Diagnosis May Hold the Key to the Successful Treatment of Nasal and Paranasal Sinus Neoplasia in the Horse. *Equine Vet. Educ.* **2011**, *23*, 441–447. [CrossRef]

27. Zakia, L.; Shaw, S.; Bonomelli, N.; O'Sullivan, S.; Zur Linden, A.; Dubois, M.; Baird, J.; Guest, B. Hematuria in a 3-Month-Old Filly with an Internal Umbilical Abscess and Internal Iliac Artery Aneurysm. *Can. Vet. J.* **2021**, *62*, 877–881.

28. Strohmayer, C.; Klang, A.; Kneissl, S. Computed Tomographic and Histopathological Characteristics of 13 Equine and 10 Feline Oral and Sinonasal Squamous Cell Carcinomas. *Front. Vet. Sci.* **2020**, *7*, 591437. [CrossRef]

29. Cilliero, I.; Williams, J.; Carstens, A.; Duncan, N.M. Three Cases of Osteoma and an Osseous Fibroma of the Paranasal Sinuses of Horses in South Africa: Clinical Communication. *J. S. Afr. Vet. Assoc.* **2008**, *79*, 185–193. [CrossRef]

30. De Zani, D.; Zani, D.D.; Borgonovo, S.; Di Giancamillo, M.; Rondena, M.; Verschooten, F. An Undifferentiated Sarcoma in the Cervical Region in a Horse: Undifferentiated Cervical Sarcoma in a Horse. *Equine Vet. Educ.* **2011**, *23*, 138–141. [CrossRef]

31. Cissell, D.D.; Wisner, E.R.; Textor, J.; Mohr, F.C.; Scrivani, P.V.; Théon, A.P. Computed Tomographic Appearance of Equine Sinonasal Neoplasia. *Vet. Radiol. Ultrasound* **2012**, *53*, 245–251. [CrossRef]

32. Henninger, W.; Mairi Frame, E.; Willmann, M.; Simhofer, H.; Malleczek, D.; Kneissl, S.M.; Mayrhofer, E. CT Features of alveolitis and sinusitis in horses. *Vet. Radiol. Ultrasound* **2003**, *44*, 269–276. [CrossRef]

33. Fenner, M.F.; Verwilghen, D.; Townsend, N.; Simhofer, H.; Schwarzer, J.; Zani, D.D.; Bienert-Zeit, A. Paranasal Sinus Cysts in the Horse: Complications Related to Their Presence and Surgical Treatment in 37 Cases. *Equine Vet. J.* **2019**, *51*, 57–63. [CrossRef]

34. Annear, M.J.; Gemensky-Metzler, A.J.; Elce, Y.A.; Stone, S.G. Exophthalmus Secondary to a Sinonasal Cyst in a Horse. *J. Am. Vet. Med. Assoc.* **2008**, *15*, 233–285. [CrossRef]

35. Ostrowska, J.; Lindström, L.; Tóth, T.; Hansson, K.; Uhlhorn, M.; Ley, C.J. Computed Tomography Characteristics of Equine Paranasal Sinus Cysts. *Equine Vet. J.* **2020**, *52*, 538–546. [CrossRef]

36. Hilton, H.; Puchalski, S.M.; Aleman, M. The Computed Tomographic Appearance of Equine Temporohyoid Osteoarthropathy. *Vet. Radiol. Ultrasound* **2009**, *50*, 151–156. [CrossRef]

37. Divers, T.J.; Ducharme, N.G.; de Lahunta, A.; Irby, N.L.; Scrivani, P.V. Temporohyoid Osteoarthropathy. *Clin. Tech. Equine Pract.* **2006**, *5*, 17–23. [CrossRef]

38. Bras, J.J.; Davis, E.; Beard, W.L. Bilateral Ceratohyoidectomy for the Resolution of Clinical Signs Associated with Temporohyoid Osteoarthropathy. *Equine Vet. Educ.* **2014**, *26*, 116–120. [CrossRef]

39. Kowalczyk, L.; Boehler, A.; Brunthaler, R.; Rathmanner, M.; Rijkenhuizen, A.B.M. Squamous Cell Carcinoma of the Paranasal Sinuses in Two Horses. *Equine Vet. Educ.* **2011**, *23*, 435–440. [CrossRef]

40. Dixon, P.M.; Barnett, T.P.; Morgan, R.E.; Reardon, R.J.M. Computed Tomographic Assessment of Individual Paranasal Sinus Compartment and Nasal Conchal Bulla Involvement in 300 Cases of Equine Sinonasal Disease. *Front. Vet. Sci.* **2020**, *7*, 797. [CrossRef]

41. Sogaro-Robinson, C.; Lacombe, V.A.; Reed, S.M.; Balkrishnan, R. Factors Predictive of Abnormal Results for Computed Tomography of the Head in Horses Affected by Neurologic Disorders: 57 Cases (2001–2007). *J. Am. Vet. Med. Assoc.* **2009**, *235*, 176–183. [CrossRef]

42. Manso-Díaz, G.; García-López, J.M.; Maranda, L.; Taeymans, O. The Role of Head Computed Tomography in Equine Practice. *Equine Vet. Educ.* **2015**, *27*, 136–145. [CrossRef]

43. Garrett, K.S.; Woodie, J.B.; Cook, J.L.; Williams, N.M. Imaging Diagnosis—Nasal Septal and Laryngeal Cyst-like Malformationsin a Thoroughbred Weanling Colt Diagnosed Using Ultrasonography and Magnetic Resonance Imaging. *Vet. Radiol. Ultrasound Off. J. Am. Coll. Vet. Radiol. Int. Vet. Radiol. Assoc.* **2010**, *51*, 504–507. [CrossRef] [PubMed]

44. Murray, R.C. *Equine MRI*, 1st ed.; John Wiley & Sons, Ltd.: Hoboken, NJ, USA, 2010; pp. 249–267.

45. Fitz, J.; Gerhards, H. Magnetic Resonance Imaging of the Thorax and Abdomen in Foals. *Pferdeheilkunde* **2005**, *21*, 115–123. [CrossRef]

46. Gutierrez-Nibeyro, S.D.; Werpy, N.M.; Gold, S.J.; Olguin, S.; Schaeffer, D.J. Standing MRI Lesions of the Distal Interphalangeal Joint and Podotrochlear Apparatus Occur with a High Frequency in Warmblood Horses. *Vet. Radiol. Ultrasound* **2020**, *61*, 336–345. [CrossRef]

47. Werpy, N.M. Magnetic Resonance Imaging of the Equine Patient: A Comparison of High- and Low-Field Systems. *Clin. Tech. Equine Pract.* **2007**, *6*, 37–45. [CrossRef]

48. Hatabu, H.; Ohno, Y.; Gefter, W.B.; Parraga, G.; Madore, B.; Lee, K.S.; Altes, T.A.; Lynch, D.A.; Mayo, J.R.; Seo, J.B.; et al. Expanding Applications of Pulmonary MRI in the Clinical Evaluation of Lung Disorders: Fleischner Society Position Paper. *Radiology* **2020**, *297*, 286–301. [CrossRef]

49. Zani, D.D.; Rabbogliatti, V.; Ravasio, G.; Pettinato, C.; Giancamillo, M.D.; Zani, D.D. Contrast Enhanced Magnetic Resonance Imaging of the Foot in Horses Using Intravenous versus Regional Intraarterial Injection of Gadolinium. *Open Vet. J.* **2018**, *8*, 471–478. [CrossRef]

50. Waselau, M.; McKnight, A.; Kasparek, A. Magnetic Resonance Imaging of Equine Stifles: Technique and Observations in 76 Clinical Cases. *Equine Vet. Educ.* **2020**, *32*, 85–91. [CrossRef]

51. Barrett, M.F.; Selberg, K.T.; Johnson, S.A.; Hersman, J.; Frisbie, D.D. High Field Magnetic Resonance Imaging Contributes to Diagnosis of Equine Distal Tarsus and Proximal Metatarsus Lesions: 103 Horses. *Vet. Radiol. Ultrasound* **2018**, *59*, 587–596. [CrossRef]

52. Ferrell, E.A.; Gavin, P.R.; Tucker, R.L.; Sellon, D.C.; Hikes, M.T. Magnetic Resonance for Evaluation of Neurologic Disease in 12 Horses. *Vet. Radiol. Ultrasound* **2002**, *43*, 510–516. [CrossRef]

53. Spoormakers, T.J.P.; Ensink, J.M.; Goehring, L.S.; Koeman, J.P.; Braake, F.T.; van der Vlugt-Meijer, R.H.; van der Belt, A.J.M. Brain Abscesses as a Metastatic Manifestation of Strangles: Symptomatology and the Use of Magnetic Resonance Imaging as a Diagnostic Aid. *Equine Vet. J.* **2010**, *35*, 146–151. [CrossRef]

54. Tessier, C.; Brühschwein, A.; Lang, J.; Konar, M.; Wilke, M.; Brehm, W.; Kircher, P. Magnetic Resonance Imaging Features of Sinonasal Disorders in Horses. *Vet. Radiol. Ultrasound.* **2013**, *54*, 54–60. [CrossRef]

55. Garrett, K.S.; Woodie, J.B.; Embertson, R.M.; Pease, A.P. Diagnosis of Laryngeal Dysplasia in Five Horses Using Magnetic Resonance Imaging and Ultrasonography. *Equine Vet. J.* **2009**, *41*, 766–771. [CrossRef]

56. Garrett, K.S. Advances in Diagnostic Imaging of the Larynx and Pharynx. *Equine Vet. Educ.* **2012**, *24*, 17–18. [CrossRef]

57. Rieger, C.; Herzog, P.; Eibel, R.; Fiegl, M.; Ostermann, H. Pulmonary MRI–a New Approach for the Evaluation of Febrile Neutropenic Patients with Malignancies. *Support. Care Cancer.* **2008**, *16*, 599–606. [CrossRef]

58. Syrjala, H.; Broas, M.; Ohtonen, P.; Jartti, A.; Pääkkö, E. Chest Magnetic Resonance Imaging for Pneumonia Diagnosis in Outpatients with Lower Respiratory Tract Infection. *Eur. Respir. J.* **2017**, *49*, 1601303. [CrossRef]

59. Eibel, R.; Herzog, P.; Dietrich, O.; Rieger, C.T.; Ostermann, H.; Reiser, M.F.; Schoenberg, S.O. Pulmonary Abnormalities in Immunocompromised Patients: Comparative Detection with Parallel Acquisition MR Imaging and Thin-Section Helical CT. *Radiology* **2006**, *241*, 880–891. [CrossRef]

60. Couetil, L.L.; Rosenthal, F.S.; DeNicola, D.B.; Chilcoat, C.D. Clinical Signs, Evaluation of Bronchoalveolar Lavage Fluid, and Assessment of Pulmonary Function in Horses with Inflammatory Respiratory Disease. *Am. J. Vet. Res.* **2001**, *62*, 538–546. [CrossRef]

61. Butler, P.J.; Woakes, A.J.; Smale, K.; Roberts, C.A.; Hillidge, C.J.; Snow, D.H.; Marlin, D.J. Respiratory and Cardiovascular Adjustments during Exercise of Increasing Intensity and during Recovery in Thoroughbred Racehorses. *J. Exp. Biol.* **1993**, *179*, 159–180. [CrossRef]

62. Hyatt, R.E.; Scanlon, P.D.; Nakamura, M. *Interpretation of Pulmonary Function Tests*; Lippincott Williams & Wilkins: Philadelphia, PA, USA, 2014; ISBN 978-1-4511-4380-5.

63. Hoffman, A.M. Clinical Application of Pulmonary Function Testing in Horses. Available online: https://www.ivis.org/library/equine-respiratory-diseases/clinical-application-of-pulmonary-function-testing-horses (accessed on 20 December 2021).
64. Burnheim, K.; Hughes, K.J.; Evans, D.L.; Raidal, S.L. Reliability of Breath by Breath Spirometry and Relative Flow-Time Indices for Pulmonary Function Testing in Horses. *BMC Vet. Res.* **2016**, *12*, 268. [CrossRef]
65. Herteman, N.; Mosing, M.; Waldmann, A.; Gerber, V.; Schoster, A. Exercise-induced Airflow Changes in Horses with Asthma Measured by Electrical Impedance Tomography. *J. Vet. Intern. Med.* **2021**, *35*, 2500–2510. [CrossRef]
66. Erck, E.; Votion, D.; Art, T.; Lekeux, P. Measurement of Respiratory Function by Impulse Oscillometry in Horses. *Equine Vet. J.* **2010**, *36*, 21–28. [CrossRef]
67. Moens, Y.P.S. Clinical Application of Continuous Spirometry with a Pitot-Based Flow Meter during Equine Anaesthesia. *Equine Vet. Educ.* **2010**, *22*, 354–360. [CrossRef]
68. Evans, D.L.; Kiddell, L.; Smith, C.L. Pulmonary Function Measurements Immediately after Exercise Are Correlated with Neutrophil Percentage in Tracheal Aspirates in Horses with Poor Racing Performance. *Res. Vet. Sci.* **2011**, *90*, 510–515. [CrossRef]
69. Herholz, C.; Straub, R.; Braendlin, C.; Imhof, A.; Lüthi, S.; Busato, A. Measurement of Tidal Breathing Flow-Volume Loop Indices in Horses Used for Different Sporting Purposes with and without Recurrent Airway Obstruction. *Vet. Rec.* **2003**, *152*, 288–292. [CrossRef]
70. Raidal, S.L.; Burnheim, K.; Evans, D.; Hughes, K.J. Effects of Sedation and Salbutamol Administration on Hyperpnoea and Tidal Breathing Spirometry in Healthy Horses. *Vet. J.* **2017**, *222*, 22–28. [CrossRef] [PubMed]
71. Mosing, M.; Marly-Voquer, C.; Macfarlane, P.; Bardell, D.; Bohm, S.; Bettschart-Wolfensberger, R.; Waldmann, A. Regional Distribution of Ventilation in Horses in Dorsal Recumbency during Spontaneous and Mechanical Ventilation Assessed by Electrical Impedance Tomography: A Case Series. *Vet. Anaesth. Analg.* **2016**, *44*, 127–132. [CrossRef] [PubMed]
72. Ambrisko, T.D.; Schramel, J.P.; Adler, A.; Kutasi, O.; Makra, Z.; Moens, Y.P.S. Assessment of Distribution of Ventilation by Electrical Impedance Tomography in Standing Horses. *Physiol. Meas.* **2015**, *37*, 175–186. [CrossRef] [PubMed]
73. Secombe, C.; Adler, A.; Hosgood, G.; Raisis, A.; Mosing, M. Can Bronchoconstriction and Bronchodilatation in Horses Be Detected Using Electrical Impedance Tomography? *J. Vet. Intern. Med.* **2021**, *35*, 2035–2044. [CrossRef] [PubMed]
74. Van Erck, E.; Votion, D.M.; Kirschvink, N.; Art, T.; Lekeux, P. Use of the Impulse Oscillometry System for Testing Pulmonary Function during Methacholine Bronchoprovocation in Horses. *Am. J. Vet. Res.* **2003**, *64*, 1414–1420. [CrossRef]
75. Van Erck, E.; Votion, D.; Kirschvink, N.; Genicot, B.; Lindsey, J.; Art, T.; Lekeux, P. Influence of Breathing Pattern and Lung Inflation on Impulse Oscillometry Measurements in Horses. *Vet. J.* **2004**, *168*, 259–269. [CrossRef]
76. Klein, C.; Smith, H.-J.; Reinhold, P. The Use of Impulse Oscillometry for Separate Analysis of Inspiratory and Expiratory Impedance Parameters in Horses: Effects of Sedation with Xylazine. *Res. Vet. Sci.* **2006**, *80*, 201–208. [CrossRef]
77. Stucchi, L.; Ferrucci, F.; Bullone, M.; Dellacà, R.L.; Lavoie, J.P. Within-Breath Oscillatory Mechanics in Horses Affected by Severe Equine Asthma in Exacerbation and in Remission of the Disease. *Animals* **2022**, *12*, 4. [CrossRef]
78. Vogelmeier, C.F.; Criner, G.J.; Martinez, F.J.; Anzueto, A.; Barnes, P.J.; Bourbeau, J.; Celli, B.R.; Chen, R.; Decramer, M.; Fabbri, L.M.; et al. Global Strategy for the Diagnosis, Management, and Prevention of Chronic Obstructive Lung Disease 2017 Report. GOLD Executive Summary. *Am. J. Respir. Crit. Care Med.* **2017**, *195*, 557–582. [CrossRef]
79. Moore, V.C. Spirometry: Step by Step. *Breathe* **2012**, *8*, 232–240. [CrossRef]
80. Leith, D.E. Comparative Mammalian Respiratory Mechanics. *Physiologist* **1976**, *19*, 485–510.
81. Jat, K.R. Spirometry in Children. *Prim. Care Respir. J.* **2013**, *22*, 221–229. [CrossRef]
82. Guthrie, A.J.; Beadle, R.E.; Bateman, R.D.; White, C.E. Characterization of Normal Tidal Breathing Flow-Volume Loops for Thoroughbred Horses. *Vet. Res. Commun.* **1995**, *19*, 331–342. [CrossRef]
83. Art, T.; Lekeux, P. Respiratory Airflow Patterns in Ponies at Rest and during Exercise. *Can. J. Vet. Res.* **1988**, *52*, 299–303.
84. Connally, B.A.; Derksen, F.J. Tidal Breathing Flow-Volume Loop Analysis as a Test of Pulmonary Function in Exercising Horses. *Am. J. Vet. Res.* **1994**, *55*, 589–594.
85. Petsche, V.M.; Derksen, F.J.; Robinson, N.E. Tidal Breathing Flow-Volume Loops in Horses with Recurrent Airway Obstruction (Heaves). *Am. J. Vet. Res.* **1994**, *55*, 885–891.
86. Mazan, M.R.; Hoffman, A.M. Effects of Aerosolized Albuterol on Physiologic Responses to Exercise in Standardbreds. *Am. J. Vet. Res.* **2001**, *62*, 1812–1817. [CrossRef]
87. Frerichs, I.; Amato, M.B.P.; van Kaam, A.H.; Tingay, D.G.; Zhao, Z.; Grychtol, B.; Bodenstein, M.; Gagnon, H.; Böhm, S.H.; Teschner, E.; et al. Chest Electrical Impedance Tomography Examination, Data Analysis, Terminology, Clinical Use and Recommendations: Consensus Statement of the TRanslational EIT developmeNt stuDy Group. *Thorax* **2017**, *72*, 83–93. [CrossRef]
88. Wang, Y.-M.; Sun, X.-M.; Zhou, Y.-M.; Chen, J.-R.; Cheng, K.-M.; Li, H.-L.; Yang, Y.-L.; Zhang, L.; Zhou, J.-X. Use of Electrical Impedance Tomography (EIT) to Estimate Global and Regional Lung Recruitment Volume (VREC) Induced by Positive End-Expiratory Pressure (PEEP): An Experiment in Pigs with Lung Injury. *Med. Sci. Monit.* **2020**, *26*, e922609. [CrossRef]
89. Dijkstra, A.M.; Brown, B.H.; Leathard, A.D.; Harris, N.D.; Barber, D.C.; Edbrooke, D.L. Review Clinical Applications of Electrical Impedance Tomography. *J. Med. Eng. Technol.* **1993**, *17*, 89–98. [CrossRef]
90. Secombe, C.; Waldmann, A.; Hosgood, G.; Mosing, M. Evaluation of Histamine-provoked Changes in Airflow Using Electrical Impedance Tomography in Horses. *Equine Vet. J.* **2019**, *52*, 556–563. [CrossRef]

91. Wey, C.; Meira, C.; Mosing, M.; Bleul, U. Development of the Postnatal Lung in Bovine Neonates Illustrated by Electrical Impedance Tomography (EIT). In Proceedings of the Conference: Workshop on Gonadal Function, Gamete Interaction and Pregnancy (GGP), Giessen, Germany, 30 October 2017.

92. Ambrisko, T.; Schramel, J.; Auer, U.; Moens, Y.; Staffieri, F. Impact of Four Different Recumbencies on the Distribution of Ventilation in Conscious or Anaesthetized Spontaneously Breathing Beagle Dogs: An Electrical Impedance Tomography Study. *PLoS ONE* **2017**, *12*, e0183340. [CrossRef]

93. Moens, Y.; Schramel, J.; Tusman, G.; Ambrisko, T.; Solà, J.; Brunner, J.; Kowalczyk, L.; Böhm, S. Variety of Non-Invasive Continuous Monitoring Methodologies Including Electrical Impedance Tomography Provides Novel Insights into the Physiology of Lung Collapse and Recruitment—Case Report of an Anaesthetized Horse. *Vet. Anaesth. Analg.* **2013**, *41*, 196–204. [CrossRef]

94. Mosing, M.; Sacks, M.; Tahas, S.; Ranninger, E.; Bohm, S.; Campagna, I.; Waldmann, A. Ventilatory Incidents Monitored by Electrical Impedance Tomography in an Anaesthetized Orangutan (Pongo Abelii). *Vet. Anaesth. Analg.* **2017**, *44*, 973–976. [CrossRef]

95. Schramel, J.; Nagel, C.; Auer, U.; Palm, F.; Aurich, C.; Moens, Y. Distribution of Ventilation in Pregnant Shetland Ponies Measured by Electrical Impedance Tomography. *Respir. Physiol. Neurobiol.* **2012**, *180*, 258–262. [CrossRef]

96. Milne, S.; Huvanandana, J.; Nguyen, C.; Duncan, J.M.; Chapman, D.G.; Tonga, K.O.; Zimmermann, S.C.; Slattery, A.; King, G.G.; Thamrin, C. Time-Based Pulmonary Features from Electrical Impedance Tomography Demonstrate Ventilation Heterogeneity in Chronic Obstructive Pulmonary Disease. *J. Appl. Physiol.* **2019**, *127*, 1441–1452. [CrossRef] [PubMed]

97. Mosing, M.; Waldmann, A.; Macfarlane, P.; Iff, S.; Auer, U.; Bohm, S.; Bettschart-Wolfensberger, R.; Bardell, D. Horses Auto-Recruit Their Lungs by Inspiratory Breath Holding Following Recovery from General Anaesthesia. *PLoS ONE* **2016**, *11*, e0158080. [CrossRef] [PubMed]

98. Auer, U.; Schramel, J.; Moens, Y.; Mosing, M.; Braun, C. Monitoring Changes in Distribution of Pulmonary Ventilation by Functional Electrical Impedance Tomography in Anaesthetized Ponies. *Vet. Anaesth. Analg.* **2018**, *46*, 200–208. [CrossRef] [PubMed]

99. Wettstein, D.; Moens, Y.; Jaeggin-Schmucker, N.; Böhm, S.H.; Rothen, H.U.; Mosing, M.; Kästner, S.B.R.; Schatzmann, U. Effects of an Alveolar Recruitment Maneuver on Cardiovascular and Respiratory Parameters during Total Intravenous Anesthesia in Ponies. *Am. J. Vet. Res.* **2006**, *67*, 152–159. [CrossRef]

100. Mosing, M.; Auer, U.; Macfarlane, P.; Bardell, D.; Schramel, J.; Bohm, S.; Bettschart-Wolfensberger, R.; Waldmann, A. Regional Ventilation Distribution and Dead Space in Anaesthetised Horses Treated with and without Continuous Positive Airway Pressure (CPAP). *Vet. Anaesth. Analg.* **2017**, *45*, 31–40. [CrossRef]

101. Balleza, M.; Casan, P.; Riu, P.J. Tidal Volume Monitoring with Electrical Impedance Tomography (EIT) on COPD Patients. Relationship Between EIT and Diffusion Lung Transfer (DL,CO). In *Proceedings of the World Congress on Medical Physics and Biomedical Engineering, Munich, Germany, 7–12 September 2009*; Dössel, O., Schlegel, W.C., Eds.; Springer: Berlin/Heidelberg, Germany, 2009; pp. 549–552.

102. Vogt, B.; Pulletz, S.; Elke, G.; Zhao, Z.; Zabel, P.; Weiler, N.; Frerichs, I. Spatial and Temporal Heterogeneity of Regional Lung Ventilation Determined by Electrical Impedance Tomography during Pulmonary Function Testing. *J. Appl. Physiol.* **2012**, *113*, 1154–1161. [CrossRef]

103. Leclere, M.; Lavoie-Lamoureux, A.; Lavoie, J.-P. Heaves, an Asthma-like Disease of Horses. *Respirology* **2011**, *16*, 1027–1046. [CrossRef]

104. Komarow, H.D.; Myles, I.A.; Uzzaman, A.; Metcalfe, D.D. Impulse Oscillometry in the Evaluation of Diseases of the Airways in Children. *Ann. Allergy Asthma Immunol.* **2011**, *106*, 191–199. [CrossRef]

105. Richard, E.A.; Fortier, G.D.; Denoix, J.-M.; Art, T.; Lekeux, P.M.; Erck, E. van Influence of Subclinical Inflammatory Airway Disease on Equine Respiratory Function Evaluated by Impulse Oscillometry. *Equine Vet. J.* **2009**, *41*, 384–389. [CrossRef]

106. Koulouris, N.G.; Hardavella, G. Physiological Techniques for Detecting Expiratory Flow Limitation during Tidal Breathing. *Eur. Respir. Rev.* **2011**, *20*, 147–155. [CrossRef]

Article

Within-Breath Oscillatory Mechanics in Horses Affected by Severe Equine Asthma in Exacerbation and in Remission of the Disease

Luca Stucchi [1], Francesco Ferrucci [1,*], Michela Bullone [2], Raffaele L. Dellacà [3] and Jean Pierre Lavoie [4]

[1] Dipartimento di Medicina Veterinaria, Università degli Studi di Milano, 26900 Lodi, Italy; luca.stucchi@unimi.it

[2] Dipartimento di Scienze Veterinarie, Università di Torino, 10095 Grugliasco, Italy; michela.bullone@unito.it

[3] Dipartimento di Elettronica, Informazione e Bioingegneria, Politecnico di Milano, 20133 Milano, Italy; raffaele.dellaca@polimi.it

[4] Département de Sciences Cliniques, Faculté de Médecine Vétérinaire, Université de Montréal, Saint-Hyacinthe, QC J2S 2M2, Canada; jean-pierre.lavoie@umontreal.ca

*   Correspondence: francesco.ferrucci@unimi.it

Simple Summary: Equine asthma shares similarities with human asthma. The aim of the study was to evaluate whether within breath analysis improved the sensitivity of oscillometry at detecting subclinical airway obstruction in horses with asthma in remission of clinical signs. From this study, we can conclude that the within-breath oscillometry is sensitive in discriminating horses with severe asthma in clinical remission of the disease from control horses. Additionally, oscillometry allowed to identify the increase in expiratory reactance similar to that due to expiratory flow limitation observed in human asthmatic patients with airway obstruction.

Abstract: Oscillometry is a technique that measures the resistance (R) and the reactance (X) of the respiratory system. In humans, analysis of inspiratory and expiratory R and X allows to identify the presence of tidal expiratory flow limitation (EFLt). The aim of this study was to describe inspiratory and expiratory R and X measured by impulse oscillometry system (IOS) in horses with severe asthma (SEA) when in clinical remission (n = 7) or in exacerbation (n = 7) of the condition. Seven healthy, age-matched control horses were also studied. Data at 3, 5, and 7 Hz with coherence > 0.85 at 3 Hz and >0.9 at 5 and 7 Hz were considered. The mean, inspiratory and expiratory R and X and the difference between inspiratory and expiratory X (ΔX) were calculated at each frequency. The data from the three groups were statistically compared. Results indicated that in horses during exacerbation of severe asthma, X during expiratory phase is more negative than during inspiration, such as in humans in presence of EFLt. The evaluation of X during inspiration is promising in discriminating between horses with SEA in remission and control horses.

Keywords: equine asthma; impulse oscillometry; airway obstruction; lung function test

Citation: Stucchi, L.; Ferrucci, F.; Bullone, M.; Dellacà, R.L.; Lavoie, J.P. Within-Breath Oscillatory Mechanics in Horses Affected by Severe Equine Asthma in Exacerbation and in Remission of the Disease. *Animals* **2022**, *12*, 4. https://doi.org/10.3390/ani12010004

Academic Editor: Diana M. Hassel

Received: 21 November 2021
Accepted: 17 December 2021
Published: 21 December 2021

Publisher's Note: MDPI stays neutral with regard to jurisdictional claims in published maps and institutional affiliations.

1. Introduction

Horses can spontaneously develop equine asthma, a non-infectious chronic lower airway disorder of adult horses, which shares several similarities with human asthma [1,2]. Based on the severity and the clinical presentation, the disease is classified as mild-moderate or severe equine asthma. Severe equine asthma (SEA) is characterized by coughing, exercise intolerance, and recurrent episodes of increased respiratory effort at rest, representing the exacerbation of the condition, alternated with periods of remission of the clinical signs [3]. The gold standard for the diagnosis of SEA is the cytological examination of bronchoalveolar lavage fluid (BALf) in horses with compatible clinical signs, that shows the presence of a marked neutrophilic inflammation [4]. Nevertheless, BALf collection is a relatively invasive procedure, that requires the sedation of the patient and the instillation of a large volume of fluid in the lungs of the horse. Moreover, during the remission of the clinical signs, BALf cytology may not allow to discriminate between healthy and affected horses [5]. In humans, the gold standard for the asthma diagnosis is the detection of alterations in pulmonary function testing [6]. In horses affected by SEA, conventional lung mechanics allows to identify the presence of airway obstruction [7–9]; however, this technique requires the use of an esophageal balloon and shows a low sensitivity for mild obstruction [10], therefore it is currently performed only in research settings. For this reason, in the last 30 years the attention of the researchers has been focused on oscillometry, a technique currently used in humans for the evaluation of asthmatic patients. As described by Dubois et al. [11], oscillometry allows to measure the mechanical properties of the lung (i.e., resistance, R, and reactance, X), evaluating the response of the respiratory system to external forcing over-imposed to spontaneous breathing. The Impulse Oscillation System (IOS) is a method based on a repetition of impulses generated from a loudspeaker and applied to the respiratory system that allows the determination of R and X across multiple frequencies [12]. Van Erck et al. [13–16] first reported on the use of IOS in horses with severe asthma, in which they evaluated the frequencies from 5 to 20 Hz. They reported that the results of IOS and conventional lung mechanics were well correlated. Moreover, IOS was more sensitive than standard mechanics during bronchoprovocation tests. Lower airway obstruction was characterized by negative frequency dependence of R, positive frequency dependence of X and negative X values throughout the frequency range. In 2006, Klein and colleagues evaluated the results of IOS at frequencies of 1, 5, and 10 Hz, that were considered the most representative of equine lower airways [17]. For the first time, the within-breath analysis was performed, and inspiratory and expiratory R and X were reported. Additionally, the within-breath analysis of asthmatic horses with subclinical inflammation, showed higher R values and lower X values when compared to those of controls [18]. Recently, other authors reported an association between R measured by IOS and histopathological findings of the airways of SEA horses [19]. To date, there is no report of IOS values of SEA when horses are in remission of the clinical signs.

In humans, the within-breath analysis of X allows to calculate the parameter Delta X (ΔX), defined as the difference between the inspiratory and expiratory reactance at each frequency. It has been shown that this parameter allows the detection of tidal expiratory flow limitation (EFLt) in COPD and asthma. This measure has an important diagnostic value [20,21], but it has not been applied for the evaluation of asthmatic horses. The aim of the present work was therefore to describe the results of within-breath analysis, including ΔX, measured by IOS in horses with severe asthma when they are in the exacerbation and in the remission phases of the disease.

2. Materials and Methods

2.1. Sample Selection

To perform the study, seven horses with SEA in exacerbation of the clinical signs (4 geldings and 3 mares, age of 11.9 ± 3.4 years), seven horses with SEA in clinical remission (1 gelding and 6 mares, aged 16.4 ± 5.0 years) and seven age matched healthy controls (7 mares, with a mean age of 13.1 ± 3.5 years) were studied. The horses with SEA were selected from a well-characterized population of asthmatic horses of the research herd of the Equine Asthma Laboratory, Faculty of Veterinary Medicine, University of Montréal. Horses in asthma exacerbation were kept in stable and fed hay; horses in remission of the clinical signs were kept at pasture 24 h/day for at least 6 months and fed pelleted hay when needed. Control horses were from the teaching herd of the Faculty of Veterinary Medicine of the University of Montréal and were considered free from respiratory diseases based on history and clinical examination. All horses had been previously trained to IOS measurement. The study was approved by the Animal Care Committee of the Université de Montréal (Protocol Rech-1324) and conducted in compliance with the guidelines of the Canadian Council on Animal Care.

2.2. IOS Measurement

Horses were restrained in stock and underwent IOS measurement by Equine IOS MasterScreen (Jaeger, Würzburg, Germany), as previously described [13]. Briefly, the system consisted in a plastic mask adapted to fit on the muzzle of the horse, sealed by a rubber tape. The mask was attached through a tube to a loudspeaker that produced the impulses, and to a pneumotachograph placed directly in front of the face mask. The pressure and flow response of the respiratory system to the impulses superimposed to the animal spontaneous breathing were measured. Prior to each experiment, the system was calibrated by means of a 2-L calibration syringe, forcing known volumes of air through the pneumotachograph. At least three measurements of 30 s each were performed, and the mean value of the three measurements was studied.

The data collected by LabManager (version 4.53, Jaeger, Würzburg, Germany) was then analyzed using Fast-Fourier transformation (FAMOS imc, Meßsysteme, Berlin, Germany). The mean total, inspiratory and expiratory R and X and the corresponding coherence (Co) of the respiratory system at all frequencies of impulses (from 0.1 to 20 Hz) was obtained. For this study, only values at 3, 5, and 7 Hz were studied, as they were the only frequencies with Co considered adequate (Co > 0.85 at 3 Hz and 0.9 at 5 and 7 Hz). Co reflects the quality of the measurement [17]. The ΔX, measured as the difference between the mean inspiratory and expiratory reactance at each frequency, was also calculated.

2.3. Statistical Analysis

The mean inspiratory and expiratory R and X and the ΔX were calculated at each frequency for the three groups and collected on an electronic spreadsheet (Microsoft Excel, Redmont, WA, USA). Data distribution was evaluated by means of Shapiro–Wilk normality test. If data were normally distributed, the comparison between the three groups was performed by means of one-way ANOVA and Dunnett's multiple comparison test. If data were not normally distributed, the comparison was performed by Kruskal–Wallis test and Dunn's multiple comparison test. Statistical analysis was performed using a statistical software (Prism Graphpad 9.1.0 for MacOs; San Diego, CA, USA). Statistical significance was set at $p < 0.05$.

3. Results

The Shapiro–Wilk normality test showed a normal distribution for all the parameters, except for inspiratory R at 3 Hz (R3i) and ΔX at 7 Hz (ΔX7). Results of IOS measurement are reported in Table 1.

Statistical comparison between groups showed significant differences between SEA horses in exacerbation and control horses for R at 3 Hz, for mean (R3, $p = 0.0002$), inspiratory (R3i, $p = 0.0011$) and expiratory (R3e, $p = 0.0008$) parameters.

For X, significant differences were present between horses in exacerbation and control horses at each frequency for mean (X3, $p < 0.0001$; X5, $p < 0.0001$; X7, $p < 0.0001$), inspiratory (X3i, $p < 0.0001$; X5i, $p < 0.0001$; X7i $p = 0.0007$), and expiratory parameters (X3e, $p < 0.0001$; X5e, $p < 0.0001$; X7e, $p < 0.0001$). Between control horses and asthmatic horses in remission, differences were present for mean X at 7 Hz (X7, $p = 0.0173$) and for inspiratory X at 3, 5 and 7 Hz (X3i, $p = 0.009$; X5i, $p = 0.0017$; X7i, $p = 0.012$).

The ΔX values were significantly higher in horses in exacerbation of severe asthma than in control horses at 3 and 5 Hz (ΔX3, $p = 0.0029$; ΔX5, $p = 0.001$), indicating a worsening of the airway obstruction during the expiratory phase of breathing.

Table 1. Results of IOS measurement in the three group. Normally distributed data are displayed as mean ± standard deviation. Non-normal distributed values are presented as median and interquartile range. (* = $p < 0.05$; ** = $p < 0.01$; *** = $p < 0.001$).

	Controls (cmH$_2$O * s/L)	Remission (cmH$_2$O * s/L)	Exacerbation (cmH$_2$O * s/L)
Mean			
R3	0.68 (±0.21)	0.87 (±0.35)	0.142 (±0.24) ***
R5	0.75 (±0.22)	0.86 (±0.32)	0.93 (±0.09)
R7	0.91 (±0.24)	0.92 (±0.34)	0.82 (±0.12)
X3	0.12 (±0.06)	−0.03 (±0.14)	−0.97 (±0.56) ***
X5	0.21 (±0.06)	−0.01 (±0.2)	−0.67 (±0.35) ***
X7	0.24 (±0.11)	−0.03 (±0.2) *	−0.49 (±0.24) ***
Inspiratory			
R3i	0.60(0.053–0.77)	0.95 (0.76–1.05)	1.2 (1.17–1.35) **
R5i	0.72 (±0.2)	0.88 (±0.32)	0.92 (±0.07)
R7i	0.87 (±0.21)	0.93 (±0.31)	0.80 (±0.19)
X3i	0.16 (±0.09)	−0.06 (±0.15) **	−0.39 (±0.14) ***
X5i	0.22 (±0.05)	−0.05 (±0.17) **	−0.31 (±0.14) ***
X7i	0.28 (±0.01)	−0.05 (±0.18) *	−0.18 (±0.28) ***
Expiratory			
R3e	0.71 (±0.24)	0.85 (±0.35)	1.4 (±0.27) ***
R5e	0.79 (±0.24)	0.85 (±0.31)	0.9 (±0.11)
R7e	0.97 (±0.27)	0.92 (±0.35)	0.8 (±0.14)
X3e	0.08 (±0.05)	−0.02 (±0.15)	−1.15 (±0.68) ***
X5e	0.19 (±0.08)	0.01 (±0.21)	−0.8 (±0.43) ***
X7e	0.2 (±0.13)	−0.02 (±0.22)	−0.54 (±0.27) ***
Delta (Δ) X			
ΔX3	0.08 (±0.07)	−0.05 (±0.1)	0.76 (±0.58) **
ΔX5	0.03 (±0.04)	−0.06 (±0.1)	0.47 (±0.32) **
ΔX7	0.05 (0.03–0.07)	−0.01 (−0.14–0.04)	0.16 (0.13–0.47)

4. Discussion

The present study represents the first report on IOS measurements and ΔX in asthmatic horses in remission of the disease. Horses in remission of severe asthma are of particular interest, because they can be used as a model for subclinical airway obstruction. A previous study, in fact, demonstrated the presence of a residual bronchoconstriction even after one year of treatment with inhaled corticosteroids or strict antigen avoidance [5].

Concerning the measurement technique, IOS generated a spectrum of frequencies ranging from 0.1 to 20 Hz. Nevertheless, it has been demonstrated that the frequencies lower than 10 Hz are the most representative of the lower airways in the equine species [17]. Moreover, in a previous study reporting data obtained in horses by means of forced oscillations (FOT), the frequencies considered as the most sensitive were 1, 2, and 3 Hz [22]. For this reason, we decided to evaluate only the results at 3, 5, and 7 Hz; we excluded the data at lower frequencies because the oscillations generated by the IOS could interfere with higher harmonics of spontaneous respiratory frequencies, and therefore the quality of data could be negatively influenced [23]. For the same reason, and as suggested previously [17], only impedance data showing high values of coherence (>0.85 at 3 Hz and >0.9 at 5 and

7 Hz) were included, in order to optimize the quality of data. Other studies on IOS values in horses with asthma did not report the coherence values [13–16,18], and therefore the comparison with our data is not possible.

In agreement with previous reports, IOS identified several differences between horses with SEA in exacerbation of the clinical signs and controls [13–16]. Differences in R were found only at the frequency of 3 Hz for the whole breath and for inspiratory and expiratory R. This result is similar to what reported in a recent study [19]. In human medicine, the increase in R at low frequencies is indicative of the presence of lower airway obstruction during clinical exacerbation of asthma [24]. The absence of differences at higher frequencies is also coherent with previous reports, as SEA horses are characterized by negative frequency dependence of R [16].

In the present study, horses in asthma exacerbation also showed significant lower values of X compared to controls, at all frequencies and for all the phases of breathing. This finding agrees with previous reports [15,19]. Negative values of X at low frequencies reflect peripheral airway obstruction [12]. In humans, X is decreased in the presence of various obstructive respiratory diseases, including asthma, chronic obstructive pulmonary disease (COPD), and emphysema [25].

Moreover, our results showed that the values of ΔX at 3 and 5 Hz in the SEA group in exacerbation were significantly higher than in controls, meaning that the expiratory X was significantly lower than the inspiratory X. Similar findings have been observed in human patients with COPD, but specifically only in patients where the airflow during expiration did not increase despite increasing efforts of the patient. This conditions, called tidal Expiratory Flow Limitation (EFLt), is due to the narrowing of some airways (chock points) consequent from the dynamic compression of peripheral airways [26]. During an IOS measurement, the oscillations cannot penetrate through the choke points and, therefore, impedance data represents the mechanical properties of the part of the lung between choke points and airway opening only. As most of lung compliance is located in the lung periphery (i.e., between chock points and alveoli), when choke points develop, the expiratory reactance drops [20]. In a similar way, exacerbation of SEA is characterized by an early peak of expiration, and a consequent decrease in the expiratory flow [27]. This also may be due to the presence of some choke points that cause a drop in expiratory X and a consequent increase in ΔX. This supports the presence of EFLt in SEA, which may contribute to pulmonary hyperinflation and exercise intolerance, as reported in humans [28].

Finally, IOS allowed to identify significantly lower X7, X3i, X5i, and X7i in SEA horses in remission compared to controls. This is the first report of the sensitivity of IOS in discriminating between healthy and horses in remission of severe asthma. This finding is surprising because negative values of inspiratory reactance are suggestive of restrictive diseases, such as interstitial lung disease, more than obstructive [29]. Moreover, horses with SEA in clinical remission of the clinical signs have normal lung function when evaluated using standard lung mechanics, despite the presence of a residual airway obstruction has been demonstrated [5]. It could be hypothesized that this residual bronchospasm does not interfere with the measurement of expiratory reactance by IOS. Nevertheless, it has been reported that asthmatic horses suffer from a chronic remodeling of the airways, that involves not only the smooth muscle mass [5], but also the vessels [30], the epithelium and the interstitial tissue [31], that is only partially improved by treatment or antigen avoidance. As the presence of pulmonary fibrosis and emphysema induces lower values of inspiratory X in humans [32], it could be speculated that the presence of a persistent subepithelial fibrosis and hyperinflation in horse with SEA in clinical remission [31] could have contributed to the decrease in inspiratory X.

5. Conclusions

The within breath analysis of IOS measurement showed some differences between control horses and SEA horses in remission, that could be a promising result for the identification of asthmatic horses in absence of clinical signs. Moreover, the parameter

ΔX suggests the presence of EFLt and dynamic airway compression in SEA horses in exacerbation of the clinical signs.

Author Contributions: Conceptualization, L.S. and J.P.L.; methodology, L.S. and J.P.L.; formal analysis, L.S., F.F., M.B., R.L.D. and. J.P.L.; investigation, L.S. and M.B.; writing—original draft preparation, L.S.; writing—review and editing, L.S., F.F., M.B., R.L.D. and J.P.L. All authors have read and agreed to the published version of the manuscript.

Funding: Funding was provided by the Canadian Institutes of Health Research (JPL, PJT-148807).

Institutional Review Board Statement: The study was approved by the Animal Care Committee of the Université de Montréal (Protocol Rech-1324) and conducted in compliance with the guidelines of the Canadian Council on Animal Care.

Informed Consent Statement: Not applicable.

Data Availability Statement: The data presented in this study are available on request from the corresponding author.

Acknowledgments: The authors thank Khristine Picotte for the technical help in IOS measurement.

Conflicts of Interest: The authors declare no conflict of interest.

References

1. Couetil, L.L.; Cardwell, J.M.; Gerber, V.; Lavoie, J.P.; Leguillette, R.; Richard, E.A. Inflammatory airway disease of horses-revised consensus statement. *J. Vet. Intern. Med.* **2016**, *30*, 503–515. [CrossRef]
2. Lange-Consiglio, A.; Stucchi, L.; Zucca, E.; Lavoie, J.P.; Cremonesi, F.; Ferrucci, F. Insights into animal models for cell-based therapies in translational studies of lung diseases: Is the horse with naturally occurring asthma the right choice? *Cytotherapy* **2019**, *21*, 525–534. [CrossRef] [PubMed]
3. Lo Feudo, C.M.; Stucchi, L.; Alberti, E.; Conturba, B.; Zucca, E.; Ferrucci, F. Intradermal Testing Results in Horses Affected by Mild-Moderate and Severe Equine Asthma. *Animals* **2021**, *11*, 2086. [CrossRef] [PubMed]
4. Couetil, L.; Cardwell, J.M.; Leguillette, R.; Mazan, M.; Richard, E.; Bienzle, D.; Bullone, M.; Gerber, V.; Ivester, K.; Lavoie, J.P.; et al. Equine Asthma: Current Understanding and Future Directions. *Front. Vet. Sci.* **2020**, *30*, 450. [CrossRef] [PubMed]
5. Leclere, M.; Lavoie-Lamoureux, A.; Joubert, P.; Relave, F.; Lanctot Setlakwe, E.; Beauchamp, G.; Couture, C.; Martin, J.G.; Lavoie, J.P. Corticosteroids and Antigen Avoidance Decrease Airway Smooth Muscle Mass in an Equine Asthma Model. *Am. J. Respir. Cell Mol. Biol.* **2012**, *47*, 589–596. [CrossRef]
6. Global Initiative for Asthma, Global Strategy for Asthma Management and Prevention. 2020. Available online: www.ginasthma.org (accessed on 15 April 2021).
7. Lavoie, J.P.; Pascoe, J.R.; Kupershoek, C.J. Effect of head and neck position on respiratory mechanics in horses sedated with xylazine. *Am. J. Vet. Res.* **1992**, *53*, 1652–1657.
8. Robinson, N.E.; Berney, C.; Eberhart, S.; deFeijter-Rupp, H.L.; Jefcoat, A.M.; Cornelisse, C.J.; Gerber, V.M.; Derksen, F.J. Coughing, mucus accumulation, airway obstruction, and airway inflammation in control horses and horses affected with recurrent airway obstruction. *Am. J. Vet. Res.* **2003**, *4*, 550–557. [CrossRef]
9. Mazan, M.R.; Deveney, E.F.; DeWitt, S.; Bedenice, D.; Hoffman, A. Energetic cost of breathing, body composition, and pulmonary function in horses with recurrent airway obstruction. *J. Appl. Physiol.* **2004**, *97*, 91–97. [CrossRef]
10. Pirrone, F.; Albertini, M.; Clement, M.G.; Lafortuna, C. Respiratory mechanics in Standardbred horses with sub-clinical inflammatory airway disease and poor athletic performance. *Vet. J.* **2007**, *173*, 144–150. [CrossRef]
11. DuBois, A.B.; Brody, A.W.; Lewis, D.H.; Burgess JR, B.F. Oscillation mechanics of lungs and chest in man. *J. Appl. Physiol.* **1956**, *8*, 587–594. [CrossRef]
12. Smith, H.J.; Reinhold, P.; Goldman, M.D. Forced Oscillation Technique and Impulse Oscillometry. *Eur. Respir. Monogr.* **2005**, *31*, 72–105.
13. Van Erck, E.; Votion, D.M.; Kirschvink, N.; Art, T.; Lekeux, P. Use of the Impulse Oscillometry System for Testing Pulmonary Function during Methacholine Bronchoprovocation in Horses. *Am. J. Vet. Res.* **2003**, *64*, 1414–1420. [CrossRef] [PubMed]
14. van Erck, E.; Votion, D.; Art, T.; Lekeux, P. Measurement of Respiratory Function by Impulse Oscillometry in Horses. *Equine Vet. J.* **2004**, *36*, 21–28. [CrossRef] [PubMed]
15. Van Erck, E.; Votion, D.; Kirschvink, N.; Genicot, B.; Lindsey, J.; Art, T.; Lekeux, P. Influence of Breathing Pattern and Lung Inflation on Impulse Oscillometry Measurements in Horses. *Vet. J.* **2004**, *168*, 259–269. [CrossRef]
16. Van Erck, E.; Votion, D.; Art, T.; Lekeux, P. Qualitative and quantitative evaluation of equine respiratory mechanics by impulse oscillometry. *Equine Vet. J.* **2006**, *38*, 52–58. [CrossRef] [PubMed]
17. Klein, C.; Smith, H.J.; Reinhold, R. The Use of Impulse Oscillometry for Separate Analysis of Inspiratory and Expiratory Impedance Parameters in Horses: Effects of Sedation with Xylazine. *Res. Vet. Sci.* **2006**, *80*, 201–208. [CrossRef]

18. Richard, E.A.; Fortier, G.D.; Denoix, J.M.; Art, T.; Lekeux, P.M.; Van Erck, E. Influence of Subclinical Inflammatory Airway Disease on Equine Respiratory Function Evaluated by Impulse Oscillometry. *Equine Vet. J.* **2009**, *41*, 384–389. [CrossRef]

19. Bullone, M.; Hélie, P.; Joubert, P.; Lavoie, J.P. Development of a Semiquantitative Histological Score for the Diagnosis of Heaves Using Endobronchial Biopsy Specimens in Horses. *J. Vet. Int. Med.* **2016**, *30*, 1739–1746. [CrossRef]

20. Dellacà, R.L.; Santus, P.; Aliverti, A.; Stevenson, N.; Centanni, S.; Macklem, P.T.; Pedotti, A.; Calverley, P.M.A. Detection of Expiratory Flow Limitation in COPD Using the Forced Oscillation Technique. *Eur. Respir. J.* **2004**, *23*, 232–240. [CrossRef] [PubMed]

21. Mikamo, M.; Fujisawa, T.; Oyama, Y.; Kono, M.; Enomoto, N.; Nakamura, Y.; Inui, N.; Sumikawa, H.; Johkoh, T.; Suda, T. Clinical Significance of Forced Oscillation Technique for Evaluation of Small Airway Disease in Interstitial Lung Diseases. *Lung* **2016**, *194*, 975–983. [CrossRef]

22. Young, S.S.; Tesarowski, D.; Viel, L. Frequency dependence of forced oscillatory respiratory mechanics in horses with heaves. *J. Appl. Physiol.* **1997**, *82*, 983–987. [CrossRef] [PubMed]

23. Peslin, R.; Fredberg, J. Oscillation mechanics of the respiratory system. In *Handbook of Physiology, Section 3: The Respiratory System*; Macklem, P., Mead, J., Eds.; American Physiological Society: Bethesda, MD, USA, 1986; Volume 3, pp. 145–166.

24. Clement, J.; Landser, F.J.; van de Woestijne, K.P. Total resistance and reactance in patients with respiratory complaints with and without airways obstruction. *Chest* **1983**, *83*, 215–220. [CrossRef] [PubMed]

25. Bickel, S.; Popler, J.; Lesnick, B.; Eid, N. Impulse Oscillometry: Interpretation and Practical Applications. *Chest* **2014**, *146*, 841–847. [CrossRef]

26. Pedersen, O.F.; Butler, J.P. Expiratory flow limitation. *Compr. Physiol.* **2011**, *1*, 1861–1882.

27. Petsche, V.M.; Derksen, F.J.; Robinson, N.E. Tidal breathing flow-volume loops in horses with recurrent airway obstruction (heaves). *Am. J. Vet. Res.* **1994**, *55*, 885–891. [PubMed]

28. Nyman, G.; Björk, M.; Funkquist, P. Gas exchange during exercise in standardbred trotters with mild bronchiolitis. *Equine Vet. J.* **1999**, *30*, 96–101. [CrossRef]

29. Sugiyama, A.; Hattori, N.; Haruta, Y.; Nakamura, I.; Nakagawa, M.; Miyamoto, S.; Onari, Y.; Iwamoto, H.; Ishikawa, N.; Fujiaka, K.; et al. Characteristics of inspiratory and expiratory reactance in interstitial lung disease. *Respir. Med.* **2013**, *107*, 875–882. [CrossRef]

30. Ceriotti, S.; Bullone, M.; Leclere, M.; Ferrucci, F.; Lavoie, J.P. Severe asthma is associated with a remodeling of the pulmonary arteries in horses. *PLoS ONE* **2020**, *15*, e0239561. [CrossRef] [PubMed]

31. Bullone, M.; Joubert, P.; Gagné, A.; Lavoie, J.P.; Hélie, P. Bronchoalveolar Lavage Fluid Neutrophilia Is Associated with the Severity of Pulmonary Lesions during Equine Asthma Exacerbations. *Equine Vet. J.* **2018**, *50*, 609–615. [CrossRef]

32. Mori, K.; Shirai, T.; Mikamo, M.; Shishido, Y.; Akita, T.; Morita, S.; Asada, K.; Fujii, M.; Hozumi, H.; Suda, T.; et al. Respiratory Mechanics Measured by Forced Oscillation Technique in Combined Pulmonary Fibrosis and Emphysema. *Respir. Physiol. Neurobiol.* **2013**, *185*, 235–240. [CrossRef]

MDPI

St. Alban-Anlage 66

4052 Basel

Switzerland

www.mdpi.com

Animals Editorial Office

E-mail: animals@mdpi.com

www.mdpi.com/journal/animals